Textbook of

Algae

1st Edition

Dr. A .K. Kushwaha

M.Sc., B.Ed., M.A., Ph.D.

Department of Botany,

Sarla Dwivedi P.G College, Akbarpur Kanpur (D) U.P, India

209101

Dr. Manoj Kumar Shukla

M.Sc. B.Ed., Ph.D. SET

Department of Botany

Chhatrasal Government. P.G College, Panna, M.P, India

488001

© Reserved

All disputes subject Kanpur Dehat (UP) Jurisdiction only.

PREFACE

Algae are photosynthetic organisms that possess photosynthetic pigments such as chlorophyll. However, they lack true roots, stems and leaves characteristic of vascular plants. Some of them are unicellular whereas others are multicellular. They may also form colonies. Most **algae** are aquatic.

This book will serve as an introduction to Algae to the beginners in the field. Actually the book is in intended to fulfil the long felt need of student of graduate and postgraduate level of all universities. The syllabi of all the universities have been kept in view during the preparation of the manuscript of this text. This work may also serve as laboratory manual. The present text provides a background of facts, terminology, general principle and occurrence, thallus structure and life cycle.

The text has been written in simple language and profusely illustrated with self-explanatory diagrams and coloured photographs. The figures are provided with detailed Legends. Some of the diagrams, of the diseased specimens have been drawn author himself, and most of the figure have been quoted from the authentic works of various authors and online websites.

I have already mentioned the book is primary and elementary text for degree and postgraduate students and not a text for researchers, and therefore, The Bibliography has been kept at minimum.

The suggestion and healthy criticism of the book will be of much value to me for the improvement of the next edition.

A.K. Kushwaha
Manoj Kumar Shukla
anilrania@gmail.com

CONTENTS

Chapter........................Page no.

1. Algae: General Characters...1-35
- Occurrence
- Range of thallus structur,
- Structure of Algal Cell
- Reproduction
- Types of life cycle.

2. Classification of algae........36-46
- Evolution of Algal Classification
- Classification Proposed by Fritsch
- Classification Proposed by Smith's
- Classification Proposed by G. F. Papenfuss (1955)
- Classification Proposed by V. J. Chapman (1962)
- Classification Proposed by Christensen (1964)
- Classification Proposed by F. E. Round (1973)
- Classification Proposed by V. J. Chapman and D. J. Chapman (1973)
- Classification Proposed by H. C. Bold and M. J. Wynne (1978)
- Classification Proposed by R. E. Lee (2008)

3. Myxophyceae (Cyanophyceae) 47-56
- General Character
- Classification, Occurrence of Cyanophyceae,
- Cell Structure,
- Heterocysts,
- Origin of Cyanophyceae,
- Economic Importance of Cyanophyceae.

4. Chroococcales 57-61
- Family. Chroococcaceae; Gloeacapsa, Microcystis,

5. Nostocales......................62-85
- Family. Oscillatoriaceae. *Oscillatoria, spirulina,*
- Family. Nostocaceae; *Nostoc, Anabena, Aulosira,*
- Family. *Scytonemataceae*; Scytonema,
- Family. Rivulariaceae; *Gloeotrichia, Rivularia.*

6. Stigonematales 86-88
- Family: Stigonemataceae; *Stigonema*

7. Chlorophyceae.................89-92
- General characteristic
- Classification and
- Alternation of Generations

8. Volvocales.....................93-112
- Family. Chlamydomonadaceae: *Chlamydomonas*
- Family. Volvocaceae; *Volvox, Pandorina, Eudorina*

9. Chlorococcales............113-124
- Family. Chlorellaceae; *Chlorella, Chlorococcum*
- Family. Hydrodictyaceae; *Hydrodictyon*
- Family. Coelastraceae; *Scenedesmus*

10. Ulotrichales 125-140
- Family. Ulvaceae; Ulva; *Enteromorpha*
- Family. Ulotrichaceae; *Ulothrix*

11. Cladophorales...........141-147
- Family. Cladophoraceae: *Cladophora*

12. Chaetophorales.........148-178
- Family. Chaetophoraceae; *Fritschiella, Chaetophora, Draparnaldia*
- Family. Coleochaetaceae; *Coleochaete*
- Family. Trentepohliaceae: *Trentepohlia*
- Family. Pleurococcaceae: *Pleurococcus*

13. Oedogoniales............179-191
- Family. Oedogoniaceae: *Oedogonium*

14. Conjugales...............192-208
- Family: Zygnemaceae: *Spirogyra, Zygnema*
- Family: Desmidiaceae: *Cosmarium*

15. Siphonales.................209-230
- Family. Caulerpaceae: *Bryopsi, Caulerpa*
- Family Codiaceae: *Codium*
- Family. Vaucheriaceae: *Vaucheria*

16. Charales.................231-247
- Family. Characeae: *Chara*

17. Bacillariophyceae 248-260
- General Characters and Classification

18. Pennales 261-265
- Family. Naviculoideae: *Navicula*

19. Phaeophyceae.............266-271
- Description and Classification

20. Ectocarpales 272-282
- Family. Ectocarpaceae: *Ectocarpus*

21. Fucales.....................283-298
- Family. Fucaceae: *Fucus*
- Family. Sargassaceae: *Sargassum*

22. Dictyotales 299-303
- Family. Dictyotaceae: *Dictyota*

23. Laminariales............304-311
- Family. Laminariaceae: *Laminaria*

24. Rhodophycae...........312-316
- Special Characteristics
- Classification
- Economic importance

25. Nemalionales 317-324
- Family. Batrachospermaceae: *Batrachospermum*

26. Ceramiales................325-331
- Family. Rhodomelaceae: *Polysiphonia*

27. Origin and Evolution of Sex in Algae 332-335

28. Economic Importance of Algae 336-343

29. Culturing of algae……..344-367

30. Toxins of Algae ………-369-368

31. M. O. P. Iyengar, Professor F.E. Fritsch…………………..370-372

32. Glossary of Terms for Algae

33. References

1. Algae: General Characters

Algae are eukaryotic organisms that have no roots, stems, or leaves but do have chlorophyll and other pigments for carrying out photosynthesis. Algae can be multicellular or unicellular. **Unicellular algae** occur most frequently in water, especially in plankton. **Phytoplankton** is the population of free-floating microorganisms composed primarily of unicellular algae. In addition, algae may occur in moist soil or on the surface of moist rocks and wood. Algae live with fungi in **lichens.**

According to the Whittaker scheme, algae are classified in seven divisions, of which five are considered to be in the Protista kingdom and two in the Plantae kingdom. The cell of an alga has eukaryotic properties, and some species have flagella with the "9-plus-2" pattern of microtubules. A nucleus is present, and multiple chromosomes are observed in mitosis. The chlorophyll and other pigments occur in **chloroplasts**, which contain membranes known as **thylakoids.**

Most algae are **photoautotrophic** and carry on photosynthesis. Some forms, however, are **chemoheterotrophic** and obtain energy from chemical reactions and nutrients from preformed organic matter. Most species are saprobes, and some are parasites.

Reproduction in algae occurs in both asexual and sexual forms. Asexual reproduction occurs through the fragmentation of colonial and filamentous algae or by spore formation (as in fungi). Spore formation takes place by mitosis. Binary fission also takes place (as in bacteria).

During sexual reproduction, algae form differentiated sex cells that fuse to produce a diploid **zygote** with two sets of chromosomes. The zygote develops into a sexual spore, which germinates when conditions are favourable to reproduce and reform the haploid organism having a single set of chromosomes. This pattern of reproduction is called **alternation of generations.**

Key Points

The study of algae is known as PHYCOLOGY. The one who study algae is called Phycologist.

General characters of algae

1. Thalloid plant body
2. In Eichler's system of classification, algae are placed in the Division Thallophyta along with Fungi and Lichens.
3. Algae are autotrophs (synthesize food using light energy)
4. Algae differ from fungi in:
5. Presence of photosynthetic pigment – chlorophyll ⊕. Mode of nutrition (autotrophs)
6. Majority of algae are in aquatic habitat (fresh water or marine), some algae are terrestrial also
7. Algae are present in all parts of the world including Arctic and Antarctic regions (universal occurrence)
8. Sex organs are unicellular or multicellular
9. Sex organs lack jacket cells around them (naked sex organs)
10. If jacket cells are present, they have different origin
11. There is a progressive complexity in the reproduction of different algal groups
12. Embryos is not formed after zygote formation
13. Show distinct alternation of generation
14. Cellular organization may be prokaryotic (blue green algae) or eukaryotic (all other algae)

OCCURRENCE

Most of the part of the land is covered over either by fresh water or sea water. Besides, several other algae are found in somewhat drier conditions. They are found on the trunks of trees, on telephone wires, on rocks, on walls, in hot springs and in several other unusual habitats. Here some of the algae have been classified according to their habitats. Special emphasis has been given on the occurrence of fresh water algae.

1. Hydrophytes:

They are more or less completely submerged or free floating on the surface of the water. The hydrophytes may be subdivided into following heads.

(i) Benthophytes: Several fresh water and marine algae are found in attached condition. The fresh water such as *Chara, Nitella, Cladophora, Gongrosira, Chaemosiphon,* etc., are found attached to some substratum in the bottom of the water. Almost all of brown algae (Phaeophyceae) are found in attached condition to some substrata in the sea.

(ii) Epactiphytes:

Such algae grow along the shores of lakes and ponds, and may be delimited from benthophytes with some difficulty. The most important fresh water forms are – *Oedogonium, Chaetophora,* some species of *Spirogyra, Mougeotia,* some diatoms, *Scytonema* and *Rivularia.*

(iii) Thermophytes:

Many algae are reported from hot springs. These algae may tolerate the temperature upto 70°C or more than that. According to Copeland, 53 genera and 153 species of Chroococcaceae may survive upto 84°C. Some Oscillatoriaceae may survive upto 85°C. This supports that Myxophyceae (blue-green algae) are primitive.

(iv) Planktophytes:

The algae which float on the surface of the water are called 'planktophytes'. They may be of two types, 'i.e., (a) euplanktophytes (b) tychoplanktophytes.

(a) Euplanktophytes:

They are never attached, and from the very beginning are free floating, e.g., diatoms, *Cosmarium, Closterium, Microcystis, Sphaeroplea, Scenedesmus, Pediastrum,*

Chlamydomonas, Volvox, other Volvocales and some members of Chroococcales. The above given forms are fresh water in habit.

(b) Tychoplanktophytes:

In the beginning, such algae are attached, but later on they become detached and free floating, e.g., some species of *Spirogyra, Zygnema, Cladophora, Oedogonium, Rhizoclonium, Mougeotia, Tribonema, Microspora, Cylindrospermum, Tetraspora, Rivularia, Nostoc, Gloeotrichia, Sargassum,* etc.

(v) Halophytes:

The algae occur in saline waters are known as 'halophytes'. The most striking examples are *Dunaliella* and *Chlamydomonas* which occur in salt lakes, the species of *Scenedesmus, Aphanocapsa, Pediastrum, Aphanothece,* and *Oscillatoria* are found in saline waters; the species of Enteromorpha are found in inland astuaries (the tidal mouth of a large river, where the tide meets the stream.); many species of Ulvales, Ulotrichales, Conjugales, and Myxophyceae are found near the sea in astuaries.

(vi) Epiphytes:

Many algae are found upon other living plants and bigger species of algae. *Aphanochaete, Bulbochaete, Oedogonium* and *Microspora,* are found as epiphytes upon larger species of *Oedogonium, Cladophora, Rhizoclonium, Vaucheria* and **Hydrodictyon** species. Co- leochaete nitellarum is epiphytic upon species of *Chara* and *Nitella.*

Some of the species of Coleochaete are epiphytic upon some grasses grown on the banks of the ponds and the hydrophytes such as—*Vallisneria, Typha, Ipomoea* and several other aquatic plants. *Chaetonema* is found epiphytic on the mucilaginous masses of *Tetraspora* and *Batrachospermum.*

(vii) Epizoophytes:

Certain algae are found on living aquatic animals such as turtles, mollusc shells, fishes, etc. Species of *Cladophora* grow upon mollusc shells. *Protoderma* and *Basicladia* occur on the **back of turtles.** *Characiopsis* and *Characium* occur on the posterior and anterior legs of Branchipus respectively.

2. Edaphophytes:

Such algae are also called terrestrial algae. They are found upon or inside the surface of the earth. They can be (i) saphophytes and (ii) cryptophytes.

(i) Saphophytes:
They are surface algae. Most of the species of Myxophyceae are found upon the surface of the soil. Besides, Mesotaenium, Botrydium, Protosiphon, Oedocladium, Vaucheria, Fritschiella and many others are met with upon the surface of the wet soil.

(ii) Cryptophytes:

Such algae are subterranean in habit and occur inside the soil. The species of Myxophyceae are found in the soil. The species of Nostoc, Anabaena and Euglena have been reported from the paddy fields, where they also fix the atmospheric nitrogen in the soil to enrich the fertility of the fields.

3. Aerophytes:

Such algae are aerial in habitat. They are found upon the trunks of trees, walls, fencing wires, rocks, and animals and so many other aerial substrata.

(i) Epiphyllophytes:

Such algae are epiphytic upon leaves of trees. Species of Trentepohlia are commonly found upon the bark of trees. They also occur upon rocks and fencing wires. They are abundantly found on the fencing wires of Calcutta botanical gardens. Phycopeltis occurs upon Rubus; Phyllosiphon on Arisaema; Rhodochytrium on Asclepias and Solidago.

Phycopeltis epiphyton , Somatochroon , Cephaleuros **Surface of leaves**
(Epiphyllophytes)

(ii) Epiphloephytes:

These algae grow on the bark of trees mixed with many mosses and liverworts. *Phormidium, Scytonema, Haplosiphon* and *Schizothrix* grow on the bark of trees mixed with liverworts.

Trentepholia , Scytonema ,

Chroococus, Pleurococcus (Epiphloeophytes)	Bark of trees

(iii) Epizoophytes:

These algae are found even on the bodies of land animals. Certain Chaetophorales are found even on the hairs of sloth.

(iv) Lithophytes:

Many algae grow on the rocks and walls. The species of Scytonema grow on the walls in rainy season and the whole wall becomes black spotted. Vaucheria, Nostoc and many other algae are also found on wet rocks.

Cyanoderma, Trichophilus, Trentopholia, Pleurococcus, Chlorococcales	Rocks and stones

4. Cryophytes:

These algae are found on ice and snow. These algal forms cause red snow, green snow, yellow snow, yellowish green snow and violet snow.

Chlamydomonas nivalis, Scotiella, Gloeocapsa	Red color
Acyclonema	Brown or purple color
Chlamydomonas yellowstonensis,	Green color
Euglena Enkistrodesmus, Mesotaenium Scitiella, Protoderma, Pleurococcus, Nostoc green	Yellow or yellowish
Raphidonema	Black colour of snow

5. Symbionts or endophytes:

Many algae grow in symbiotic association of other plants. The most striking example of symbiosis are lichens, here the algae are found in symbiotic association of fungi. Various Myxophyceae, e.g., *Chroococcus, Nostoc, Microcystis, Gloeocapsa, Scytonema, Rivularia,* etc., have been separated from lichens. Some green algae, e.g., *Coccomyxa, Chlorella, Protococcus, Palmella,* etc., are also found as symbionts in lichens.

Besides, several algae are endophytes in the tissue of other plants. *Anabaena azollae* is found inside the leaves of Azolla (a pteridophyte). *Anabaena cycadae* is found in the corralloid roots of Cycas. Nostoc has been reported from the tissues of *Anthoceros* and *Notothylas*. *Nostoc* is found in the leaves of *Sphagnum* (Bryophyta) and several angiosperms. *Chlorochytrium* is endophytic inside *Lemna, Ceratophyllum* and certain mosses.

6. Endozoophytes:

Algae growing inside the body of vertebrates or aquatic animals are called endozoic algae for e.g.

Algae	Host
Zoochlorella viridis	*Hydra*
Zooxanthella	Fesh water sponges
Oscillatoria, Simonsiella, Anabaeniolum	Several vertebrates

7. Parasites:

Members of algae are known to live as parasite and semiparasite on other algae as well as higher plants, where they cause severe damage, for e.g.

1. *Cephaleuros virescence* (Chlorophyceae) grows on Tea plants (Causes Red rust of tea)
2. *Cephaleuros virescence* (Chlorophyceae) grows on *Coffea arabica.,*
3. *Rhododendron, Magnolia* and *Piper nigrum* causing Red Rust.
4. *Rhodochytrium* (Chlorophyceae) grows on ragweed (*Ambrosia*) leaves.
5. *Phyllosiphon* (Chlorophyceae) grows on the leaves of *Arisarum vulgare*.
6. *Polysiphonia fastigata* (Rhodophyceae) grows on *Ascophyllum nodusum* as semiparasite.
7. *Ceratocolax* (Rhodophyceae) grows in *Phyllophora thallus*.

8. Fluviatile algae:

Such algae are found in rapidly flowing waters; *Ulothrix* occurs in mountain falls. *Stigeoclonium, Batrachospermum* are reported from the swift running streams of Dehradun and other hilly tracts.

Range of thallus structure

Range of algal thallus varies from unicellular to multicellular forms or microscopic to macroscopic structures, with their size ranging from a few microns to some meters. *Micromonas pusilla* is known to be smallest algae which is unicellular and is 1 μm (0.00004 in.), on the other hand giant kelps has longest thalli that reaches up to 60 m (200 ft) in length. The unicellular forms may remain solitary as a single unit which are capable of completing their life cycle by providing all physiological, biochemical, genetical requirements and may be motile or non-motile. When these unicellular forms are held together in a common gelatinous matrix, they constitute colonial forms, which are considered intermediate between unicellular and multicellular structures. The other intermediate stages considered in thallus organization of algae are palmella, dendroid, palmelloid, coccoid, fi lamentous, siphonaceous, heterotrichus, uniaxial, multiaxial etc. In colonial forms the individual cells are independent in both structure and function. The multicellular form ranges from microscopic to macroscopic, where some of the macroscopic forms reported from phaeophyceae and rhodophyceae grow upto few meters. The multicellular forms may be parenchymatous or sometimes the thallus is differentiated (Round 1973).

The range of thallus organization in algae may be classified as follows:

1. Unicellular

Motile and non-motile

2. Aggregates

Palmelloid and Dendroid

3. Colonial

(a) Colony motile

(b) Colony non-motile

4. Filamentous

(a) Un-branched

(b) Branched

(i) Simple

(ii) Heterotrichous

(iii) Pseudoparenchymatous.

5. Siphonaceous.

6. Parenchymatous.

1. Unicellular algae:- The unicellular algae are all sizes and shapes. They range from small spherical cells to large irregular shaped cells. The unicellular forms may be spherical, oblong, pear shaped or sometimes elongated bearing flagella (*Chlamydomonas, Euglena*) etc. These are of following two types –

(a) Non-motile:- Non-motile unicells, without flagella. Many unicellular algae do not possess any outgrowth for
   locomotion, e.g., *Chlorella*

(b) Motile:- Motile unicells, with flagella. Many unicellular motile forms may be spherical, oblong, pear-shaped or
   sometimes elongated and bears flagella for locomotion, eg., *Chlamydomonas, Euglena,* etc.

2. Aggregates

Aggregates are formed by the collection of single cells to make thallus. Unlike the coenobium the aggregation of cells does not have fixed number of cells shape or size.

The cells are aggregated into more or less irregular colony like mass. When the cells divide, the daughter cells remain in same gelatinous mass. Thus there is increase in the number of cells after division. The aggregates can be palmelloid, dendroid and rhizopodial in form.

(a) Palmelloid: Colonial members in which "non – motile" cells remain embedded in an amorphous gelatinous or mucilaginous matrix. In this type neither the number, nor the shape and size of cells is constant. The cells are aggregated in a common mucilaginous envelope. All the cells are independent of each other and fulfil the function of an individual. Most normally fl agellate or coccoid unicellular algae may enter (often temporary) palmella stage, a condition where the flagella are lost and the individuals undergo successive vegetative divisions while embedded in a common gelatinous matrix, named after the volvocale an (Chlorophyceae) genus *Palmella*. This term may be strictly applied to those algae where the cells will readily revert to a motile condition or may be expanded to include all algae where the palmelloid habit is more permanent. Example: *Palmella*, *Microcystis*.

(b) Dendroid: The cells are united in a branching manner by localized production of mucilage at the base of each cell. The whole colony looks like a tree in habit.
Examples: *Chrysodendron, Ecballocyctis, Dinobryon.*

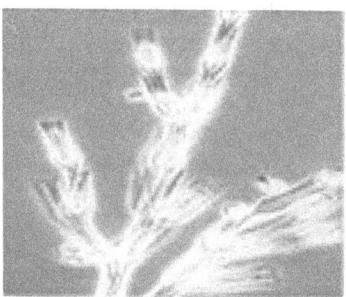

Fig.1. Dendroid

(c) Rhizopodial: The cells of rhizopodial colonies are united through rhizopodia, as in *Chrysidiastrum* (Chrysophyta).

3. Colonial Forms

A further evolution of the unicellular types from occasional and indefinite type of colony like structures—with independent individual cells inside it to a well-defined colony prasinocladus with interlinks among the cells results in a true colonial habit.

Here varying numbers of unicells aggregate together in different ways, often within a mucous envelope. Colonial forms are seen among Chlorophyceae. Chrysophyceae, Bacillariophyceae, Dinophyceae, Xanthophyceae etc. The colony may be (a) non-motile or (b) motile

(a) Non-motile Colony:- Number of cells in these colonies is indefinite and they are non motile (*Hydrodictyon, Pediastrum*).

(b) Motile Colony:- In these algae, definite number of cells are found and these are motile, such as *Scenedesmus, Gonium* (small colony), *Volvox* (big colony).

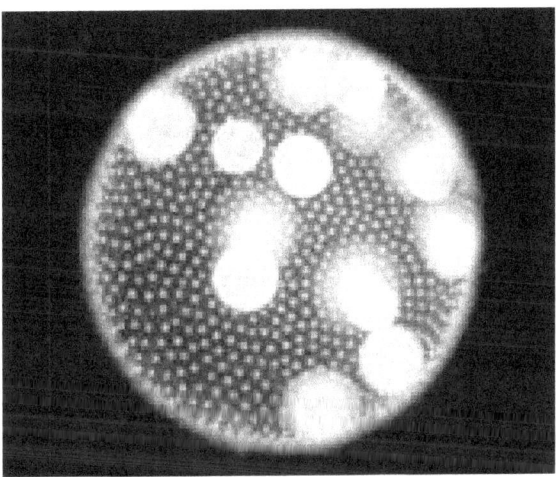

Fig.2. *Volvox* (big colony)

Motile Coenobial Colony:

Definite number of motile cells are embedded in a gelatinous matrix with their flagella protruded out, or are held together by cytoplasmic connection. The cells may be compact or loosely arranged. Thus a colony is formed of definite number of cells arranged in a specific manner forming what is called a coenobium (pi. coenobia)

4. Filamentous:- Filamentous algae (floating or attached). The simplest filamentous algae consist of a thallus, of asingle chain of cells. This is the result of cell division in one plane only. The filaments may be unbrached filamentous forms.

Such types of thalli are found in many algae and consist of a straight row of cells (*Spirogyra, Ulothrix, Oedogonium*).

(a) Unbranched filaments: Cells arranged one above the other without any branching points.e.g. – *Spirogyra, Zygnema* (free floating), *Nostoc, Anabaena, Spirulina*

Ulothrix, Oedogonium (attached to the substratum by a basal specialized cell)

Fig.3. *Spirogyra*

Branched filaments, Filaments with branching often dichotomous branching. e.g., –*Cladophora, Pithophora, Bulbochaete*

(a)

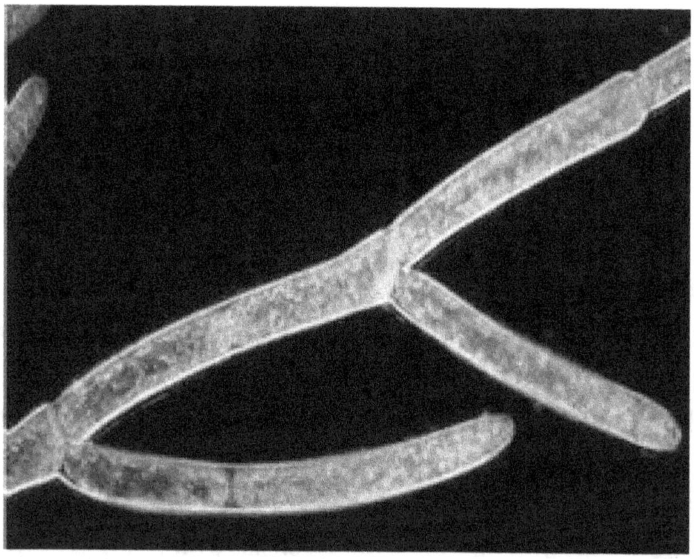

Fig.4. *Cladophora*

(i) **Psuedo-branched filaments**: Appear like branched under microscope. Actually, branching is due to the close association of unbranched individual filaments e.g., *Scytonema, Ulothrix*

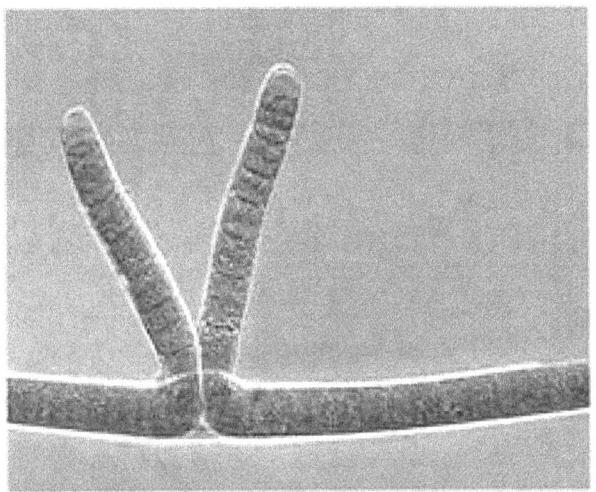

Fig.5. *Scytonema*

(ii) Heterotrichous forms (heteros =different): Some cells in the filaments divide several times in different planes resulting in two parts. Thallus consists of two parts, i.e. horizontally running main trichome and a vertical erect trichome e.g., *Draparnaldiopsis, Chara*

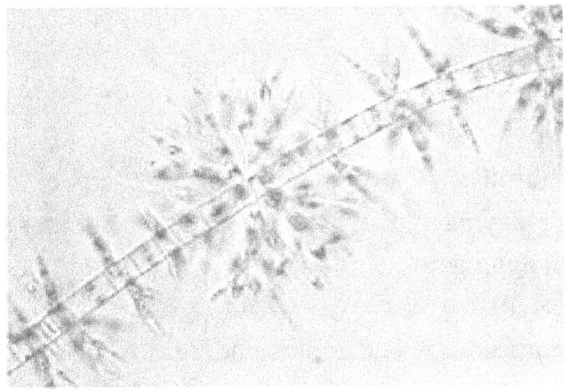

Fig.7. *Draparnaldiopsis*

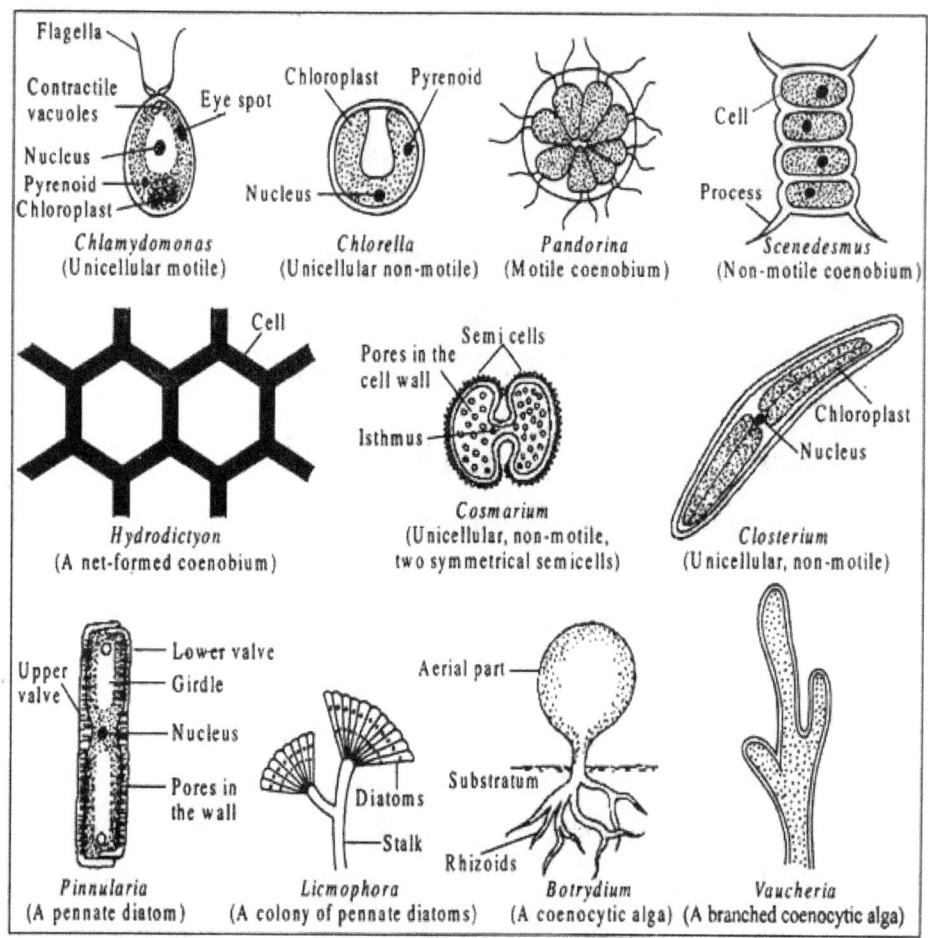

Fig.6. Different thallus structure

(iii) **Psuedoparenchymatous:-** Psuedoparenchymatous algae have thalli that superficially resemble parenchyma, but which are actually composed of appressed filaments or amorphous cell aggregates. e.g., *Monostroma*.

(5) **Siphonous or Siphonaceous:-** The thallus undergoes repeated nuclear division without the accompanying formation
of cell walls. As a result tubular structure with the multinuclear cytoplasm lining is formed known as coenocytic as in
siphonales e.g., *Vaucheria, Caulerpa*.

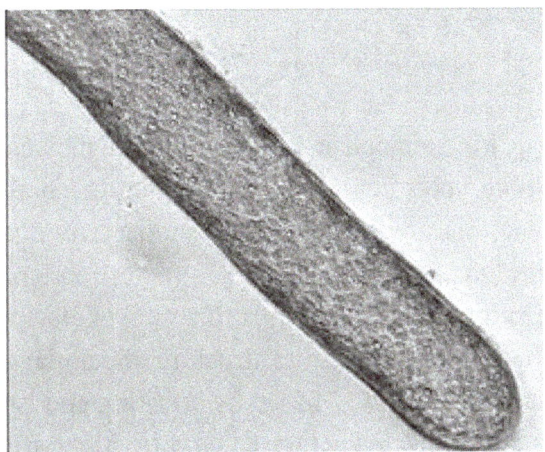

Fig.8. *Vaucheria*

(6) **Parenchymatous:-** Parenchymatous thallus organization also is a modification of the filamentous habit, with cell division in more than one plane. Depending upon the nature of cell division, the parenchymatous thalli may be 'leaf-like' or foliose tubular or highly developed structure. Flat, foliose or tubular thalli are formed by the division of the cells two or three planes. Common examples of flat and foliose structures in Viva (Chlorophyceae), Punctaria (Phaeophyceae) and Porphyra (Rhodophyceae). The example of tubular structure is *Enteromorpha* (Chlorophyceae). In some phaeophyceae e.g., in lictocarpales and Sphacelariales, the parenchymatous form develops by abundant separation of primary filament. In Phaeophyceae (e.g., *Sargassum*) cells of the thallus are differentiated into central medulla, middle cortex and outer merislodenn.

Structure of Algal Cell

Algae exhibit 2 different basic kinds of cell structure; therefore they can be separated into 2 groups - prokaryotes. Prokaryotes contain so called blue-green algae classed as Cyamophyceae or Myxophyceae, but now termed as Cyanobacteria due to their cells are prokaryote type. Eukaryotic algae fairly varied in cell structure and morphology that is taken in account for categorization. In recent years, utilization of electron microscopy has brought much novel information regarding ultra structure of cellular components of algae. Chemical composition and functions are found by breaking and isolating each of its organelles separately. This study discloses that eukaryotic algae illustrate several features which resemble higher plant groups.

Prokaryotic Algal Cell:

Cyanobacteria closely look like bacteria in their ultrastructure. Though, cyanobacteria are not flagellated. Specific features of the cellular component are given below:

1) Cell Wall and Cell Sheath:

Cell of cyanobacteria are enveloped by the gelatinous heath and also have separate cell wall outside plasma membrane. This can be eliminated by digesting it with enzyme-lysozyme. It chemical examination illustrates that it is composed of mucopolysaccharide (peptidoglycan) like that of bacterial cell wall. It has complex structure, composed of polymer of N-aceylmuramic acid and Nacetylglucosamine which are cross linked by peptides and other compounds. Wall in fact, illustrate at least 4 layers and outermost may include lipo-polysaccharides and proteins. In several cyanobacteria cell wall is enveloped by gelatinous mucilage. It may be thin and colorless as in planktonic forms. In subaerial forms sheath is thick, firm and colored yellow or orange brown and multilayered. Few aquatic forms like scytonema Petalonema possibly have multilayered and colored sheath.

2) Photosynthetic Lamellae:

Cyanobacteria contain no chloroplasts but only pigmented membranes that inhabit peripheral region of cells known as chromatoplasm. In this area photosynthetic lamellae or thylakoids are present. Lamellae are folded double membranes in which photosynthetic pigments-chlorophyll a, and many kinds of carotenoid are embedded. On surface of thylakoids are found rows of granules known as phycobilisomes which have phycocymanin, allophycocyanin and at times also phycoerything, features of cyanobacteria. It has been discovered that theylakoids also have enzymes needed for respiration.

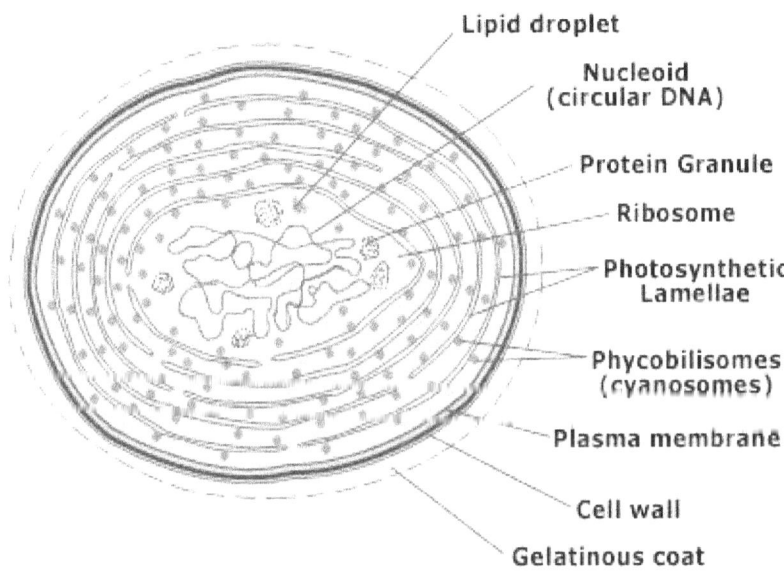

Fig.10. Diagrammatic representation of *Prokaryotic Algal Cell*

3) Granular Inclusions of Cytoplasm:

Ultrastructure of cyanobacteria sytoplasm illustrates many kinds of gramules. Between thylakoids glycogen is discovered in form of granules of various sizes. Protein granules known as cyanophycin granules composed of polymer of 2 aminoacids aspartic acid and arginine are for storage of nitrogen. Another kind of granule common in algae developing in waters rich in phosphate, is polyphosphate, storage form of phosphate. Few algae also have granules of polybatalydroxrate as big crystals. Another exclusive granule discovered in cyanobacteria are polyhedra crystalline bodies called as carboxysomes. They are composed of ribulose-biphosphate carcoxylase (Rubisco) enzyme is needed in photosynthetic fixation are of 70s kind unlike 80s type found in eukaryotes.

4) Gas Vesicles:

Several planktonic cyanobacteria such as Microvysis have in their cell lengthened, cylindrical vesicle individually or in bundles called as gas vesicles. They make cells float on surface of water. When, gas escapes they collapse turn out to be flat and cells sink to bottom. Wall of vesicle is composed of single layer of protein molecules and permeable to gases but not to water.

5) Nucleoplasm:

Central portion of cell generally referred as nucleoplasm has genetic material DNA, corresponding to nucleus of eukaryotes. It seems as the network of fibrils, and like that of bacteria it is long thread in shape of ring, usually referred to as circular chromosome. There may be numerous copies of it in cell. Histone proteins discovered in eukaryotic cells are not related with DNA of cyanobacteria.

6) Plasmids:

Similar in bacteria, DNA is also discovered in cells of cyanobacteria as small covalently related circular molecule called as plasmid that has genes which make organism resistant to antibiotics. Plasmids are not permanent characteristic of cells, they may be lost and regained further and they can also grow inside host cells.

Specialized Cells of Cyanobacteria:

These are thick cells discovered in filamentous cyanobacteria illustrate two other kinds of structures, heterocysts and akinetes. These are explained below:

1) *Heterocysts*:

These are thick walled cells discovered in filamentous cyanobacteria either in between vegetative cells (intercalary or at ends (terminal) of filament. Many significant function of heterocysts is fixation of atmospheric nitrogen as they have essential enzyme system, nitrogenase.

Structure of Heterocyst:

Unlike the vegetative cell, heterocyst contains thick wall with 3 layers that are structurally dissimilar. Inner most layer has certain glycoligids that make heterocyst resistant to oxygen, otherwise O2 lives in action of nitrogenase and avoids nitrogen fixation. Heterocysts are joined with adjacent cells through fine protoplasmis strands plasmodesmata at poles and also with large glossy granules - polar granules composed of cyanophycin. Heterocysts also have photosynthetic lamellae, but these are less dense which in vegetative cells. The lamellae contain chlorophyll a and carotenoids. However, phycocyanin is lost when a vegetative cell changes into a heterocyst. Therefore, mature heterocysts cannot fix carbon dioxides, so O2 is not liberated in light. Polyphosphate and glycogen granules, carboxysomes and gas vesicles are entirely absent in the cytoplasm of the hetercocyst.

2) *Akinetes:*

These are thick walled cells also known as spores, meant for perenation. All the vegetative cells of a filament or only a few cells like those adjacent to a heterocyst may develop into spores. Akinetes have thick walls and they are generally light brown, deep brown or black in colour. The contents of the cell are highly granular with glycogen but polyphosphate is lacking.

Akinetes can withstand prolonged desiccation and under suitable conditions geminate giving rise to new filaments.

Eukaryotic Algal Cell:

Eukaryotic algae include many divisions each comprising its own cell structure and other specific characters. Though, basic characteristics common to all groups are - mitochondria, existence of membrane bound nuleus, plastids, chromosomes, golgo bodies, and 80s type of ribosomes. Besides cell division by mitosis, several groups illustrate sexual method of reproduction having fusion of gametes and meiosis (reduction division). The following account provides significant characteristics of algal cells of different groups.

Cell Wall:

Algal cell wall is primarily composed of cellulose. Other extra compounds may be added to it during growth. In brown algae hemicelluloses, fucin, alginic acid, fucoidin are also present. In diatoms wall material is mainly silica.

Cells of Division Chrsophyta do not have proper cell wall. They are enclosed by scales of silica (like Mallomonas). In coccolithophorides elaborate scales have calcium carbonate (calcite). Cell wall of red algae has polysylphate esters carbohydrates additionally to cellulose and pectin. Calcium carbonate deposits are discovered over surface of algae belonging to various groups of several marine seaweed, called as calcarious algae, for instance, Neomeris, Udotia (green algae). Corallina (red alga), Padina (brown alga) and fresh water alga Chara.

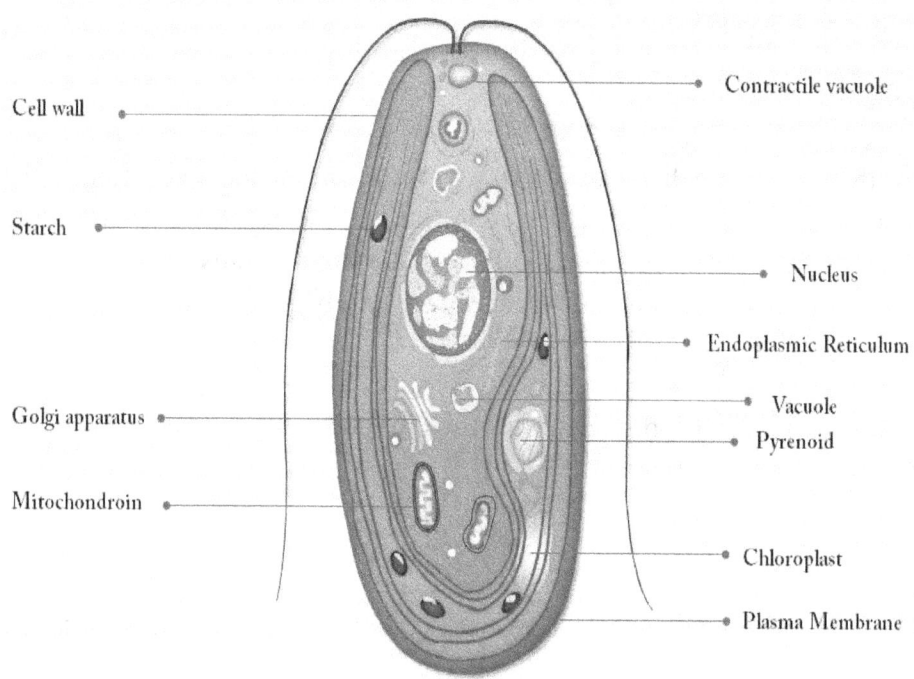

Fig.11. Diagrammatic representation of Eukaryotic *Algal Cell*

Plastids:

Every photosynthetic algae illustrate plastids - chloroplasts whose basic structure is related to chloroplasts of higher plants. Shape and location of chloroplasts in algae differs from species to species. When situated at centre of cell, they are known as axile, and when situated near periphery is known as parietal. Their number also following shapes of chloroplasts can be effortlessly recognized: spiral band (Spirogyra), girdle like (Ulothrix), cup like (Chlamydomonas), and stellate (star-shaped-Zygenema).

Ultrastrcuture:

Ultrastructure of algal chloroplast resembles that of higher plants, it is enveloped by double membrane. The number of thylakoid lamellae are spread into matrix -stroma. Lamellae are composed of lipoprotein complexes interspersed with molecules of chlorophylls and carotenoids. When phycobilins are present as in case of red algae, they are present in form of granules called as phycobilisomes, joined to membrane surface in linear rows. Stroma of chloroplast has many enzymes joined with photosynthetic carbon fixation. Arrangement of thylakoids in chloroplasts differs in different algae. They may be extremely closely stacked to form garna (sing, granum), as in green, brown algae and enlemophytes. In red algae they are extensively divided from each other.

One significant feature of chloroplast is presence of circular or like DNA. Plastids of Acetabularia, diatoms, Englena, chlamydomonas, members of Chrysophyceae, Xanthophyceae, phaeophyceae all have circular DNA. Chloroplasts give rise to novel plastids by easy division. Chloroplasts have ribosomes of 70s type that are present in cytoplasm. They also have complete machinery for protein synthesis. Ribosomes of 70s type are feature of prokaryotes such as cyanobacteria. Due to this fact it is thought that chloroplasts of eukaryotes were certainly cyanobacteria that became endosymbiontic during course of evolution.

Pyrenoids:

Plastids of several green algae have major proteinaceous granules known as pyrnoids around which starch is deposited. In several cases photosynthetic thylakoids are traversing matrix of pyrnoid or at least related with it. When chloroplasts separate, pyrenoids also split to give rise to novel pyrenoids. Several algae have only one nucleus per cell. Though, green algae like *Chladophord Camlerpa* and *Vaucheria* (Xantheophyceae) have more than one nucleus (multinucleate) similar to eukaryotic plant and animal nuclei, algal nucleus is enveloped by the distinct double membrane punctured by pores. During interphase (not separating, resting nuclets) uncoiled, fine chromatin threads are visible in nucleus.

Several algal nuclei have globular nucleoli, one or more in number, at times attached to specific region of chromosome nucleolus organizer. Nucleolus may disintegrate and vanish during cell division but reappear during interphase. It is now know that nucleolus is involved in synthesis of cytoplasmic ribosomes. Structure of nuleus in algal groups Euglenophyta and Dinophyta is fairly unique and is dissimilar from all other eukaryotes. During interpphase, nucleus inside its membrane illustrates not uncoiled chromatin fibres but very condensed chromosomes further, unlike in other organisms, they do not have histone proteins. Number of chromosomes present in every genus or species of ala has no relation with systematic position. Smallest number recorded is n=2 and highest may be 600 or more. Size of individual chromosomes is also variable. Large chromosomes are discovered in *Oedogonium, Cladophoraand Chara*.

Other Organelles of the Eukaryotic cells:

1) Mitochondria:

Number of mitochondria in algal cells differs from one as in few flagellates to several in other algae. Their size and shape also differs widely. Ultrastructure illustrates double membrane, inner one folded inwardly forming cristae protruding into lumen. Novel mitochondria arise by division of mitochondria present in parent cell, much like platids. It is thought that mitochondria from endosymbiotic bacteria adapted to intracellular existence inside ancestral host eukaryotic cells. Like chloroplasts they also have circular DNA, RNA, 70s ribosomes machinery for protein synthesis.

2) Golgi bodies:

These are also called as dictyosomes and are extensively found in algal cells. They are composed of 2-20 lamellae or membranes set in stack. They play the significant role in formation of cell wall material as in case of red algae. In several algae they are related with secretory function.

3) Flagella:

The motile members of algae, zoospores and gametes have one or more flagella which are organs of motion. Flagella are absent in members of Cyanophyceae and Rhodophyceae. Flagella are uniform or thread-like protoplasmic appendages.

All flagella are uniform in their internal structure. Each flagellum is made of two central tubules surrounded by nine peripheral tubules. The structure of 9 + 2 is surrounded by a membrane. In different algal groups flagella differ in number, size, location and types.

The flagella can be 2, 4 or indefinite in number. All flagella of one algae can be equal in size i.e., isokontic (Fig. 12 A, B) or unequal in size i.e., heterokontic. The flagella can be apical, sub-apical and lateral in position (Fig. 12 E).

The flagella can be of following types in algae:

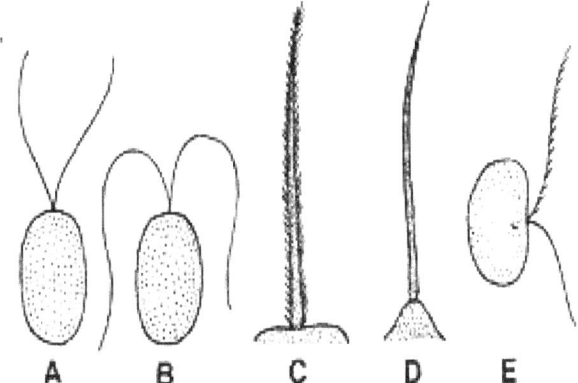

Fig. 12. (A-E) Types of flagella A, B Two equal acronematic flagella C-Pantonematic. D. Single acronematic flagellum E. Heterokontic lateral flagella.

(a) Whiplash or acronematic type:

These flagella do not have hair like appendages and their surface is smooth (Fig. 1 A, B, D, Fig. 2A, B).

(b) Tinsel or pleuronematic type:

These flagella have hair like appendages on their surface. These appendages are called mastigonemes or flimmers.

These can be of following types on the basis of arrangement of mastigonemes.

(i) Pantonematic:

The mastigonemes are arranged in two opposite rows or radially (Fig. 12C, Fig. 12 A, B).

(ii) Pantacronematic:

It is a pantonematic flagellum with terminal fibril (Fig. 12C).

(iii) Stichonematic:

Mastigonemes are present on one side of flagellum (Fig. ID).

4) Eyespots:

Motile cells of algae belonging to phaeophyta, Euglenophyta chlorophyta. Chrysophyta have orange - red colored eyespots. In few algae eyespot may form the part of chloroplast and it is situated at base of flagellum, but in Euglena it is fairly distinct and away from chloroplasts. Common kind of eyespot as located in green algae like chlamydomonas seems to have row of orange colored lipid gramules as part of thylakoids at anterior portion of chloroplast. Granules are discovered to have carotenoids, B-carotene being most important.

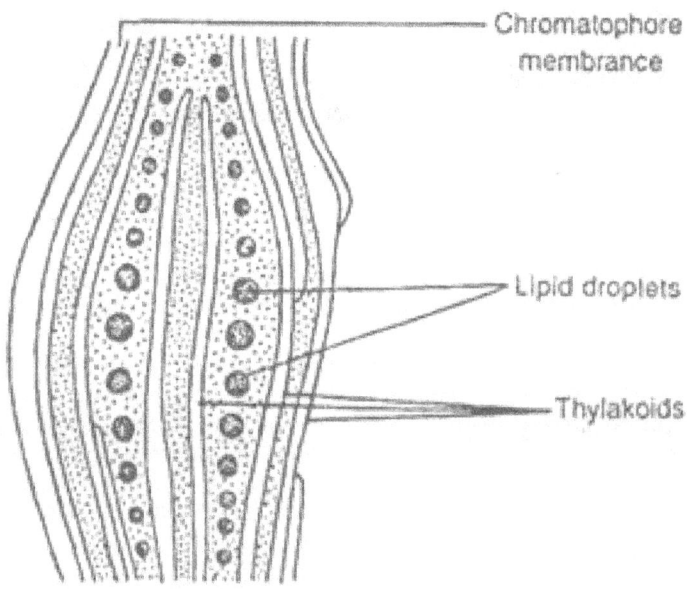

Fig.13. An eyespot of *Chlamydomonas*

The pigments that provide the actual colour of the thallus are of various types:
i. Chlorophylls:
There are five types of chlorophylls found in algae, Chi a, b, c, d, and e. Of them, chlorophyll a is present in all groups of algae. Chlorophyll b is found only in Chlorophyceae, Chlorophyll c in Phaeophyceae, Cryptophyceae, Bacillariophyceae and Chrysophyceae, Chlorophyll d in some red algae, and chlorophyll e in certain Xanthophyceae.
  ii. Caroteinoids:
(2). Carotenoids:
  (i) Carotenoids are fat soluble yellow pigments present in algae
  (ii) Carotenoids are found in close association with chlorophylls.

(iii) They protect chlorophylls from photo-damage (solarization)
(iv) Chemically carotenoids are tetraterpenoids
(v) Carotenoids are present in almost all algal groups.
(vi) Carotenoids with beta-ionone ring have Vitamin-A like activity
(vii) All carotenoids are strong antioxidants

Two types of carotenoids are found in algae.

(1). Carotenes:
(i) Carotenes are yellow coloured pigments
(ii) They are unsaturated fat soluble hydrocarbons
(iii) They do not contain oxygen
(iv) They absorb blue and green light and transmit yellow and red light.
(v) Examples: α-carotene, β-carotene, and Lycopene

(2). Xanthophylls:

Ø Xanthophylls are also called as carotenols
Ø They are oxygen derivatives of carotenes.
Ø Example: lutein and zeaxanthin (both are responsible for the yellow colour of egg yolk)

(3). Phycobilins:

(i) Phycobilins are water soluble pigments
(ii) Phycobilins are always bonded with some water soluble proteins called
(iii) They are blue and red in colour
(iv) They are present in Cyanophyceae and Red algae
(v) Phycobilins are usually found in organisms living in deep water for the efficient absorption of light
(vi) All phycobilins are strongly fluorescent. They emit orange or red light after fluorescence.

Two classes of phycobilins are present in algae.

(1). Phycocyanin:

(i) Phycocyanin are blue coloured pigments
(ii) They are blue green algae pigments
(iii) Phycocyanins are also present in red algae
(iv) They absorb green, yellow and red light and transmit blue colour.

(v) Phycocyanins are the principal pigment of Cyanophyceae.

(2). Phycoerythrin:

(i) Phycoerythrin are red coloured pigments
(ii) Phycoerythrin are red algae pigments
(iii) They absorb blue green, green and yellow light and transmit red light.
(iv) Phycoerythrin present abundantly in members of Rhodophyceae (red algae)

Distribution pattern of different pigments in different algal groups

(i) All photosynthetic algae contain chlorophylls, carotenoids and Xanthophylls.
(ii) Members of Cyanophyceae (Myxophyceae = blue green algae) and Rhodophyceae (red algae) contain large amount phycobilins, particularly phycocyanins.

Distributions of individual pigments in specific algal group are summarized in the following table:

Division	Common Name	Botanical Name	Major Pigment
Chlorophyta	Green algae	Chlorella sp.	Chlorophyll b
Charophyta	Charophytes	Spirogyra	Chlorophyll b
Euglenophyta	Euglenoids	Euglena gracilis	Chlorophyll b
Phaeophyta	Brown algae	Fucus vesiculosus	Chlorophyll $c_1 + c_2$, Fucoxanthin
Chrysophyta	Yellow-brown or golden brown algae	Dunaliella salina	Chlorophyll $c_1 + c_2$, Fucoxanthin
Pyrrhophyta	Dinoflagellates	Amphidinium carterae	Chlorophyll c_2, Peridinin
Cryptophyta	Cryptomonads	Cryptomonas sp.	Chlorophyll c_2, Phycobilins
Rhodophyta	Red algae	Porphyridium cruentum	Phycoerythrin, Phycocyanin
Cyanophyta	Blue-green algae	Spirulina platensis	Phycoerythrin, Phycocyanin

REPRODUCTION

There are three common methods of reproduction found in algae – (i) vegetative, (ii) asexual, and (iii) sexual. In addition to these methods, several perennating bodies also develop which face the adverse conditions.

1. Vegetative reproduction:

This may be of several types.

(i) By cell division:

The mother cells divide and the daughter cells are produced, which become new plants. This is exclusive type of reproduction in Pleurococcus, some desmids, diatoms, *Euglena*, etc.

(ii) Fragmentation:

The plant body breaks into several parts or fragments and each such fragment develops into an individual. This type of vegetative reproduction is commonly met within filamentous forms, e.g., *Ulothrix, Spirogyra*, etc. The fragmentation of colonies also takes place in several blue green algae, e.g., *Aphanocapsa, Aphanothece, Nostoc*, etc.

(iii) Hormogone formation:

When the trichomes break in small pieces of two or more cells, such pieces are called 'hormogones'. Each hormogone develops into a new plant, e.g., *Oscillatoria, Nostoc, etc.*

(iv) Hormospores or hormocysts:

They are thick-walled hormogones, and produced in somewhat drier conditions.

(v) By adventitious thalli:

Certain special structures of thalli are formed which help in vegetative reproduction. The well-known propagula of *Bryopsis*, *Sphacelaria* and *Nereocystis* are good examples.

(vi) By primary secondary protonema:

Such thread-like vegetative bodies develop in the case of *Chara*, which help in reproduction.

(vii) Tubers:

Usually these bodies are rounded and filled up with abundance of starch. Each body may give rise to a new plant, e.g., *Chara*.

(viii) Starch or amylum stars:

Such special star-shaped, starch filled bodies give rise to new plants frequently reported from *Chara*.

(ix) Bulbils:

Small bud-like structures. Usually develop on the rhizoids of Chara are called bulbils. Each such bulbil may develop into a new plant.

(x) Akinetes:

In most of the Chlorophyceae members, the Akinetes are developed. Usually the protoplast of each cell converts in a single akinete. Sometimes they are formed in chains. Each akinete may develop into a new plant, e.g., Oedogonium, Ulothrix, etc.

2. Asexual reproduction:

Usually the protoplast of a cell divides into several protoplasts and thereafter they escape from the mother and develop into new plants. (See Fig. 3.7).

(i) By zoospores:

The zoospores are formed from certain older cells of the filaments. The cytoplasm divides to form zoospores which are escaped from the mother cell. They are always formed in favourable conditions. The zoospores are always motile. They may be (i) biflagellate, (ii) tetraflagellate, (iii) stephanokontean type of zoospores, e.g., Oedogoniales and (iv) compound zoospores, e.g., Vaucheriaceae.

(ii) By aplanospores:

When motile phase of zoospores is eliminated, the bodies are called aplanospores. The aplanospores develop in unfavourable conditions. Each such spore is surrounded by a wall.

(iii) By hypnospores:

Actually they are very thick-walled aplanospores and develop only in adverse conditions, e.g., Pediastrum, Vaucheria.

(iv) Palmella stage:

Here the successive generations of divided cells are gelatinized and a thick mucilaginous envelope develops, e.g., *Chlamydomonas, Ulothrix*, etc.

(v) Autospores:

They are just like aplanospores except that they are smaller in size. They resemble in shape to mother cell except in size. Each autospore gives rise to a new plant. Such autospores are reported from many Chlorococcales.

(vi) Endospores:

In many blue green algae and Bacillariophyceae, the endospores are formed within the cells. On the approach of favourable conditions, each endospore develops in a new individual.

(vii) Auxospores:

In many members of Bacillariophyceae, such auxospores are produced. Each develops in a new plant.

(viii) Carpospores:

They are found in the carposporophytes of red algae (Rhodophyceae). Each such spore develops in a new individual.

(ix) Neutral spores:

These spores are not formed within the sporangia. They are found in Rhodophyceae.

(x) Monospores:

These spores develop within monosporangia. Each spore gives rise to a new plant, e.g., many members of Rhodophyceae (Bangia, Porphyra, Porphyridium, etc).

(xi) Paraspores:

Such spores are reported from many members of Rhodophyceae. Each spore develops into a new plant.

(xii) Statospores:

They are found in Xanthophyceae and Bacillariophyceae where they act as perennating bodies.

(xiii) Daughter colonies:

In many Volvocales and Chlorococcales, the daughter colonies are developed asexually, e.g., Volvox, Hydrodictyon, Pediastrum, etc.

(xiv) Gongrosira stage of Vaucheria:

In the aseptate filaments of Vaucheria, the protoplast divides into several parts, several hypnospores or cysts are produced and the whole filament looks like an algal form 'Gongrosira'.

(xv) Microspores:

They are produced in many Bacillariophyceae.

3. Sexual reproduction:

It is greatly advanced method of reproduction and not known in Myxophyceae (blue green algae). There are two main types, i.e., (i) isogamy and (ii) heterogamy.

(i) Isogamy:

The fusion of similar motile gametes is found in many species. Usually the gametes taking part in fusion come from two different individuals or filaments, sometimes these gametes come from two different cells of the same filament. Thousands of gametes come and aggregate in clumps.

(ii) Heterogamy:

The fusion of dissimilar gametes is called heterogamy. There are variations of it.

(a) Anisogamy:

The motile gametes taking part in fusion may either differ in size (morphological anisogamy) or physiological behaviour (physiological anisogamy).

(b) Oogamy:

In this case, the male antherozoid fuses with the female egg. This fusion may be of primitive type as found in Cylindrocapsa, or advanced type as in Oedogonium, Vaucheria, Chara, Polysiphonia, etc.

(iii) Aplanogamy or conjugation:

It implies the fusion of two non-flagellate amoeboid gametes (aplanogametes). They are morphologically similar but physiologically dissimilar, e.g., order Conjugales.

In fresh water algae, the sexual reproduction is best means of perennation because it is followed by the formation of thick-walled zygote or oospore.

Conditions for sexual reproduction:

(a) The sexual reproduction takes place after considerable accumulation of food material and the climax of vegetative activity is over.

(b) The bright light is the major factor for the production of the gametes.

(c) A suitable pH value is required.

(d) The optimum temperature is necessary.

Parthenogenesis:

The female gametes convert into zygotes without fusion. The resultants are called azygospores or parthenospores and the phenomenon 'parthenogenesis', e.g., Spirogyra, Oedogonium and many others.

Autogamy:

In this phenomenon, the fusion of the daughter protoplasts or of the divided nuclei of a cell without liberation takes place. This process is known in many diatoms and colourless dinoflagellates.

TYPES OF LIFE CYCLE

FE Fritsch (1935) has recognized five main types of life cycles in algae, such as –

1. **Haplontic:-** In this type the main vegetative body is haploid (n), i.e., gametophytic in nature and bears haploid gametes. The gemetes unite through syngamy and results in the formation of diploid zygote (2n). Next the zygote undergoes meiosis immediately after a period of rest and forms haploid spores. These spores ultimately germinate into haploid gametophytic plant. Example – *Oedogonium, Spirogyra, Ulothrix, Chlamydomonas*, etc.

2. **Diplontic:-** In this type the plant body is diploid (2n), i.e., the sporophytic in nature and bears diploid gametangia. Meiosis takes place at the time of formation of gametes (n). Fusion of these haploid take place very soon which result in the formation of diploid zygote, which develops into diploid sporophytic generation again. Example – *Fucus, Sargassum, Diatoms*, etc.

3. **Diplohaplontic:-** In this type of life cycle two distinct individuals are present – one haploid (n) i.e., the *gametophyte* and the other diploid (2n) i.e., the *sporophyte*. In this type both the generations alternate with each other showing distinct alternation of generations among plants.

Isomorphic Diplohaplontic life cycle – Here both the diploid and haploid individuals (vegetative plants) are morphologically similar i.e., isomorphic. They come alternately in the life-cycle and that type of life cycle is called, *isomorphic diplohaplontic life cycle*. The gemetophytic plant (n) bears haploid sex organs from which haploid gametes are formed. Gametes fuse to form diploid zygote (2n) which directly develops into diploid sporophytic plant (2n). The sporophytic plant bears sporangia from which haploid spores are developed by meiotic division. The haploid spores on germination produce haploid gametophytic plant body again. Example – *Cutleria*.

Heteromorphic Diplohaplontic life cycle – Here both the diploid and haploid individuals (vegetative plants) are morphologically dissimilar i.e., heteromorphic. They come alternately in the life-cycle and that type of life cycle is called, heteromorphic diplohaplontic life cycle. The sporophytic plant is large and it bears sporangia, within which haploid spores (n) are developed by meiotic division. Such spores on germination produce haploid gametophytic plant (n). Haploid gametophytic plant bears gametangia from which haploid gametes are produced. Gametes on fertilization form diploid zygote

(2n), which directly germinates to produce diploid sporophytic plant body. Example – *Laminaria*.

4. Haplobiontic:- In this type of life cycle two haploid phases are present. One is represented by gametophytic plant including gametangia and gametes (n) and the other is represented by gonioblast filament, carposporangia (n), carpospores (n) and Chantrantia stage. Here the diploid (2n) stage is restricted in the zygote only. This type is found in Bactrachospermum, Nemalion, etc.

5. Haplodiplobiontic:- This type of life cycle is found in Polysiphonia. Here the haploid (n) phase is represented by male and female gametophytic plants, sex organs and gametes (n). The diploid phases are two – first diploid phase (n) is represented by zygote, gonioblast filaments, carpogonia and carpospores (2n). All structures of the first diploid phase are together known as carposporophyte (2n) which depends on gametophyte. The second diploid phase is represented by tetrasporophyte plant bearing tetrasporangia (2n). Due to the presence of one haploid and two diploid phases, this type is known as haplodiplobiontic. Here the life cycle is triphase

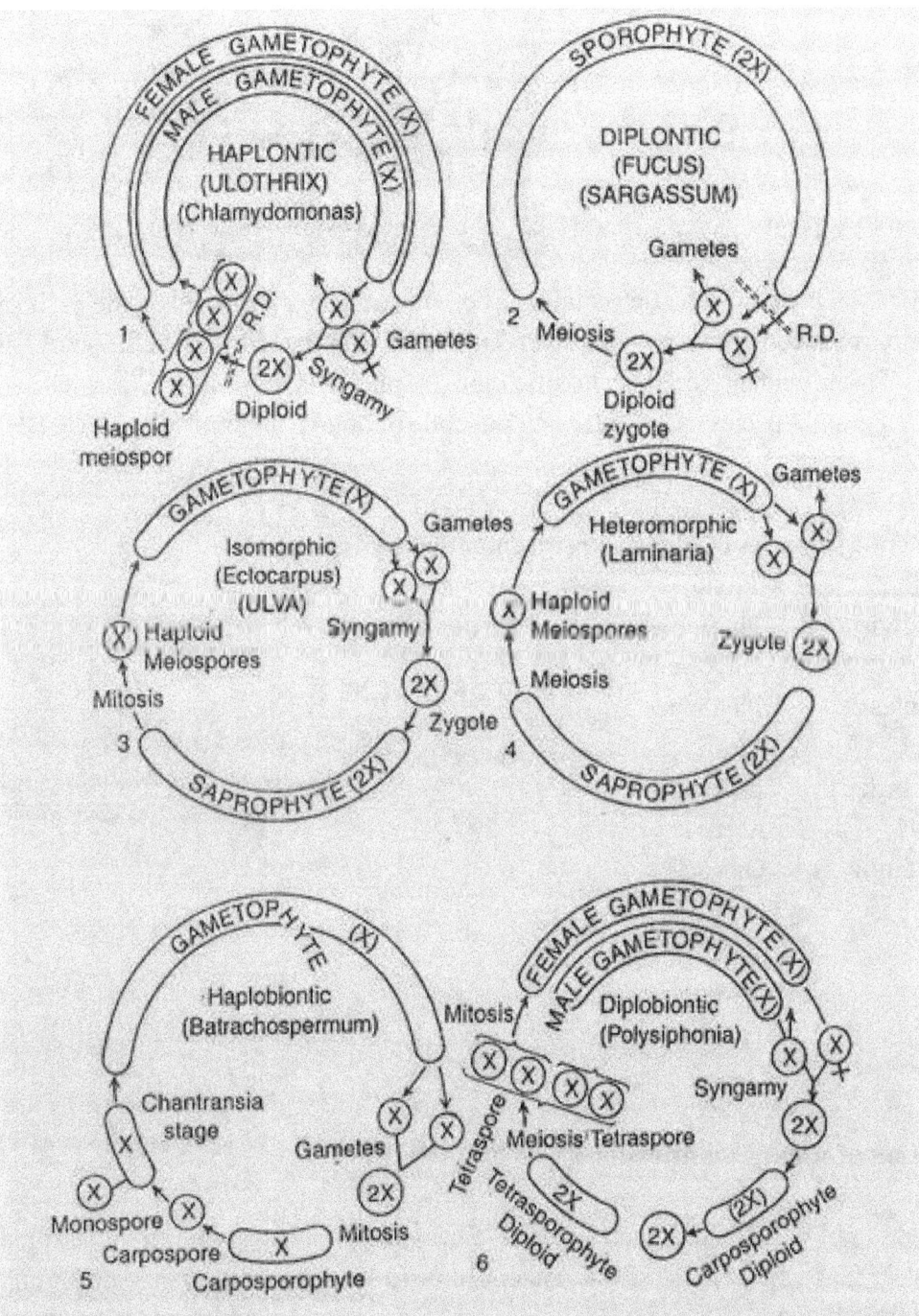

Fig.14. Life-cycle patterns in algae. 1. haplontic type; 2. diplontic type; 3. isomorphic type; 4. heteromorphic type; 5. haplobiontic type; 6. diplobiontic type.

****

2.

CLASSIFICATION OF ALGAE

Classification is the systematic grouping of organisms into categories on the basis of relationships between them, where the relationship can be either evolutionary or structural. The hierarchy for the classification of plants is Division, Class, Order, Family, Genus and Species as per International code of botanical nomenclature (ICBN) (Sharma 2011).

Different categories of algae as recommended by ICBN are

Division:	Phyta
Sub – division:	Phytina
Class:	Phyceae
Sub – class:	Phycidae
Order:	Ales
Sub – order:	Inales
Family:	Aceae
Sub – family:	Oideae
Tribe:	Ease
Genus:	
Species:	
Variety:	
Form:	

Evolution of Algal Classification

The history of classification dates back to Carolous Linnaeous, who first classified plants into 25 classes based on "sexual system" considering the number of stamens and carpels in their flowers. Out of his 25 classes, in "Cryptogamia" which contains plants with "concealed reproductive organs" Linnaeus, proposed 14 algal genera of which only 4, *Conferva, Ulva, Fucus* and *Chara* are now considered as algae (Dixon1973). W. H. Harvey is considered as one of the first algologist who proposed the first descriptive algal classifi cation. Since W. H. Harvey several classifications have been proposed based on a variety of characters including morphological, physiological, biochemical and more recently the molecular characters have also been considered. The main characters which are being widely used for algal classification are: I. Photosynthetic Pigments: Chlorophylls, Carotenoids (Carotenes and Xanthophylls), Phycobilins II. Biochemical nature of food reserve. III. Cell wall composition IV. Flagella the major classification proposed by different algologists for algae are:

1. Classification Proposed by by Fritsch(1935)

Algae possess diverse characters in their pigments, nature of reserve food, nature of cilia etc. According to these morphological and physiological differences they are classified by many people. Fritsch (1935) classified the whole of the algae into eleven classes on the basis of type of pigments, nature of reserve food material, mode of reproduction etc. They are Chlorophyceae, Xanthophyceae, Chrysophyceae, Bacillariophyceae, Cryptophyceae, Dinophyceae, Chloromonodineae, Euglinineae, Phaeophyceae, Rhodophyceae and Myxophyceae (Cyanophyceae). The classification is published in his book titled "The Structure and Reproduction of Algae".

1. Class: Chlorophyceae (Green Algae)

   - Occurrence: Most forms are fresh water and a few are marine.
   - Pigments: Chief pigments are chlorophyll a an b and carotenoids (yellow pigments)
   - Reserve food: Starch
   - Structure: Unicellular motile to heterotrichous filaments. Cell wall consists of Cellulose. Pyrenoidsare commonly surrounded by starch sheath. Motile cells have equal flagella.
   - Reproduction: Sexual reproduction ranges from isogamous to advanced oogamous type. e.g., Chlamydomonas, Volvox, Chlorella, Scenedesmus ,Pediastrum.

2. Class: Xanthophyceae (Yellow green algae)
   - Occurrence: Most forms are fresh water but a few are marine.
   - Pigments: Yellow xanthophyll is found abundantly.
   - Reserve food: oil
   - Structure: Unicellular motile to simple filamentous. Cell wall rich in pectic compounds and composed of two equal pieces overlapping at their edges. Motile cells have two very unequal flagella.
   - Pyrenoids absent.
   - Reproduction: Sexual reproduction is rare and always isogamous. e.g.: *Vaucheria*

3. Class: Chrysophyceae

- *Occurrence: Most forms occur in cold fresh water but a few are marine.*
- *Pigments: Chromatophores are brown or orange colored. Phycochrysin serves as chief accessory pigments.*
- *Reserve food: Fat and leucosin.*
- *Structure: Plants are unicellular motile to branched filamentous. Flagella are unequal attached at front end. Cells commonly contain one or two parietal chrmoatophores.*
- Reproduction: Sexual reproduction seldom occurs but is of isogamous type.

4. Class: Bacillariophyceae (Diatoms)

- Occurrence: In all kind of fresh water, sea, soil and terrestrial habitats.
- Pigments: Chromatophores are yellow or golden brown.
- Nature of accessory pigments is not very definite.
- Reserve food: Fat and volutin.
- Structure: All the members are unicellular or colonial. Cell wall is partly composed of silica and partly of pectic substances. It consists of two halves and each has two or more pieces. Cell wall is richly ornamental
- Reproduction: Forms are diploid. Sexual reproduction is special type, occurs by fusion of protoplasts of the ordinary individuals. Example: *Pinnularia*

5. Class: Cryptophyceae

- Occurrence: Both in marine and fresh water
- Pigments: Chromatophores show diverse pigmentation. It may be some shades of brown. Chromatophores are usually parietal.
- Reserve food: Solid carbohydrates or in some cases starch. Structure: Represented by motile cells and most advanced forms are coccoid, flagella are slightly unequal.
- Reproduction: Isogamous in the reported cases. Example: Chroomonas

6. Class: Dinophyceae

- Occurrence: Plants occur widely as sea water planktons. A few may be fresh water forms.
- Pigments: Starch and oil
- Reserve food: Chromaophores are dark yellow, brown, etc., and contain a number of special pigments. Structure: plants are unicellular motile to branched filamentous.

- Reproduction: Sexual reproduction is of isogamous type. it is rare and not very definite. Example: *Dinoflagellate, Ceratium*

7. **Class: Chloromonadineae**

    - Occurrence: All plants are fresh water forms.
    - Pigments: Chromatophores are bright green in colour and contain an excess of xanthophyll.
    - Reserve food: Oil
    - Structure: The plants are motile, flagellate with two almost equal flagella.

8. **Class: Euglenineae**

    - Occurrence: Only fresh water forms are known Pigments: Chromatophores are pure green. Each cell has several chromatophores.
    - Reserve food: Polysaccharide and Paramylon
    - Structure: Motile flagellates, flagella may be one or two arising from the base of canal like invagination at the front end.
    - Complex vacuolar system and a large and prominent nucleus.
    - Reproduction: Sexual reproduction is not substantially known. It is isogamous type. Example: *Euglena*

9. **Class: Phaeophyceae (Brown algae)**

    - Occurrence: Mostly marine
    - Pigments: chl a, c, carotenes, xanthophylls, not chl b
    - Reserve food: Mannitol as well as laminarin and fats Structure: The plants may be simple filamentous to bulky parenchymatous forms. Several plants attain giant size, external and internal differentiation.
    - Reproduction: Sexual reproduction ranges isogamous to oogamous. Motile gametes have two laterally attached flagella. Varied types of alternation of generation. Example: *Ectocarpus, Sargassum*

10. **Class: Rhodophyceae (Red algae)**

    - Occurrence: Few forms are fresh water and others are marine. Pigments: Chromatophores are res blue containing pigments like red phycoerythrin and blue phycocyanin, Chl-a,d, carotenes.
    - Reserve food: Floridean starch

> Structure: Simple filamentous to attaining considerable complexity of structure. Motile structures are not known.

> Reproduction: Sexual reproduction is advanced oogamous type. The male organ produces non motile gametes and the female organ has a long receptive neck. After sexual reproduction special spores (carpospores) are produced Example: *Batrachospermum, Polysiphonia*

11. Class: Myxophyceae (Cyanophyceae or Blue green algae)

> Occurrence: Found in sea and fresh water, Pigments: Chlorophyll, carotenes, xanthophylls, and phycocyain and phycoerythrin. The ratio of last two pigments exhibits colour variation, commonly blue green.

> Reserve food: Sugars and Glycogen

> Structure: Simple type of cell to filamentous, some of the filamentous forms show false or true branching, very rudimentary nucleus, no proper chromatophores, the photosynthetic pigments being diffused throughout the peripheral position. No motile stages.

> Reproduction: There is no sexual reproduction. Example: Oscillatoria, Nostoc

2. Classification Proposed by Smith's

In 1955 C. M. Smith has classified algae into seven divisions, each of which contains one or more classes.

A brief account of this classification is given below:

i. Cyanophyta:

Cyanophyta or blue green algae-inhabitants of moist soil and rocks, fresh water or saline water; unicellular or colonial; plastids not well-defined; presence of nucleoplasm in the centre of the cell and chromoplasm in the periphery; nucleus is primitive; phycocyanin is the chief pigment but sometimes phycoerythrin may be present; sexual reproduction and flagellated cells are always absent; reproduction is vegetative or asexual by resting spores; glycogen is the reserve food.

Examples:

Oscillatoria, Nostoc, Gleocapsa, etc.

ii. Chlorophyta or Green Algae:

They are found in fresh water or saline water; unicellular, multicellular or colonial; cells flagellated or non-flagellated; reproduction is asexual or sexual; presence of chlorophyll, carotin and xanthophyll like higher plants; reserve food is starch.

Examples:

Spirogyra, Oedogonium, Ulothrix, Vaucheria, Chara, etc.

iii. Euglenophyta:

Found in fresh water or stagnant water or on moist soil; cells flagellated or naked; coloured plastids prominent; cell divisions longitudinal; reserve food is paramyleum (a kind of carbohydrate).

Example:

Euglena

iv. Pyrophyta:

Chiefly marine; unicellular; presence of two unequal flagella; yellowish green or yellowish brown pigments present; sexual reproduction is rare; reserve food is starch or oil.

v. Chrysophyta:

Chiefly consists of Diatoms; terrestrial, fresh water or saline water; flagellated or non-flagellated; unicellular or multicellular, colonial or filamentous; cell wall made of pectin with deposition of silica particles on the surface; of the pigments carotin and Xanthophyll are present in abundance; special method of sexual reproduction; oil and leucosin (a complex carbohydrate) are reserve food.

The division consists of three classes, of which Bacillariophyceae is the main plants consisting of this class are known as Diatoms or golden brown algae; the yellow pigment diatomin is always present with chlorophyll.

vi. Phaeophyta or Brown Algae:

Chiefly marine; plant body large and complicated by the presence of pigment like fucoxanthin and phycophene; both sexual and asexual reproduction take place; sexual reproduction is isogamous and oogamous carbohydrate and oil are reserve food.

Examples:

Ectocarpus, Fucus, etc.

vii. Rhodophyta or Red Algae:

Chiefly marine; plant body is very complex; main pigment is phycoerythrin; sexual reproduction is oogamous; male gametes known as spermatia, are non-flagellated; reserve food is Florideam starch.

Examples:

Batrachospermum, Polysiphonia, etc.

3. Classification Proposed by G. F. Papenfuss (1955)

G. E. Papenfuss proposed algal classification based on phylogenetic relationship. He recognized 7 division and 12 classes as described in the following table. The blue green algae were kept together in a separate phylum Schizophya along with bacteria.

S. No.	Divisions	Class
1.	Chlorophycophyta	Chlorophyceae
2.	Charophycophyta	Charophyceae
3.	Euglenophycophyta	Euglenophyceae
4.	Chrysophycophyta	(a) Xanthophyceae
		(b) Chrysophyceae
		(c) Bacillariophyceae
5.	Pyrrophycophyta	(a) Dinophyceae
		(b) Cryptophyceae
		(c) Chloromonadophyceae
6.	Phaeophycophyta	Phaeophyceae
7.	Rhodophycophyta	Rhodophyceae

4. Classification Proposed by V. J. Chapman (1962)

Chapman, considered pigments, morphological characters, biochemical differences and also phylogenetic relationships with in different algae for its classification. He divided algae into four different divisions and further into classes as given in table:

S. No.	Divisions	Class
1.	Euphycophyta	(a) Chlorophyceae
		(b) Phaeophyceae
		(c) Rhodophyceae
2.	Myxophycophyta	(a) Myxophyceae
3.	Chrysophycophyta	(a) Chrysophyceae
		(b) Xanthophyceae
		(c) Bacillariophyceae
4.	Pyrrophycophyta	(a) Cryptophyceae
		(b) Dinophyceae

5. Classification Proposed by Christensen (1964)

Christensen (1964), divided algae on the basis of prokaryotic and eukaryotic features of cell into Prokaryota and Eukaryota.

			Division	
Algae	Prokaryota		(a) Cyanophyta	
	Eukaryota	(i) Aconta (Motile stages absent)	(a) Rhodophyta	
		(ii) Contophora (Motile stages present)	(a) Chlorophyta	Chlorophyll a and b predominates
			(b) Charophyta	
			(c) Euglenophyta	
			(d) Xanthophyta	Carotenoids predominates, chlorophyll a present, chlorophyll b absent
			(e) Chrysophyta	
			(f) Bacillariophyta	
			(g) Pyrrophyta	
			(h) Cryptophyta	
			(i) Phaeophyta	

6. Classification Proposed by F. E. Round (1973)

F. E. round divided algae again on the basis of presence or absence of true nucleus, membrane bound organelles and phylogenetic relationship etc. He classified algae into 12 phyla and further into classes. The classification proposed by Round is

S. No.	Group	Phylum
1.	Prokaryota	Cyanophyta
2.	Eukaryota	Euglenophyta
		Chlorophyta
		Charophyta
		Prasinophyta
		Xanthophyta
		Haptophyta
		Dinophyta
		Bacillariophyta
		Chrysophyta
		Phaeophyta
		Rhodophyta

7. Classification Proposed by V. J. Chapman and D. J. Chapman (1973)

They classifi ed algae into Prokaryota and Eukaryota which were further divided into divisions and classes as:

S. No.		Division	Class
1.	Prokaryota	I. Cyanophyta	(a) Cyanophyceae
2.	Eukaryota	II. Rhodophyta	(a) Rhodophyceae
		III. Chlorophyta	(a) Chlorophyceae
			(b) Prasinophyceae
			(c) Charophyceae
		IV. Euglenophyta	(a) Euglenophyceae
		V. Chloromonadophyta	(a) Chloromonadophyceae
		VI. Xanthophyta	(a) Xanthophyceae
		VII. Bacillariophyta	(a) Bacillariophyceae
		VIII. Chrysophyta	(a) Chrysophyceae
			(b) Haptophyceae
		IX. Phaeophyta	(a) Phaeophyceae
		X. Pyrrophyta	(a) Dinophyceae
			(b) Desmophyceae
		XI. Cryptophyta	Cryptophyceae

8. Classification Proposed by H. C. Bold and M. J. Wynne (1978)

Bold and Wynne followed the classification proposed by Papenfuss and they accepted the use of "phyco" before "phyta" in algal divisions. They divided algae into nine divisions as:

S. No.	Divisions
1.	Cyanochloronta
2.	Chlorophycophyta
3.	Charophyta
4.	Euglenophycophyta
5.	Phaeophycophyta
6.	Chrysophycophyta
7.	Pyrrophycophyta
8.	Cryptophycophyta
9.	Rhodophycophyta

9. Classification Proposed by R. E. Lee (2008)

Lee classified algae in two groups Prokaryota and Eukaryota which were further divided into divisions. Prokaryota has just one division Cyanophyta, whereas Eukaryota were further divided on the basis of nature of chloroplast membrane.

S. No.	Groups		Divisions	Class
	Prokaryota	I	(i) Cyanophyta	Cyanophyceae
	Eukaryota	II. Chloroplast surrounded by the two membranes of the chloroplast envelope	(b) Glaucophyta	
			(c) Rhodophyta	
			(d) Chlorophyta	
		III. Chloroplast surrounded by one membrane of chloroplast endoplasmic reticulum	(a) Euglenophyta	
			(b) (Euglenoids)	
			(c) Dinophyta	
			(d) (Dinoflagellates)	
		IV. Chloroplast surrounded by two membranes of chloroplast endoplasmic reticulum envelope	(a) Cryptophyta (cryptophytes)	
			(b) Prymnesiophyta (haptophytes)	Prymnesiophyceae
			(c) Heterokontophyta (heterokonts)	Chrysophyceae
				Synurophyceae
				Dictyophyceae
				Pelagophyceae
				Bacillariophyceae
				Raphidophyceae
				Xanthophyceae
				Eustigmatophyceae
				Phaeophyceae

****

3.

Myxophyceae (Cyanophyceae): General Character and Classification

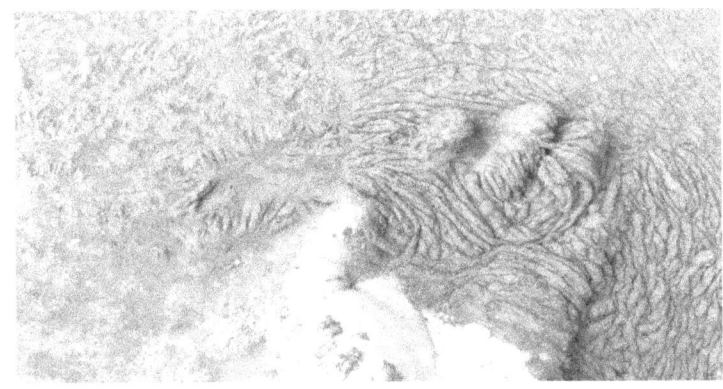

The term Cyanophyceae acknowledges that these prokaryotic algae are more closely related to the prokaryotic bacteria than to eukaryotic algae. For the last quarter century, cyanobacteria were thought to have evolved about 3.5 billion years ago. These reports were based on interpretation of microfossils, difficult at best with such small organisms. It now appears that these investigators selected specimens that fit the assumptions of the authors, with most phycologists now rejecting their claims. Based on other reports, the actual time of evolution of Cyanophyceae is thought to be closer to 2.7 billion years ago (Buick, 1992; Brasier et al., 2002, Dalton, 2002). Cyanobacteria have chlorophyll a (some also have chlorophyll b or d), phycobiliproteins, glycogen as a storage product, and cell walls containing amino sugars and amino acids. At one time, the occurrence of chlorophyll b in cyanobacteria was used as a criterion to place the organisms in a separate group, the Prochloro phyta. Modern nucleic-acid sequencing, however, has shown that chlorophyll b evolved a number of times within the cyanobacteria and the term Prochlorophyta has been discarded (Palenik and Haselkorn, 1992; Urback et al., 1992).

It is a primitive group of algae, consists of 150 genera and about 2,500 species. In India, the division is represented by 98 genera and about 833 species. Members of the class Myxophyceae (Cyanophyceae) are commonly known as blue green algae. The name blue green algae is given because of the presence of a dominant pigment c-phycocyanin, the blue green pigment.

In addition, other pigments like chlorophyll a (green), c-phycoerythrin (red), β-carotene and different xanthophylls are also present. The members of this class are the simplest living autotrophic prokaryotes.

General Character

1. The individual cells are prokaryotic in nature. The nucleus is incipient type and they lack membrane bound organelles.

2. Both vegetative and reproductive cells are non-flagellate.

3. Cell wall is made up of microfibrils and is differentiated into four (4) layers. The cell wall composed of mucopeptide, along with carbohydrates, amino acids and fatty acids.

4. Locomotion is generally absent, but when occurs, it is of gliding or jerky type.

5. The principal pigments are chlorophylls a (green), c-phycocyanin (blue) and c-phyco- erythrin (red). In addition, other pigments like β-carotene and different xanthophylls like myxoxanthin and myxoxanthophyll are also present.

6. Membrane bound chromatophore are absent. Pigments are found embedded in thylakoids.

7. The reserve foods are cyanophycean starch and cyanophycean granules (protein).

8. Many filamentous members possess specialized cells of disputed function (supposed to be the centre of N_2 fixation) known as heterocysts.

9. Reproduction takes place by vegetative and asexual methods. Vegetative reproduction takes place by cell division, fragmentation etc. Asexual reproduction takes place by endospores, exospores, akinetes, nannospores etc.

10. Sexual reproduction is completely absent. Genetic recombination is reported in 2 cases.

Occurrence of Cyanophyceae:

Members of Cyanophyceae are available in different habitats. Most of the species are fresh water (e.g., Oscillatoria, Rivularia), a few are marine (e.g., Trichodesmium, Darmocarpa), and some species of Oscillatoria and Nostoc are grown on terrestrial habitat.

Species of some members like Anabaena grow as endophytes in thallus of Anthoceros (Bryophyta) and in leaves of Azolla (Pteridophyta) and Nostoc in the root of Cycas (Gymnosperm).

Species of Nostoc, Scytonema, Gloeocapsa, and Chroococcus grow symbiotically with different fungi and form lichen. Some members like Nostoc, Anabaena etc. can fix atmospheric nitrogen and increase soil fertility.

Cell Structure

Cell Structure The cyanobacteria have prokaryotic organization. They lack membrane bound organelles. The internal membranes which separate DNA from cytoplasm, photosynthetic and respiratory organelles are absent (Fig. 3).

1. Sheath Outside the cell wall is present mucilaginous sheath which consist of three layers of microfibrils reticulately arranged within an amorphous matrix. The sheath retains absorbed water which is useful during the period of desiccation and for sloughing off parasite.

2. Cell Wall Inside the sheath is present a double layered cell wall, a very rigid structure that gives shape to the cell. The inner layer is made of mucopeptide and muramic acid. Electron microscopy studies showed that the wall consist of four layers. Each layer is 10 nm thick. The second (L II) layer has mainly peptidoglycan and first (L-I) lie next to plasmalemma. In structure and function the cell wall is similar to the gram negative bacteria. The cell wall in both is composed of mucopeptide together with carbohydrates, amino acids and fatty acids.

3. Plasma Membrane Beneath the cell wall is present plasmalemma. This membrane invaginates inside the cells, which are considered sites for biochemical reactions and functions normally associated with mitochondria, endoplasmic reticulum and golgi bodies in Eukaryotic cells.

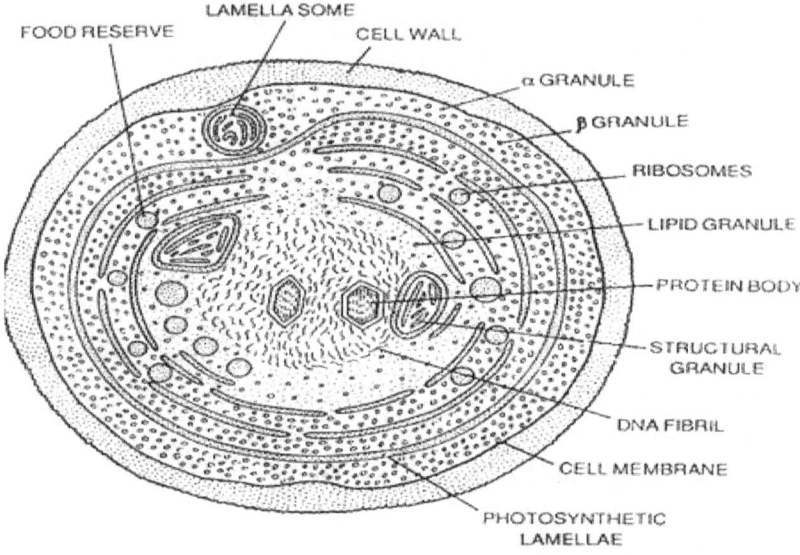

5. Prokaryotic cellular organization in Myxophyceae. Line diagram of an electron micrograph of typical Myxophycean cell.

Fig.1.

4. Photosynthetic Apparatus Membrane bound plastids are absent, instead pigments are found embedded within lamellae composed of two membranes joined at the ends. The structures are known as thylakoids. They appear as elongated, fl attened sacs consisting of two unit membranes, each about 75 A^0 thick. Adjacent thylakoids are separated from each other by a space of 50 nm, occupied by contiguous rows of discoidal phycobilisomes that transfer light energy to phytosystem II reaction centers like chlorophyll b, c, and d. They are able to dissolve in water and balanced in solution and poor light sensitive. The basic subunit of a phycobilisome consists of apoproteins α and β, each of which is attached to a chromophore (Anderson and Toole 1998; Samsonoff and MacColl 2001). In the core of the phycobilisome α and β are attached to allophycocyanins, which are adjacent to chlorophyll in the energy Cyanobacteria 60 transmission way. In the outer rods, α and β are attached to phycocyanin. The α and β molecules are assembled into hexamers (α1 and β1) cylindrical in shape and are joined together by linker polypeptides. The hexamers that make up the core of the phycobilisomes are assembled in pairs, wit-h the hexamers of the rods radiating from the core. The linker polypeptides are basic whereas the hexamers are acidic; this proposes that electrostatic interactions are important in assembling phycobiliproteins. There are high-molecular- weight polypeptides that anchor the phycobilisomes to the area of the thylakoid membrane that contains the reaction centre and associated chlorophylls (Fig. 1). Structurally phycobilin, are chromoproteins in which the prosthetic group (nonprotein part of the moleculeor chromophores is a tetrapyrole bile pigment) known as phycobilins and is tightly bound to the apoprotein (protein part of the molecule) moiety by covalent linkage and its diffi cult to separate the pigments from apoprotein. The cyanophyceae have four phycobiliproteins: c-phycocyanin (λmax 620 nm) and allophycocyanin (λmax 650 nm), c-phycoerythrin (λmax at 565 nm), and phycoerythrocyanin (λmax at 568 nm). All cyanobacteria contain the fi rst two, whereas c-phycoerythrin and phycoerythrocyanin occur in few species.

Intra Cytoplasmic Inclusions Inside the plasma membrane various kinds of inclusion have been found in cyanobacteria like.

I. Cyanophycin granules – Are enormous and frequently found near the crosswalls of fi lamentous forms and are composed of a polymer (arginine and asparagine) which utilize as nitrogen replacement.

II. II. Polyhedral bodies (now called carboxysomes) – Contain important enzymes, ribulose diphosphate carboxylase.

III. III. Polyphosphate bodies (volutin) – Sources of phosphate for the production of nucleic acids, phospho lipids and ATP.

IV. IV. Polyglucoside bodies – Similar to α granule which are known to store carbon and energy in the form of carbohydrate. They are known to function in blasting the cell and play a role in buoyancy regulation.

V. V. Poly - beta - hydroxybutyric acid – is biopolymer found exclusively in prokaryotes in which it act as a storage and is also used as proplastic e.g., Spirulina and Nostoc muscorum

5. Gas Vacuoles Cyanobacterial gas vacuoles are not tonoplast bound, instead they are made of protein cylinders which are hollow packets. These vacuoles contain metabolic gases and function as gas vesicles that help cells move upward, towards the light, where photosynthesis can occur more readily. When many vesicles are present, cyanobacterial cells tend to fl oat. If photosynthesis increases cell contain higher concentration of sugars that increases the turgidity and collapsing of the gas vesicle. Collapse of gas vesicle causes cyanobacteria to sink. In deeper water respiration is performed by cellular carbohydrates allowing gas vesicle to re-form as a result buoyancy cycle continues and this can be interrupted by cyanobacterial blooms at the surface (Fig. 1). These thick blooms may be unable to get suffi cient resources to produces sugars, which allow gas vesicles to remain intact.

6. Nucleoplasm Nucleolus and nuclear envelope are absent (no true nucleus), instead the DNA present in the cytosol (cytosol = liquid component of the cytoplasm). Since DNA is not associated with protein material (histone or protamines), organized chromosomes are not found as in eukaryotic cells. RNA is present in addition to DNA.

Heterocyst

A specialized cell found in nitrogen-fixing cyanobacteria. Heterocysts are enlarged cells with thick cell walls and they lack chlorophyll, giving them a colourless appearance. They are the site of nitrogen fixation, for which they produce the enzyme nitrogenase. The lack of photosynthetic activity (which would produce oxygen), together with the thick cell wall, are thought to maintain the anoxic conditions in heterocysts that are essential for the activity of nitrogenase. Heterocysts are connected by plasmodesmata to surrounding cells, on which they depend for nutrients.

Structure

Heterocysts are thick walled, pale yellow and barrel-shaped structures. The wall is differentiated into an outer fibrous, a middle homogeneous and an inner thin lamellar layer. It is characterised by the presence of pores either at one or both the poles. The wall layers become thicker in the polar region, and the pores are plugged with a refractive material, called polar granules (nodules) (Fig. 2). Heterocysts contain dense and homogeneous cytoplasm. They contain mostly carotenoids, hence yellowish in colour. In the absence of phycobilins (which are the principal light absorbing pigments in oxygen-evolving photosynthetic reactions) and chlorophyll a, there is no CO_2 fixation or O_2 production in heterocysts.

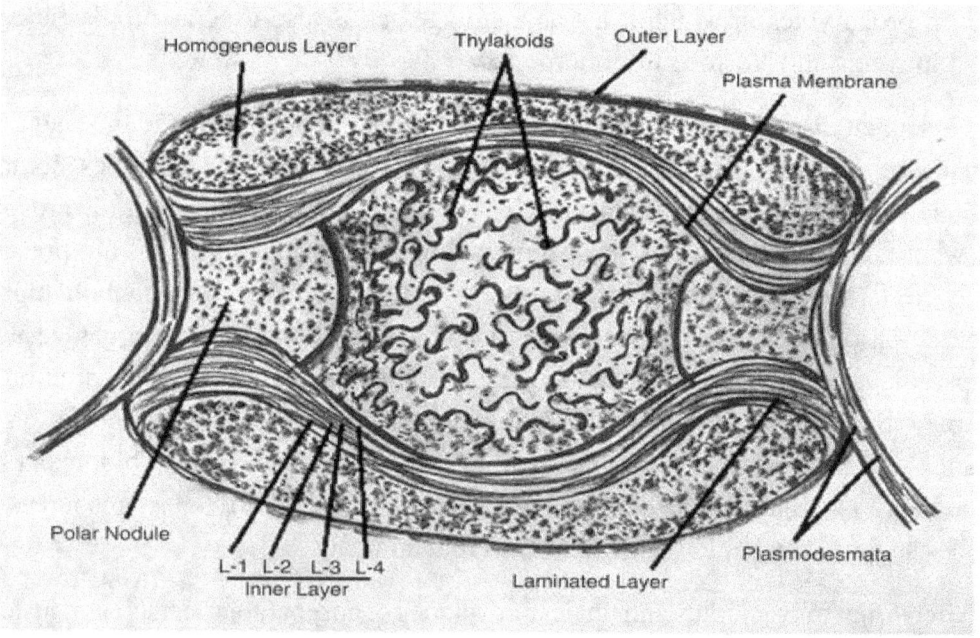

Fig.2. Heterocysts

The photosynthetic lamellae (thylakoids) are tightly packed, with reduced interlamellar spaces. They are concentrated along the periphery in a complex reticulate manner. These lamellae contain two lipids-glycolipids and acyl lipids, which are not found in vegetative cells. Three types of granules (structural granules, crystalline granules, small granules) are associated with the peripheral lamellae, but they decrease in number and eventually disappear as the heterocyst matures. The mature heterocyst may remain attached to the filament for varying periods.

Development

Heterocysts develop from vegetative cells. One or both the daughter cells of a recently divided vegetative cell may behave as proheterocyst. A proheterocyst undergoes a number of changes to form heterocyst. It enlarges in size and forms a multilayered envelop outside the cell wall. The wall becomes differentiated into an outer fibrous, a middle homogeneous and an inner lamellated layer. The outer fibrous layer is uniformly thick, whereas the middle and inner layers are thicker at the polar ends.

The pores appear one or both poles, and through these pores cytoplasmic continuity is maintained with adjacent vegetative cells. There is gradual dissolution of storage granules and breakdown of photosynthetic thylakoids. Sometimes the lamellae may form swollen vesicles and they become concentrated along the periphery and at the pores. In a heterocyst the polar pores become plugged with a refractive material (also known as polar nodules due to which their cytoplasmic connections with adjacent cells are broken.

The frequency of heterocyst production is affected by the following external factors:

(1) The blue and green light inhibit heterocyst development, whereas the white and red light support it.

(2) The concentration of phosphate salts in the medium stimulates heterocyst development, whereas the absence of Mg and Fe ions in the medium inhibits heterocyst development.

(3) The presence of combined nitrogen in the medium inhibits heterocyst development. The formation of heterocyst is inversely related to the amount of nitrogen in the medium.

Function of Heterocyst

Early workers considered heterocyst as dead cell, thus having no functions. According to (Brand 1903) if heterocyst is removed from the trichome, sporulation is inhibited. The vegetative cells adjacent to heterocyst sporulate earlier than those away from it. Canabaeus (1929) found a sort of connection between the formation of gas vacuole and heterocyst. According to Fogg (1944) these are the sites for salt accumulation. Fritsch (1951) suggested that it also stimulates growth and cell division. Fay et al. (1968) considered heterocyst as N_2-fixing organ particularly under aerobic conditions. Its thick wall restricted O_2 entry in to the heterocysts and protect nitrogenase inaction in presence of oxygen. Important heterocystous filamentous nitrogen fixing forms are *Anabaena, Aulosira, Nostoc, Tolypothrix, Cylindrospermum, Mastigocladus, Anabaenopsis, Calothrix and Scytonema*. Whereas few non heterocystous forms underneath anaerobic conditions also fix atmospheric nitrogen (e.g., *Plectonema boryanum*). Nitrogenase reduces nitrogen gas into ammonia which in turn is metabolized to glutamine in the presence of enzyme glutamine synthetase (GS). Now this glutamine is then transported from heterocyst to the adjacent vegetative cell, where it is converted into glutamate by the enzyme glutamine oxoglutarate aminotransferase (GOGAT). Both glutamine and glutamate, by various transamination reactions, form other amino acids. The fixed nitrogen may be utilized in number of ways like it is assimilated by the cyanobacteria themselves. Soluble nitrogenous compounds are liberated from healthy cells into the culture medium and after death the nitrogenous compounds are broken down into ammonia, which is eventually converted into nitrate by nitrifying bacteria. During development of heterocyst three DNA excision events take place that allows expression of nitrogen fixing genes.

Origin of Cyanophyceae

This group is considered to be the most primitive because of the presence of some important features.

These are:

a. Presence of unorganised nucleus,

b. Absence of chromatophores,

c. Absence of flagella, and

d. Lack of sexual reproduction.

They are found in all habitats where life is possible and distributed throughout the world. Fossil records indicate that they have originated in early Pre-Cambrian period. But their ancestry is not known. Absence of flagella and the prokaryotic nature of cells lead to believe that possibly they have originated from unicellular aflagellate cells. Presence of most of the members in terrestrial habitat leads to believe by most of the investigators that the Cyanophyceae have originated from terrestrial members.

Affinities of Cyanophyceae:

The members of Cyanophyceae show some relationship with both bacteria and Rhodophyceae.

Similarities of Cyanophyceae with Bacteria:
1. Cell structure is prokaryotic in both the group, having unorganised nucleus and devoid of membrane bound organelle.
2. The capsule of bacteria (if present) and mucilaginous sheath of blue green algal cells are made up of fine fibrils.
3. Cell wall composed of mucopeptide (murein).
4. *Oscillatoria* (blue green alga) shows similarity with *Beggiatoa* (sulphur bacterium), both in shape and movement.
5. Both are sensitive to antibiotics.
6. Both the groups show similarity in many metabolic processes like nitrogen and sulphur metabolism.
7. Absence of sexual reproduction.
8. Genetic recombination has been reported in *Anacystis nidulans*, a member of Cyanophyceae, showing similarity with bacteria.

Similarities of Cyanophyceae with Rhodophyceae (Red Algae):
1. Both the groups resemble in the absence of motile cells.

2. The Cyanophycean pigments, c-phycocyanin (blue) and c-phycoerythrin (red) are chemically similar to the Rhodophycean pigment r-phycocyanin and r-phycoerythrin.

3. Stignema and some other members of Cyanophyceae have pit connections, and show relationship by having similar structures as found in the members of Rhodophyceae.

Economic Importance of Cyanophyceae:

The Cyanophycean members show both beneficial and harmful activities.

Beneficial Activities:

1. Nostoc commune is boiled and used as soup in China.

2. Few species of *Nostoc, Anabaena, Scytonema* form a thick substratum over the soil resulting a reclamation of land.

3. About twenty two (22) filamentous members of Cyanophyceae like *Nostoc, Anabaena, Aulosira, Anabinopsis, Calothrix, Scytonema* etc. can fix atmospheric nitrogen and form nitrogenous compounds. These compounds are further absorbed by the plant for their metabolic activity and increase yield. All the above members have heterocyst. But certain non-heterocystous members like Plectonema boryanum are able to fix atmospheric nitrogen in anaerobic condition.

Harmful Activities:

1. Some members of Cyanophyceae cause damage of building plasters, stones etc. It can be avoided by spraying $CuSO_4$ and sodium arcenate.

2. Some members like *Microcystis*, Anabaena, form water blooms and can grow well in O_2 deficient water. Continuous respiration by submerged plants and animals during night time (when photosynthesis does not take place) causes the depletion of O_2 to almost zero level. At that condition mortality of both animals and other submerged plants takes place due to suffocation.

3. Blue green algae contaminate the water of reservoirs. They develop a foul odour in water and make it unhygienic for human being and cause several diseases.

Different diseases like gastric troubles may appear by drinking the water contaminated with *Microcystis* and *Anabaena*.

****

4.

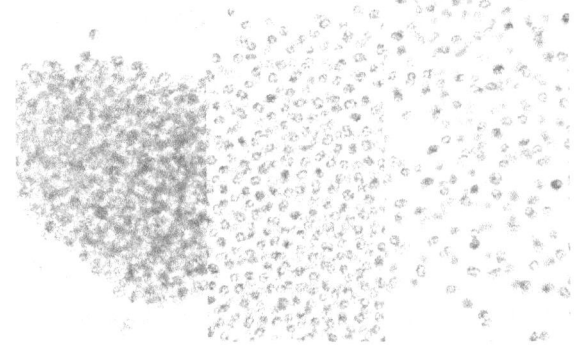

Chroococcales:
Gloeacapsa, Microcystis

Chroococcales is a group of Bacteria. There are 70 species of Chroococcales, in 55 genera and 8 family. This group has been around since the Paleoarchean Era. Chroococcales includes groups like Microcystaceae, Chroococcaceae, and Aphanothecaceae. They are a sessile organism. Almost all the members of the order are fresh water forms which are unicellular and are often organised into regular colonies, but they never show a trichome organisation. They reproduce chiefly by fission or by endospore formation. Heterocysts and hormogonia are absent.

The order divided two families.

Family Chroococcaceae Rabenhorst, 1863
Family Entophysalidaceae Geitler, 1925

Family Chroococcaceae

Members of the family are unicellular and free-living forms. They occur either singly or in colonies. The individual cells are usually spherical, ovoid, ellipsoidal or Cylindrical. Multiplication sakes place by fission or sometimes by nannocytes The common examples of the family are *Gloeacapsa, Microcyatis* and *Synechococcas*

Gloeacapsa

Systematic Position

Class : Cyanophyceae

Order : Chroococcales

Family : Chroococcaceae

Genus : *Gloeocapsa*

Occurrence: With about 23 species, Gloeocapsa is a free-floating colonial form of freshwater as well as saltwater. It is found in diverse habitats. Some species are sub-aerial or terrestrial and are found on wet rocks, moist soils, moist walls and wet brick pieces. Gloeocapsa arenarea grows on moist stones. Some species are found as phycobiont in the thalli of lichens. Common Indian Species: *Gloeocapsa decorticans, G punctata and G rupestris,* etc.

Thallus structure: Glococapsa grows as colonial form. The colony may be transparent ,hyaline, stratified and bright in colour. Various species of Gloeocapsa impart different colours, such as yellow, brown, red or violet due to the presence of certain (gloeocapsin) pigments. In a few forms, cells usually aggregate to form a 2 to 8-celled palmelloid colony (Fig. 4.c). Cell structure: Cells are spherical, cylindrical or ellipsoidal in shape. They are surrounded by a lamellated sheath which is usually coloured. The cells show typical cyanobacterial *structure,*

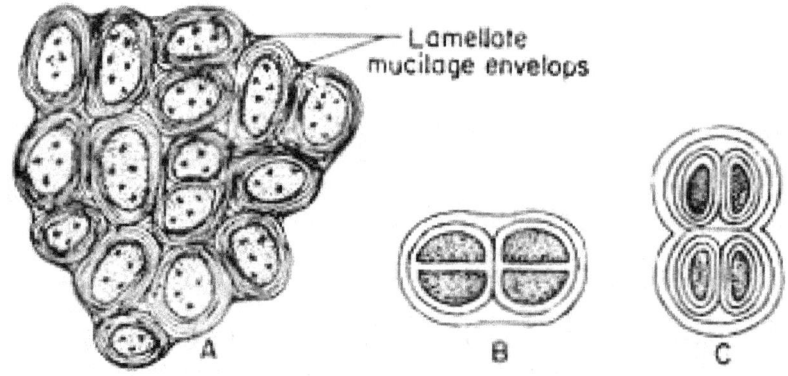

Fig.1. Glococapsa Sp. A. Cells in Group. B & C. Cells within envelope

Reproduction: Gloeocapsa reproduces by the following

Vegetative method:

(i) **Fragmentation:** The colony, after gaining certain size, breaks up into fragments, each developing into a new colony.

(ii) **Cell division (Fission):** The cell clongates and divides by simple fission in three planes at right angles to one another. The process is repeated till the formation of a large number of cells. These cells remain embedded within the parent cell sheath. New colonies are released by the breaking of this sheath. Occasionally, in some species, reproduction has been observed by nanocyte formation.

Points of interest: According to Wyatt and Silvey (1969), Gloeocapsa alpicola fixes elementary nitrogen.

Microcystis

Systematic Position:

Class : Cyanophyceae

Order : Chroococcales

Family : Chroococcaceae

Genus : *Microcystis*

Occurrence:

Microcystis (Gr. Micro = small; kystis = bladder) is a colonial fresh water alga. It is represented by 18 species. All species are free floating, and form extensive water blooms. This alga secretes poisonous substances. These substances destruct the liver and inhibit the growth of other algae.

M. aeurginosa is responsible for the production of a toxin (fast death factor) which is toxic to many animals which drink water containing this alga. M. toxica is very poisonous and is responsible for the death of thousands of sheep's and catties.

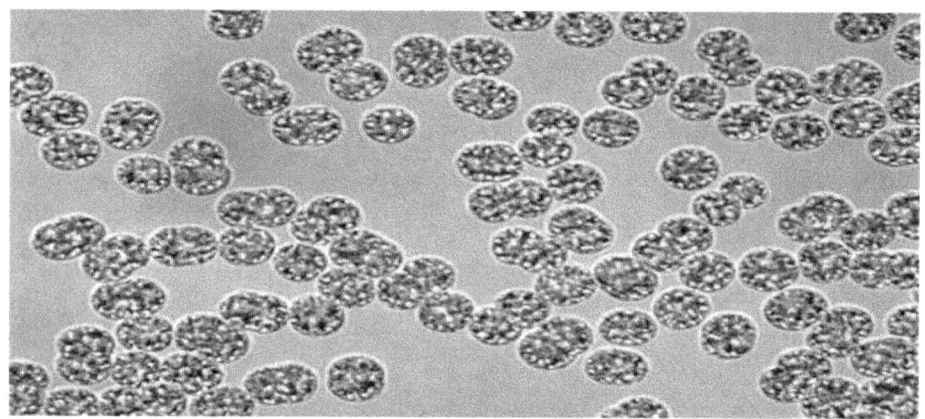

Fig. 2. Microcystis culture unicells

Thallus Structure

The colonies of Microcystis may be round and definite in shape or irregular or sometimes clathrate, depending upon the species. There are hundreds or thousands of small, marble-like cells in each colony. The spherical or elongated cells in a colony float freely in a colourless, homogeneous mucilaginous matrix. They lack individual sheath, and exhibit typical cyanophycean cell structure. Numerous pseudovacuoles are frequently present in a cell.

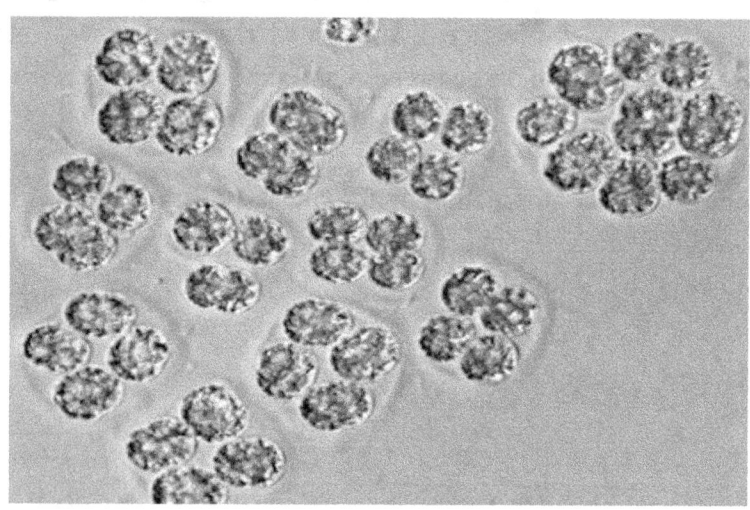

Fig.3. *Microcystis viridis Colony*

Location: water reservoir Wuppertalsperre Germany

Physical characteristics

As the etymological derivation implies, *Microcystis* is characterized by small cells (a few micrometers in diameter), possessing gas filled vesicles (also lacking individual sheaths). The cells are usually organized into colonies (macroscopic aggregations of which are visible with the naked eye) that begin in a spherical shape, losing coherence to become perforated or irregularly shaped over time. These colonies are bound by a thick mucilage composed of complex polysaccharide compounds, including xylose, mannose, glucose, fucose, galactose, rhamnose, among other compounds.

The coloration of the protoplast is a light blue-green, appearing dark or brown due to optical effects of gas-filled vesicles.

Reproduction

Microcystis reproduces only vegetatively and sexual reproduction is completely absent. The vegetative propagation takes place either by fission or by fragmentation of the colony. The fission is simple cell division which takes place in all planes. In *M. flosaquae,* **nannocytes,** the modified endospores are formed by repeated divisions of cell contents.

****

5.

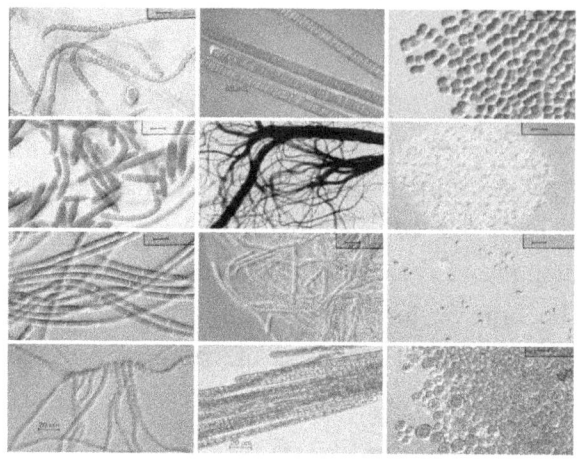

Nostocales: *Oscillatoria, spirulina, Nostoc, Anabena, Aulosira, Scytonema, Gloeotrichia, Rivularia.*

Nostocales is the largest order of the class Cyanophyceae. The thalli are mainly filamentous, mostly un-branched and sometimes show false branching. Heterocysts are commonly present but some families of the order are without heterocysts. Multiplication takes place by hormogonia, akinetes, exospores, endospores or hormospores.

Desikachary (1959) divided order Nostocales into five families: Oscillatoriaceae, Microchaetaceae, Nostocaceae, Scytonemataceae and Rivulariaceae.

Family: Oscillatoriaceae

Trichome un-branched, cylindrical cells 2-3 time wider than longer, cytoplasm granular with vacuoles, sheath absent or extremely thin. Fresh plants exhibit oscillating movement.

Oscillatoria

Systematic Position
Division: Thallophyta
Class: Cyanophyceae
Order: Nostocales
Family: Oscillatoriaceae
Genus: *Oscillatoria*

Occurrence

Oscillatoria is a genus of filamentous cyanobacteria. It has more than 100 species. It is named for the oscillation in its movement. Filaments in the colonies can slide back and forth against each other. Thus the whole mass is reoriented to its light source. It is very common in moist places rich in decay organic mailer. It is commonly found in watering-troughs waters like streams, roadside ditches, drains and sewers. It is mainly blue-green or brown-green. It forms thin blue green mucilaginous coating on the surface of flowing water. It's one specie is found in hot springs. Some species like *0. .formosa* and O. *princeps* are symbiotic. They form association with the nitrogen fixing bacteria. Some common species are *0. fomosa, 0. prolificn* and *0. formosa*.

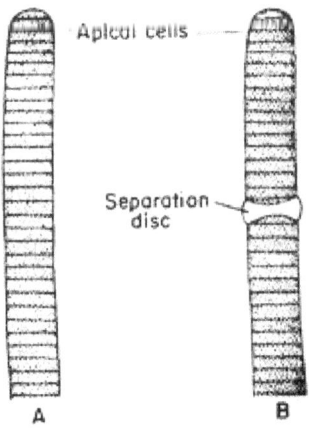

Fig. 1. *Oscillatoria* **filaments**

Vegetative structure

Its body is composed of **single row** of cells. These cells form trichomes. Its trichomes are unbranched filaments. They are covered by very thin mucilaginous sheath. All cells of a trichome are similar in shape except apical cells. The apical cells are convex at the tip. All other cells arc broader and cylindrical. In some species, the apical cells may end in subacute point. In some cases, it may have cap or **calyptra** at the tip. Some species have narrow trichome. They have cylidrical cells with their length equal or **greater than the breath.**

Cell structure

Cell wall is made of mucopeptide. Ultra structurally it consists mainly a 2000 A° structural layer external to plasma membrane. Outside the structural layer is 160 A° thick another layer and there is a third 90° A layer loosely wrapped around the two.

The structural layer has s series of 700 A° wide pores which terminate at the 160 A° layer. Under an ordinary microscope the protoplasm is distinguishable into a peripheral chromoplasm and a central colourless centroplasin or central body.

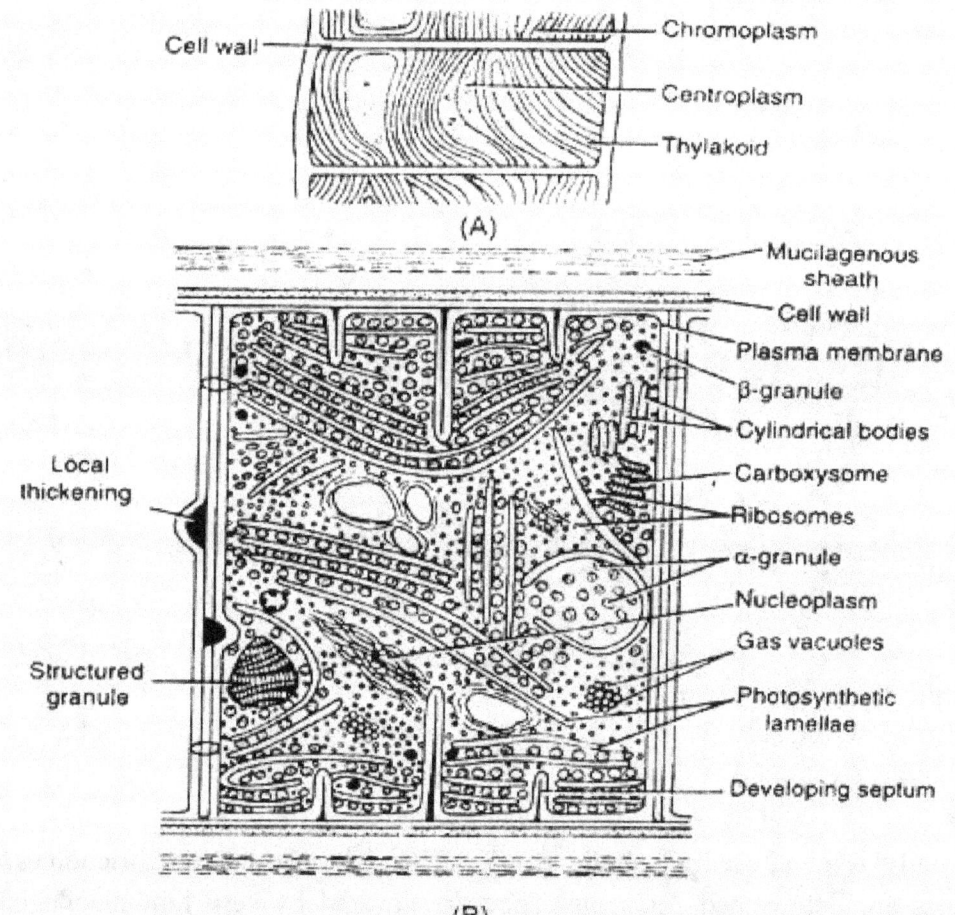

Fig. 2. (A–B). *Oscillatoria*. (A) Single cell, (B) Ultrastructure of cell.

Ultrastructure of cell shows that the chromoplasm contains photosynthetic lamellae or single thylakoid which often run parallel to one another. The thylakoids contain photosynthetic pigments like chlorophyll a, carotenes, xanthophyll's and phycobilins (C-phycocyanin, allophycocyanin, c-phycoerythrin).

Phycobilins occur in minute vesicles called phycobilisomes. The centroplasm represents the incipient nucleus called gonophore. It is represented by DNA fibrils. The cell contains many ribosomes but mitochondria, plastids, ER and Golgi bodies are absent. Reserve food material is in the form of cyanophycean starch, lipid, globules and cyanophycin. The protoplasm also contains two types of granules α and β α granules contain proteins and polysaccharides while β granules have lipid. Planktonic species of Oscillatoria possess gas vacuoles or pseudovaculoles which are devoid of any membrane. It is made of a number of **'hexagonal'** structures called **'gas vesicles'** (Fig. 2B). The trichome shows polarity. It possesses characteristic apical cell. It may be cap like (capitate) or covered by a thick membrane called calyptra. The apical cell may also be conical, dome shaped, acuminate, oval, flattened convex or coiled and accordingly to the shape of the cap cell, the species are identified (Fig. 3 A-H).

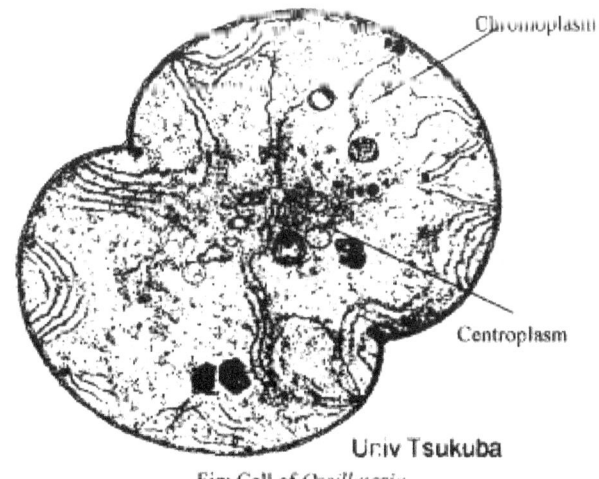

Fig: Cell of *Oscillatoria*

Fig.3.

Reproduction

Aseuxal and sexual reproduction is absent in Oscillatoria. Oscillatoria reproduces by fragmentation. Oscillatoria forms long filaments of cells which can break into fragments called **hormogonia.** Each filament does not grow indefinitely in length, but regularly breaks up into short fragments, each containing 2, 3 or several cells.

Before separation, the hormogones are delimited in the filament by the formation of double concave disks (separation disks) of gelatinous consistency between the hormogones. Each hormogone may develop into a new filament by division.

The most characteristic feature of Oscillatoria is the fact that the filaments, particularly the hormogones, are capable of swaying or oscillating movements to which the plant owes its name.

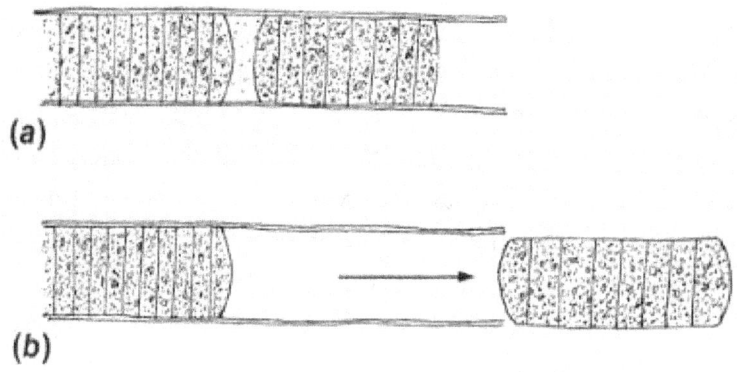

Fig.4. (a),(b) Formation of a hormogonium

Movement in Oscillatoria:

The name Oscillatoria (oscillare, to swing) is given to this alga due to the peculiar movement shown by the. trichome. It is called **'oscillatory movement'**. These are the jerky, pendulum-like movements of the apical region of the trichome.

Some other movements shown by the trichomes of Oscillatoria are:

Gliding or creeping movement:

The trichome moves forward and backward along its long axis.

Bending movement:

The tip of the trichome shows bending.

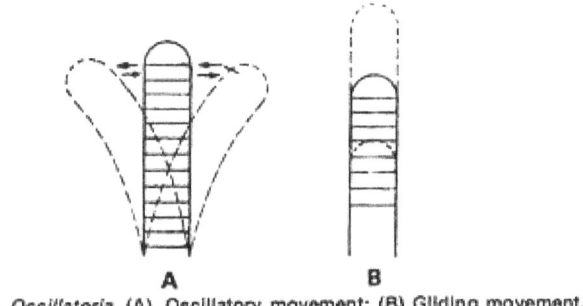

(A, B). *Oscillatoria*. (A). Oscillatory movement; (B) Gliding movement.

Fig.5.

Spirulina

Systematic Position

Class: Cyanophyceae
Order: Nostocales
Family: Oscillatoriaceae
Genus: *Spirulina*

Occurrence

Spirulina occurs in a wide variety of fresh water habitats, often found densely entangled with the *Oscillatoria* filaments. Some species are found in brackish water, and can tolerate very high salt concentrations. For example, *S. subsalsa* can grow and multiply in solutions that are more concentrated than 3M sodium chloride Asolution. Some common species of *Spirulina* are *S. albida, S. major, S. princeps, S. subsalsa and S. versicolor.* Species of *Spirulina,* are rich in proteins, essential vitamins and ungaturated fatty acids. They are receiving considerable attention for their use as a supplement to diet.

Ultra Structure

Transmission Electron Microscope observations show for *Spirulina* prokaryotic organization, capsule, pluri-stratified cell wall, photosynthetic or thylakoid lamella system, ribosomes and fibrils of DNA region and numerous inclusions. The capsule has fibrillar structure and covers each filament protecting it. The irregular presence of capsule around the filaments in *S. platensis* is a differentiating morphological characteristic to compare with S. maxima. Trichome width varies from 6 to 12 μm, and is composed of cylindrical cells. The helix diameter varies from 30 to 70 μm; the trichome length is about 500 μm, although in some cases when stirring of culture is deficient the length of filament reaches approximately 1 mm. It is very important to explain that the helical shape of *Spirulina* in liquid culture is changed to spiral shape in solid media. These changes are due to hydratation or dehydratation of oligopeptides in the peptidoglycan layer.

Spirulina cell wall is formed by four numbered layers, from the inner most outward as: LI, LII, LIII and LIV. All these layers are very weak, except layer LII made up of peptidoglycan, substance that gives the wall its rigidity. The LI layer contains b-1, 2-glucan, a polysaccharide not very digestible by human beings. However, the low concentration (<1%) of this layer, thickness its (12 nm), and the protein and lipopolysaccharide nature of the LII layer are favorite reasons for the easy human digestion of Spirulina.

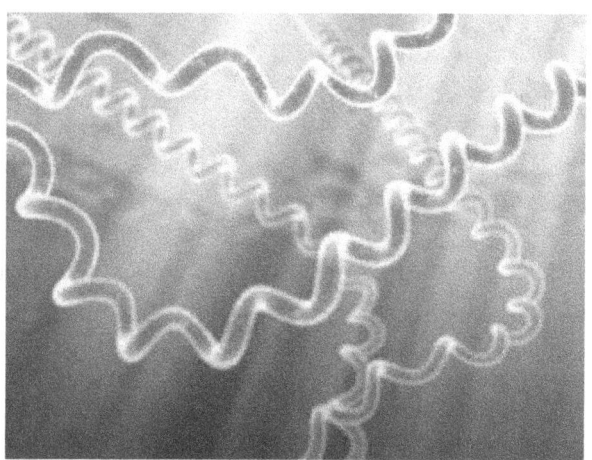

Fig.6. *Spirulina:* **Trichome structure**

In this microorganism chlorophyll a, carotenes and phycobilisomes, which contain phycocyanin (blue pigment), are located in the thylakoid system or photosynthetic lamellas. The inter-thylakoid space is limited by the presence of electronically transparent protein gas vesicles, with the cylindrical form that give Spirulina its floating capacity. Ribosomes and fibrils of DNA region are generally of central localization.

Spirulina contains numerous characteristic peripheral inclusions associated to thylakoids. Those are: cyanophycin granules, polyhedral bodies, polyglucan granules, lipid granules, and polyphosphate granules. The cyanophycin granules, or reserve granules, are important due to their chemical nature and a series of pigments. The polyhedral bodies or carboxysomes mainly contain the enzyme ribulose 1, 5-diphosphate carboxylase that allows the fixation of CO_2 in photosynthetic organisms and probably carry out a reserve function. The polyglucan granules or glycogen granules or a-granules are glucose polymers, small, circular and widely diffused in the interthylacoidal space. The lipid granules, b-granules or osmophile granules form the reservation deposit, constituted by poly-b-hydroxybutyrate (PHB), found only in prokaryotes. PHB acts as a carbon and energy reserve.

Spirulina as a unique source of protein

Spirulina is a very special alga that doesn't even need to undergo any extraction, since its initial protein content is as high as 70 percent! If you are able to obtain fresh *spirulina* per kilo without it having to undergo through drying and repacking you'll have it for a very affordable price, almost comparable to beef or egg protein. Aside from easily digestible protein, *spirulina* also contains many pigments, which are responsible for its intense color. The main pigment of *spirulina* is chlorophyll, and it is densely packed with it. Containing 20 times more chlorophyll than the green-leafy vegetables, *spirulina* even colors the stool, making it dark green. Hence, these who lack vegetables could benefit from this *spirulina* protein, otherwise it will make an excess.

Reproduction:

In *spirulina* Reproduction takes place asexually by Binary Fission. ' Sexual reproduction is absent.

Family: Nostocaceae

1. Many trichomes are present in a common mucilage.
2. All the cells are arranged in a single row, and they are similar in shape and structure.
3. Hormogonia are present.
4. The heterocysts are present in most of the genera.

5. Akinetes are present, either singly or in chain.

6. In between the individual cells of the trichome is present a well-marked constriction, which provides the trichomes a moniliform appearance.

Some important genera of Nostocaceae are *Nostoc, Anabaena. Richelia. Anahaenopsis, Cylindrospermum. Raphidiopsis, Aulosira* and *Nodularia.*

Nostoc

Systematic Position

Class: Cyanophyceae

Order: Nostocales

Family: Nostocaceae

Genus: *Nostoc*

Occurrence

Nostoc is a common freshwater or terrestrial alga, occurring abundantly in rice fields. The terrestrial species grow commonly on moist soils, mixed with many small plants such as mosses, liverworts, lichens, etc. Some species are found on moist rocks in hills. The freshwater species are found free-floating in pools, ponds, ditches and other similar temporary or permanent water reservoirs.

The plants are found in the form of ball-like colonies of pin-head size. But sometimes the colony may be as large as 30 cm in diameter. *Nostoc punctiforme* occurs endophytically within the thallus of *Anthoceros,* coralloid roots of *Cycas,* roots of *Zamia* and underground stem of *Gunnera manicata*. Some species (e.g. *N. sphaericum* and *N. collema*) are the common phycobionts of *lichens.*

Desikachary (1959) described 23 species of *Nostoc* in India, of which some of the common Indian species are *N. punctifonne. N. endophytum, N. rivulare, N. elhpsarporum.* N. *muscorum, N. calcicola, N. sphaericum* and *N hatei.*

Thallus Structure of Nostoc

The body of Nostoc is called ***thallus***. It is a simple body and consists of many threads like structures, called ***filament***. Each filament is composed of many rounded cells which are arranged in a chain like manner or in beaded form. All the cells are similar. The whole filament is covered by a gelatinous sheath.

n the filament, some cells are larger in size and thick-walled, called *heterocysts*. These cells perform two functions, **reproduction** and **Nitrogen Fixation**. In Nitrogen fixation, Nitrogen is converted into simpler Nitrogen compounds, which are used by the plant.

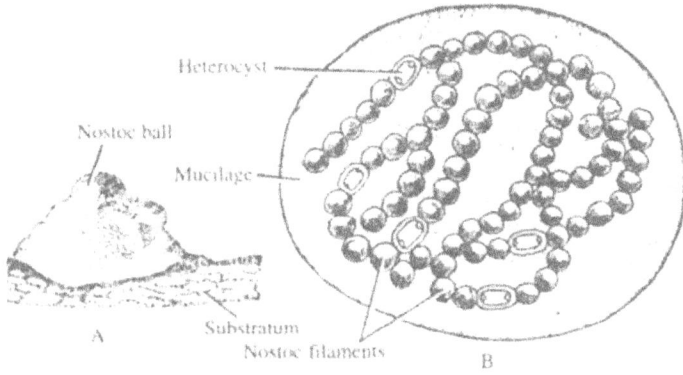

A, A gelatinous mass of *Nostoc*; B, A part of colony under light microscope.
Fig.7.

Structure of a cell of Nostoc

Each cell of *Nostoc* is rounded in shape. It is covered by an outer cell wall. The protoplasm of the cell consists of two parts:

A) Chromoplasm

It is the outer colored part along the cell wall, which contains blue-green pigments (Phycocyanin). It also prepares food material due to the presence of Chlorophyll pigments.

B) Centroplasm or Central Body

It is the inner colorless part of the cell, which contains stored food granules. It also acts as a nucleus. The cell does not contain a true nucleus. The central body performs the function of the nucleus. It has no nuclear membrane or nucleolus, so it is called *incipient nucleus*. i.e.(Incomplete nucleus).

Reproduction in Nostoc

Sexual reproduction is absent in Nostoc. In Nostoc, the reproduction takes place by the following methods:

1. by Hormogonia OR by Fragmentation

Hormogonia are formed by Fragmentation. The filament breaks at different points and each broken filament is called **hormogonia**. The filament breaks due to decay and death of ordinary cells. Heterocyst may form the breaking point. Each hormogonium grows into a new filament by repeated cell division.

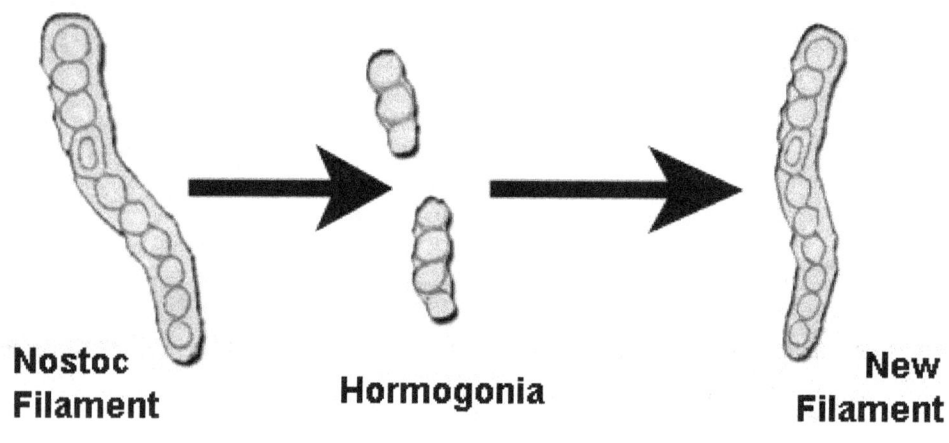

Fig.8. Hormogonia formation

2. by Arthrospores OR by Akinetes Formation

During unfavorable conditions, some cells of filament become enlarge and they are covered by a thick wall, they are called *Akinetes* or *Arthrospores* or *Resting Spores*.

They also store food material. They germinate during favorable conditions into new filaments.

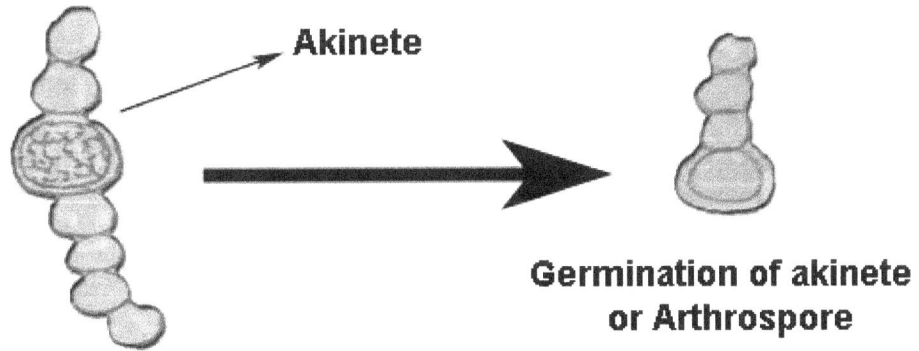

Fig.9. Akinetes Formation

3. by Heterocyst

At the time of reproduction, the heterocysts are separated from *Nostoc filament*. These heterocysts are changed into a normal reproductive cell. By the process of cell division, it is converted into a new filament.

Nutrition in *Nostoc*

Nostoc contains chlorophyll, so it can manufacture its own food i.e. it is ***Autotroph*** in nature.

Anabaena

Systematic Position

Class: Cyanophyceae

Order: Nostocales

Family: Nostocaceae

Genus: *Anabena*

Occurrence

Anabaena is a genus of filamentous cyanobacteria, or blue-green algae. It found as plankton. It is known for its nitrogen fixing abilities. They form symbiotic relationships with certain plants, such as the mosquito fern. Some species of *Anabaena* are **endophytes.** They live in the roots of *Cycas* and *Azolla.* Anabaena is found in all types of water. **Blooms** or massive growths can occur in waters with a lot of nutrients. These blooms discolour the water and give it a bad odour when the cells die and decay. They are one of four genera of cyanobacteria that produce **neurotoxins.** These toxins are harmful to local wildlife, as well as farm animals and pets. Production of these neurotoxins is part of its symbiotic relationships. It protects the host plant from grazing pressure.

Thallus Structure

It has filamentous structure. Its filament resembles the filament of *Nostoc.* Sometimes it becomes difficult to differentiate between trichomes of *Nostoc* and *Anabaena.* There is only one difference. The filaments of *Nostoc* are covered by mucilage and form a colony. It is absent in *Anabaena.* The filament of *Anabaena* consists of string of **beaded** cells. Several intercalary heterocysts are present in the trichome. Heterocysts are of same shape as of vegetative cell. The filaments are ordinarily straight. But they may be circinate or irregular. Filaments occur singly within a sheath. Sheaths are always hyaline and watery gelatinous.

Structure of cells

- The cells are spherical or barrel shaped. They are rarely cylindrical and never discoid. The majority of the cells of a colony are similar in size. Its cells have following components:
- Each cell has outer cell wall. This wall consists of three layers. The inner layer is thin cellular layer, medium is pectic layer and outer is mucilage layer.
- Protoplasm is composed of 'Soo riafts. The peripheral part is called chromoplasm. It contains pigment. Hence it is coloured. The central colourless part of protoplasm contains nucleus like material called **central body** or **chromatin granules.**
- Heterocyst are of same shape as of vegetative cell.
- Golgi bodies, endoplasmic reticulum and mitochondria are absent in their cells.

Nitrogen fixation by Anabaena

During times of low environmental nitrogen, about one cell out of every ten will differentiate into a heterocyst. Heterocyst then supply neighbouring cells with fixed nitrogen in return for the products of photosynthesis. Such nitrogen fixing cell now cannot perform photosynthesis. This separation of functions is essential. The nitrogen fixing enzyme in heterocysts is **nitrogenase.** It is unstable in the presence of oxygen.

Nitrogenases are kept isolated from oxygen. Therefore, heterocysts have developed elements to maintain a low level of oxygen within the cell. The developing heterocyst builds three additional layers outside the cell wall. These layers prevent the entrance of oxygen into the cell. It gives heterocyst its characteristic enlarged and rounded appearance. Due to these adaptations, the rate of oxygen diffusion into heterocysts is 100 times lower than of vegetative cells. One layer creates an envelope polysaccharide layer. The nitrogen is fixed in this oxygen-restricted envelope. To lower the amount of oxygen within the cell, the presence of photosystem II is eliminated.

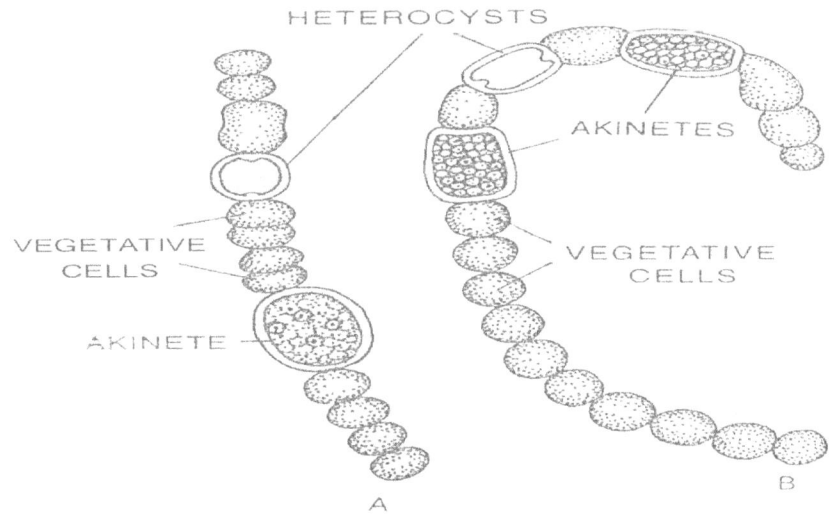

Fig: A part of filament bearing Heterocysts and Akinetes.

Fig.10.

Reproduction

Anabaena reproduces entirely asexually by the following methods –

1. Hormogonia

Hormogone formation is very common in Anabaena. In such case the trichome ruptures at places where heterocyst and the vegetative cells adjoin. In this way short, short segments of living cells are formed called the hormogonia. The hormogonia slip out of the gelatinous matrix and establish new colonies by division. The terminal cells of the hormogonia differentiate into heterocysts. The intercalary cells then divide in the plane parallel to the axis of the trichome forming a packet of cells. This is called aseriate stage.

2. Akinetes

Under certain conditions any cell or some vegetative cells of the trichome become enlarged and secretes a thick, highly resistant wall around it. They get filled with the reserved food materials. Such specially modified vegetative cellsare called akinetes or resting spores. These are well adapted to survive the unfavourable conditions like water shortage and unsuitable temperature. The resting spores survives during the unfavourable conditions and with the return of the favourable conditions, they germinate into new plants.

Note-Anabaena is not just a free living organism, it also has many symbiotic relationships with other organisms

Aulosira

Systematic position:

Class:	Cyanobacteria (Cyanophyceae)
Order:	Nostocales
Family:	Nostocaceae
Genus:	*Aulosira*

Occurrence

With its eight species, *Aulosira* is commonly found in freshwater bodies, such as ponds, tanks and ditches. *A bombayensis* is thermophilic, found in hot springs. *A pseudoramosa* is lithophytic and found on the walls of residential houses. *A fertilissima* was reported as the most important nitrogen-fixing species in paddy fields of UP and Bihar (Singh, 1946). This species is dominant for a few months and forms a papery growth on the soil surface.

Common Indian Species: *Aulosira bombayensis, A prolifica, A. aenigmatica, A. laxa* and *A. fertilissima.*

The plant body: Uniseriate trichomes enclosed in firm individual sheaths, occurring singly or sometimes in clusters, but never in mats. Occasional heterocysts occur at regular intervals along the trichome and are slightly larger than the vegetative cells. The trichomes are with sheath and indefinite, bearing intercalary heterocysts. Akinetes are developed in series and may be near the heterocyst or away from it.

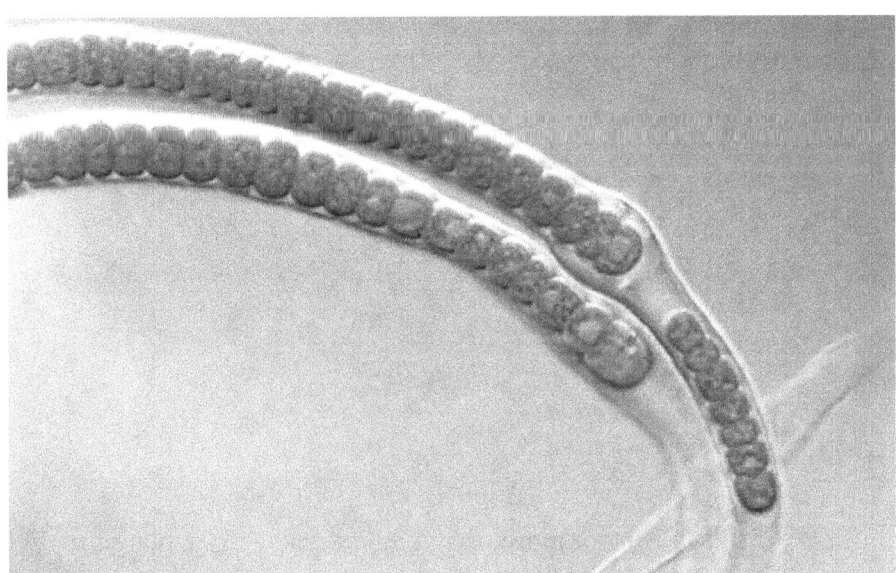

Fig.11. Aulosira Trichome

Cell structure: Each cell shows the typical cyanobacterial character.

Reproduction: *Aulosira* reproduces by means of fragmentation, akinetes, homogones and heterocysts.

1. *Aulosira fertilissima* is the main nitrogen-fixer in paddy fields of India and its inocculants are being promoted for use in rice cultivation.
2. *Aulosira fertilissima* is used bioremediation of organophosphorus pesticides.

Family: Scytonemataceae

The **Scytonemataceae** are a family of filamentous, heterocystous cyanobacteria within the order Nostocales. The family is known from freshwater, marine, and terrestrial environments, where it grows in colonies attached to the substrate. Akinetes are not known, and the members of the family are known to reproduce with nonheterocystous hormogonia.

Scytonema

Systematic position:

Class:	Cyanophyceae
Order:	Nostocales
Family:	Scytonemataceae
Genus:	*Scytonema*

Occurrence

Scytonema, with its about 40 species, grows commonly in aquatic as well as terrestrial habitats. Sometimes, it occurs as an epiphyte on wet bark of trees during rains. It also grows on wet walls and rocks, looking like black cottony patches.

Common Indian Species: *Scytonema gofmanni, S syengari, S ocellatum, S pascheri,* etc.

Thallus Structure (Fig. 12):

The trichomes of Scytonema consist of more or less cylindrical cells and are usually of uniform diameter throughout. Sometimes slight or marked constrictions appear at the transverse walls. The trichomes are surrounded by sheaths of extremely fine texture and may be hyaline or yellowish or brownish in colour. The sheath may be homogeneous throughout or stratified, and stratifications are either parallel or oblique. The filaments show characteristic false branching. The trichome, during the process of multiplication, becomes fragmented, either due to disintegration of the intercalary cells or due to formation of separating disks or heterocysts. These fragments are immobile and begin to grow within the original sheath. The branching results due to interruption of such newly formed trichomes at certain points. As a result, portions of one or both the trichomes perforate the firm sheath and grow out as laterals, which secrete a distinct sheath of their own. These false branches usually appear in pairs and occupy approximately the middle portion between two heterocysts. The heterocysts, intercalary in origin, occur singly or in series of twos or threes. They are more or less of the same size as the vegetative cells.

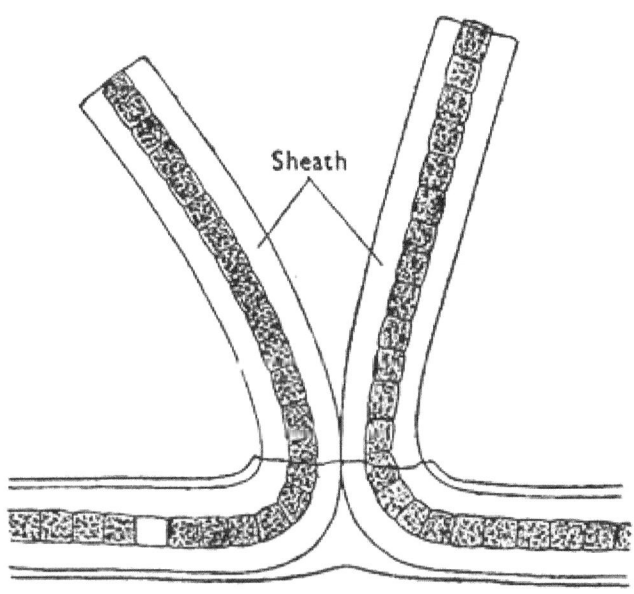

Fig.12. *Scytonema*

Branching

A characteristic feature of *Scytonema* is the presence of false branches, which are formed by the breaking of the filament. When a filament breaks, one or both the broken ends protrude out of the mucilagenous sheath and grout like branches. If both the broken ends come out of the sheath and grow like branches, it is called **geminate branching,** and if only one, then it is said to be **false branching.**

The false branches are formed by the following methods.

1. By loop formation. Certain parts of the filament increase in length due to rapid cell division. Such localized growth results in loop formation that protrudes through the sheath. Eventually the loop breaks in the middle, cell division continues and two false branches are formed (Fig. 13 A).

2. **By formation of a separation disc.** Sometimes one of the intercalary cells becomes dark coloured due to the diffusion of chromatophores in the protoplast, and its cell wall becomes thin. This cell forms a biconvex separation disc due to the pressure of the adjacent cells. The disc ultimately degenerates, resulting in the breaking of the trichome. One of the broken ends protrudes through the sheath as false branch (Fig. 13 D).

3. **By rupture of filament near the intercalary hetgrocysts.** In trichomes with intercalak· heterocysts a **zone** of **weak** adhesion lies where **the vegetative** cell and the, heterocyst about each other. If the trichome breaks on one side of the heterocyst, a singlebranch is formed and if it breaks on both the sides of the heterocyst, the branching is geminate

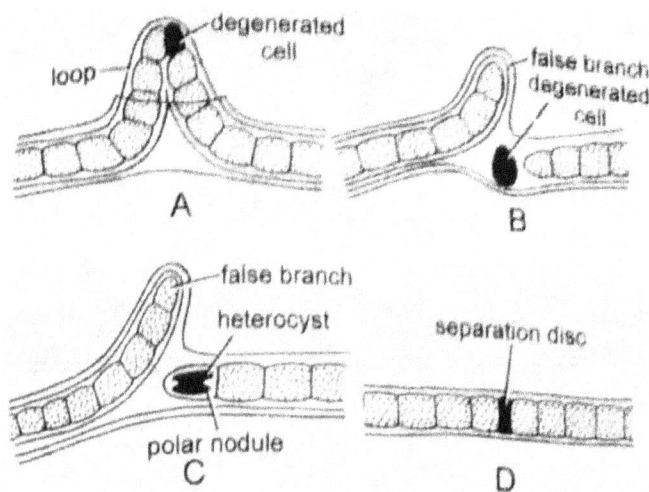

Fig. 13 A-D. *Scytonema:* **Modes of branching; A. Loop formation. B. Degeneration of intercalary cell, C. Breaking of filament near the intercalary heterocyst, D. Formation of separation disc.**

4. **By degeneration of one or more intercalary cells.** This is a very common method of the formation of false branches. One or more intercalary cells degenerate and leave gaps in the trichome. One or both the broken ends of the trichome may protrude through the sheath and develop into branches (Fig.13B).

Growth

In *Scytonema* growth occurs by repeated transverse divisions in the apical or sub-apical cell. In terrestrial species, the apical cell remains active for some time and then degenerates. Thereafter, the sub-apical cell functions as apical cell. Besides the apical and sub-apical cells, some intercalary cells also divide transversely and add to the growth of the trichome.

Reproduction

Multiplication of trichomes in *Scytonema* takes place by the formation of **hormogonia** or **akinetes.**

1. By hormogonia. Many factors, such as the presence of separation discs and heterocysts, and disintegration of intercalary cells help in the fragmentation of the trichome. These fragments are called **hormogonia,** and each hormogonium is capable of growing into a new trichome.

2. Akinetes are rarely formed in this alga.

Family Rivulariaceae

This family includes freshwater forms with uniscriate tapering trichomes. The sheaths surrounding the trichomes are homogeneous or lamellated, and have a firm texture. The trichomes usually fm'm spherical, hemispherical or irregular colonies. Except few non heterocystous genera like *Amphithrix* and *Calothrix*, most of the members of this family are heterocystous. The false branches, which arise near the tapering ends, are formed by the breaking of the trichome just below an intercalary heterocyst.

Gloeotrichia

Systematic position

Class: Cyanophyceae

Order: Nostocales

Family: Rivulariaceae

Genus: *Gloeotrichia*

Occurrence

Gloeotrichia occurs free-floating in freshwater, such as in ponds and pools. The colonies may be found attached to submerge plants or wet rocks.

Common Indian Species: *Gloeotrichia ghosei, G indica, G pilgeri, G pisum,* etc.

Thallus Structure:

Thallus is colonial. Colony appears as spherical to hemispherical and gelatinous. Each colony is exceedingly firm in consistency and contains numerous radiating filaments with repeated false branching in which, each branch terminates into a multicellular colourless hair.

One or two heterocysts are located at the broad basal end of the trichome. The lower and broader portion of the filament is covered by a sheath. Cells contain granular, undifferentiated and vacuolated cytoplasm (Fig 14).

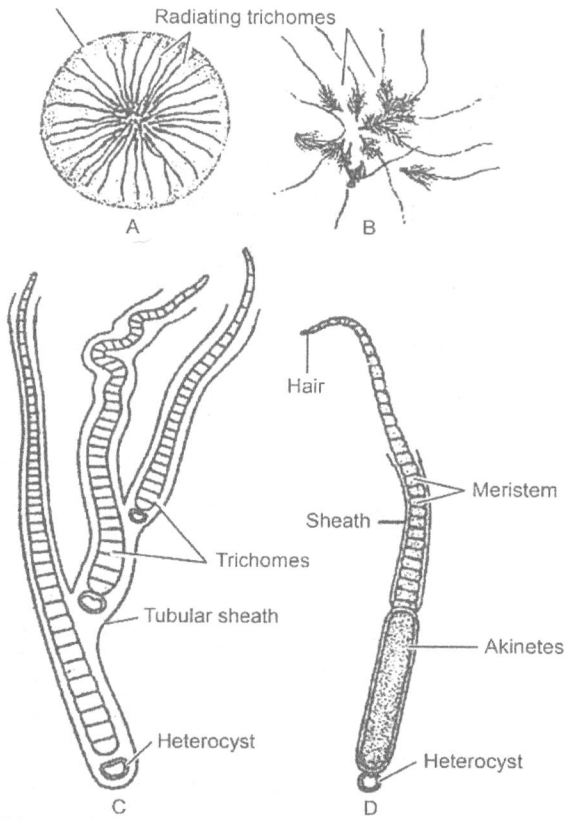

Fig.14. *Gloeotrichia*: A. Colony under low power, B. Radiating Trichomes, C. Trichomes in Gelatinous Matrix, D. Single Trichome

Reproduction

Hormogonium:

It is formed from the upper part of the trichome. It is composed of a few cells.

Akinete:

It is an elongated and thick-walled body. It is situated just above the rounded basal heterocyst.

Identification:

Thallus blue-green in colour, cells devoid of conspicuous nucleus, absence of any sex organ, presence of gelatinous sheath around the cells, (in most cases) absence of organised cell organelles like plastids.

Rivularia

Systematic position

Class:	Cyanophyceae
Order:	Nostocales
Family:	Rivulariaceae
Genus:	*Rivularia*

Occurrence

Rivularia is a freshwater form found free-floating on water surface of ponds and pools. Some species grow on submerged stones and stems and leaves of aquatic plants. *Rivularia bullata* grows on damp soils as well as in marine habitats, whereas *R polyotis* is **found** only in marine habitats.

Common Indian Species: Rivularia hansgirgi, R beccariana, R manginii, R. globiceps, etc.

Thallus Structure:

Fig.15. *Rivularia*

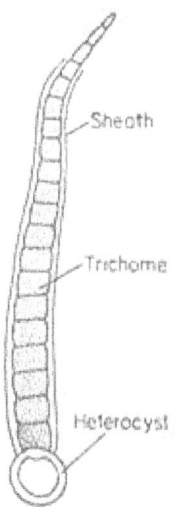

Thallus is un-branched, filamentous, but showing false branching. Filaments are colonised in more or less radial or parallel order in hemispherical or spherical mucilaginous colony. Trichomes mostly ending in a hair. Filaments often show trichothallic growth and are completely included within the gelatinous sheath. Filament contains a meristematic zone in intercalary position. It is composed of several cells (Fig 15).

Reproduction

Heterocyst:

It is basal or intercalary in position.

Hormogonium:

It is formed singly or in a group, particularly towards the base of the meristematic zone.

Akinete:

Absent

Identification:

Thallus blue-green in colour, cells devoid of conspicuous nucleus, absence of any sex organ, presence of gelatinous sheath around the cells, (in most cases) absence of organised cell organelles like plastids.

****

6.

Stigonematales: *Stigonema*

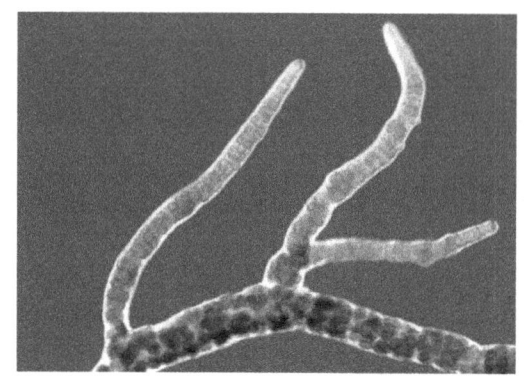

Filament branched irregularly and curved at places, sheath around trichome distinct.

Family: Stigonemataceae

Filaments of older region multi-axial – two to many rows of cells in each trichome, heterocyst roundish, intercalary or lateral in position; hormogonium formed at the tip of young branches.

Stigonema

Systematic position:

Class:	Cyanophyceae
Order:	Stigonematales
Family:	Stigonemataceae
Genus:	*Stigonema*

Occurrence:

Stigonema about 44 species which occurs in freshwater habitats throughout the world, either attached to the sub-stratum in submerged conditions or on the soil. *Stigonema alpinum* grows sub-aerophytically on the bark of trees in the alpine zone of high mountains. *S minuta* has been reported from soil crusts of the Antarctica.

Common Indian Species: Stigonema dendroideum, S panniforme, S minuta, S tuifaceum, S ocellatum, etc.

Thallus Structure:

Plant body is branched, filamentous. Branching is irregular and variously curved. Filament in older region is composed of two to many rows of cells. Sheath around each trichome is thicker in older trichomes

Cell structure: Each cell of the filament is spherical, sub-spherical or rectangular in shape and is sometimes broader than long. Each cell has typical cyanobacterial structure.

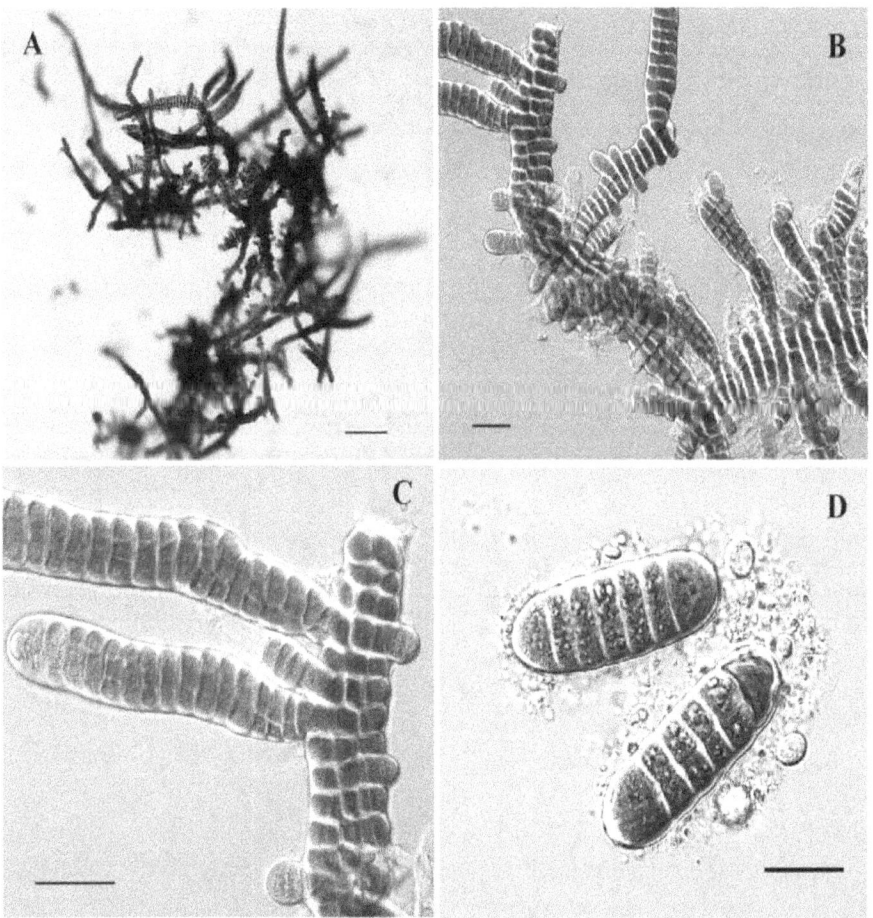

Fig. Stigonema dinghuense. A. Habit of thallus, B. Mature filaments. C. Lateral branches. D. Hormogonia. Scale bars: 200 μm (A), 50 μm (B, C), and 20 μm (D).

Reproduction

Heterocyst: It is intercalary or lateral in position. It is roundish in shape.

Hormogonium: It is formed at the end of young branches, two to few cells in length.

Identification:

Thallus blue-green in colour, cells devoid of conspicuous nucleus, absence of any sex organ, presence of gelatinous sheath around the cells, (in most cases) absence of organised cell organelles like plastids

****

7.

Chlorophyceae: General characteristic, Classification and Alternation of Generations

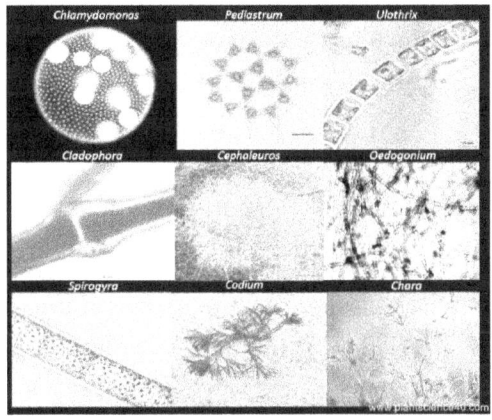

Chlorophyceae (from the Greek word chloros, meaning "green") make up an extremely large and important class of green algae. Members may be unicellular, colonial, or filamentous. Cells of unicellular and colonial chlorophyceans may have two or more flagella.

There are about 2,650 living species of chlorophyceans. The main features of the class (and most plants) are the use of starch as the principal food reserve and the green chloroplasts with chlorophylls a and b. In spite of plant characteristics, this algal group is not directly related to early land plants.

Chlorophyceans are almost entirely restricted to freshwater and terrestrial habitats. Some members of this class have adapted to life on snow as snow algae. Snow algae cause snow to appear red-burgundy or orange in color because of high levels of unusual carotenoid pigments within the algal cells.

General characteristics of green Algae: Chlorophyceae is characterized by the presence of chlorophyll a and b, starch as reserve food, pyrenoids which are surrounded by starch sheath, cellulosic cell wall, sexual reproduction is isogamous to oogamous. The thallus ranges from single unicellular motile form to complex heterotrichous form. Most of the green algae are aquatic, about 90% of which are fresh water. Chlorophyceae have been divided into nine orders: Volvocales, Chlorococcales, Ulotrichales, Cladophorales, Chaetophorales, Oedogoniales, Conjugales, Siphonalesales and Charales. All the members of Oedogoniales and Conjugales are fresh water. Members of the two groups Siphonales and Ulvaceae, are marine. Members of Volvocales, Cladophorales and Chaetophorales are found both in the fresh water and marine waters. Vaucheria and Ulothrix are usually found in damp soil, in the form of sheets. Sexual reproduction occurs by isogamy, anisogamy and oogamy. The sex organs are usually unicellular. Zygote represents the diploid phase. The life cycle in different genera may be haplontic (e.g Chlamydomonas), diplontic (e.g Codium), isomorphic (e.g. Cladophra), or heteromorphic (e.g. Urospora).

Characteristics of Class Chlorophyceae:

- Fresh water or marine algae, with unicellular or multicellular body.
- Cells are eukaryotic, containing cellulose in the cell wall.
- Chief pigments include chlorophyll a and b, alpha, beta and gamma carotenes, lycopene, hexanthin, lentin, violaxanthin and astaxanthin.
- Reserve food includes starch and very rarely oils. Starch is elaborated in pyrenoids
- Sexual reproduction is isogamous, anisogamous or oogamous.
- Zoospore formation is common.
- Male gametes are flagellate and hence motile.
- Flagella are isokont (identical)
- Life cycle is mostly haplontic.

Classification of Chlorophyceae (Green Algae):

Chlorophyceae are divided into nine Orders which may be distinguished as follows:

1. Volvocales:

Unicellular or colonial; motile throughout life or form sedentary colonies which readily revert to a motile condition; reproduce asexually as well as sexually; mainly freshwater.

Examples: *Chlamydomonas, Sphaerella, Pandorina, Eudorina Volvox*, etc.

2. Chlorococcales:

Unicellular or colonial; non-motile in the vegetative condition, reproduced by zoospores or aplanospore; almost exclusive fresh water.

Examples: *Hydrodictyon, Pediastrum*, etc.

3. Ulotrichales:

Filamentous, simple and unbranched or cellular expanse with small cells; chloroplasts parietal, axial or stellate; sometimes filaments simple and with large multinucleate cell; mostly isogamous; mostly species freshwater, some are marine.

Examples: *Ulothrix, Ulva*, etc.

4. Cladophorales:

Simple or branched, filamentous with cells containing two to many nuclei and usually with elaborate large chloroplasts; mostly isogamous; freshwater and marine.

Example: *Cladophora*.

5. Chaetophorales:

Filamentous and sharply differentiated into prostrate and erect portions (heterotrichous); erect portion often reduced and the prostrate portion often forming discoid expanse; hairs of diverse type are often present; mostly isogamous and freshwater.

Examples: *Chaetophora, Enteromorpha, Draparnaldiopsis, Trentepohlia. Coleophaete, Protococcus,* etc.

6. Oedogoniales:

Simple or branched, filamentous; zoospores multiflagellate; oogamous; cell divisions characterized by intercalation of strips of membrane between two parts of the mother cell; entirely freshwater.

Example: *Oedogonium.*

7. Conjugales:

Unicellular or colonial (generally filamentous) with elaborate chloroplasts; motile gametes unknown; reproduction by vegetative cell division or by conjugation of amoeboid gametes; exclusively freshwater.

Examples: *Spirogyra, Zygnema, Cosmarium, Closterium*, etc.

8. Siphonales:

Filamentous; without septa or elaborately differentiated; all parts coenocytic; chloroplasts numerous and discoid; sexual reproduction mainly isogamous, sometimes oogamous, unknown in many cases; mostly marine.

Examples: *Vaucheria, Caulerpa, Bryopsis*, etc.

9. Charales:

Thallus well-differentiated into nodes and internodes; internodes sometimes corticated; branches of limited growth in whorls; chloroplasts numerous and discoid; cells usually uninucleate; reproduction vegetative and sexual with elaborate oogonia and antheridia; germination of zygote indirect; fresh water and brackish water.

Examples: *Chara, Nitella,* etc.

Alternation of Generations of Chlorophyceae:

Some Chlorophyceae show regular alternation of similar haploid and diploid generations, but in great majority the plant body is haploid with the zygote representing the only diploid phase. In the simplest case, as illustrated by *Chlamydomonas,* the life cycle consists of alternation of unicellular haploid and diploid phases.

This alternation is, however, not obligatory in the sense that the haploid phase will always give rise to the diploid phase, because there may be a formation of haploid individuals in succession before the formation of a diploid phase.

On the contrary, it is obligaloid one. This represents, therefore, the most primitive type of life cycle. Starting from this primitive condition, one finds an evolution of multicellular thalli in either phases (haploid, as in Spirogyra, *Oedogonium,* etc., or diploid, as in Codium) of the life cycle. In such cases, there is an alternation of a multicellular haploid generation with a unicellular diploid phase, or a multicellular diploid generation with a unicellular haploid phase.

Lastly, there may be an alternation between multicellular haploid and diploid generations (as in Cladophoraceae, Ulvaceae, Chaetophoraceae). The two generations may be quite identical (isomorphic) with each other, or they are morphologically different (heteromorphic).

****

8.

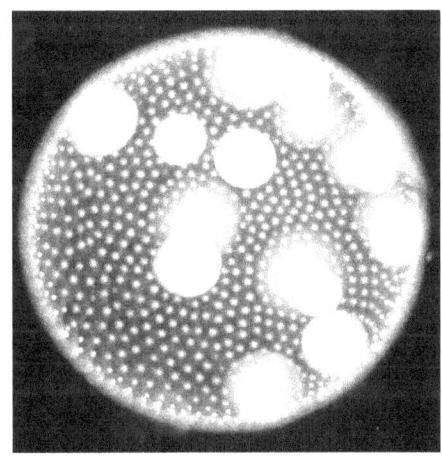

Volvocales: *Chlamydomona, Volvox*

The order Volvocales include about 60 genera and 500 species which are generally grouped in six families. The order includes the simplest members of Chlorophyceae which are either unicellular, e.g. *Chlamydomonas* or colonial, e.g., *Volvox, Pandorina, Endorina*,etc.

Distinguishing Features of the Order

(1) The plant body may be unicellular (*Chlamydomonas*) or colonial *(Pandorina, Eudorin* and *Volvox)*

2) The members are either motile throughout their life (Chlamydomonadinae) or forming sedentary colonies of palmelloid forms (*Tetrasporinea*) or dendroid forms *(Chlorodendrineae)*

(3) Species are usually fresh water

(4) Chloroplasts are usually cup or basin shaped with a single pyrenoid.

(5) Asexual reproduction by biflagellate zoospores, aplanospores or palmella stages.

(6) Sexual reproduction varies from Isogamy to oogamy.

Family: Chlamydomonadaceae:

This family includes algae possessing unicellular biflagellate vegetative cells. Asexual reproduction is by zoospores and hypnospores and sexual reproduction ranges from—isogamy, anisogamy to simple and advanced oogamy.

Chlamydomonas

Systematic position

Class: Chlorophyceae
Order: Volvocales
Family: Chlamydomonadaceae
Genus: *Chlamydomonas*

Occurrance

Chlamydomonas is common freshwater, motile unicellular alga present in stagnant rain water, ponds, pools, ditches and on moist soils. It grows mostly in waters rich in ammonium compounds and organic matter. *Chlamydomonas* halophile has been found in salt water. *Chlamydomonas yellowstonensis* is reported from snow in Yellow Stone, National Park of America. *C. nivalis* is reported from the red snow in Alpine and Arctic regions. There are about 325 species of *Chlamydomonas*.

Thallus Structure

It is a unicellular green alga with varying:

(i) Shapes: i.e., spherical, ovoid, subcylindrical or ellipsoidal. In many species, an anterior papilla-like outgrowth is present.

(ii) Protoplast: It is surrounded by a definite layer of cellulose wall and some species have a gelatinous pectic sheath outside the cellulose layer.

(iii) Chloroplast: Most of the species have a single cup shaped chloroplast. In *C. reticulata*, chloroplast is reticulate whereas in *C. arachne* and *C, steinii*, it is stellate, i.e., star-shaped.

(iv) Pyrenoid: Generally a single pyrenoid is present in a chloroplast but some of the species have two (*C. debaryana*) or several (*C. gigantea, C. sphagnicola*) or may be absent (*C. reticulata*). Pyrenoid is made of protein surrounded by starch plate or starch grains and it is the centre of starch formation.

(v) Flagella: At the anterior end arise two flagella, which come out through the same or separate canals. The flagellum arises from a tiny basal granule, the blepharoplast. The flagella are acronematic (whiplash) and of equal length, connected together by a transverse fibre called paradesmose which is attached to intranuclear centrosome of the nucleus by rhizoplast.

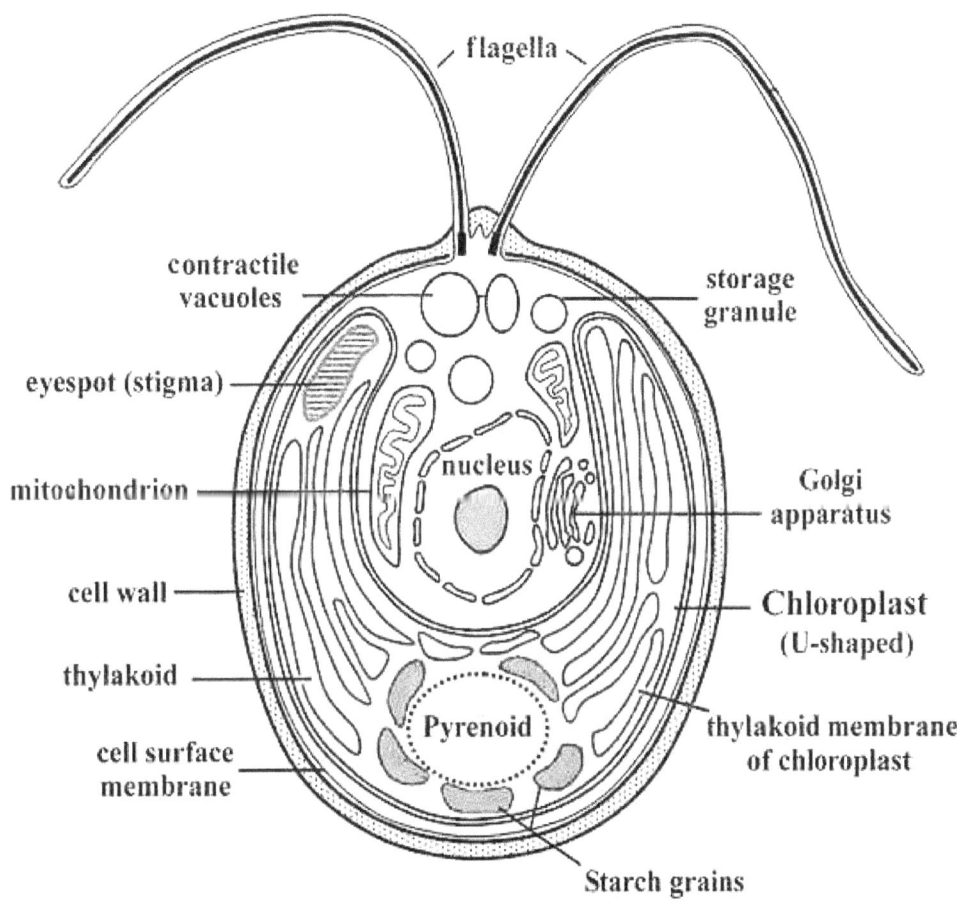

Fig.1. Typical structure of *Chlamydomonas* cell

In some species of *Chlamydomonas* e.g., *C. nasuta*, a sensitive neuro-motor apparatus is present. It controls movement of thallus in response to light, chemical and other stimuli.

The neuro-motor apparatus consists of two basal granules or blepharoplasts from which the flagella originate, a transverse cytoplasmic fibre paradesmos which connects two blepheroplasts, a cytoplasmic fibre rhizoplast connecting one blepheroplast with the centrosome and a small delicate fibre connecting centrosome with nucleolus.

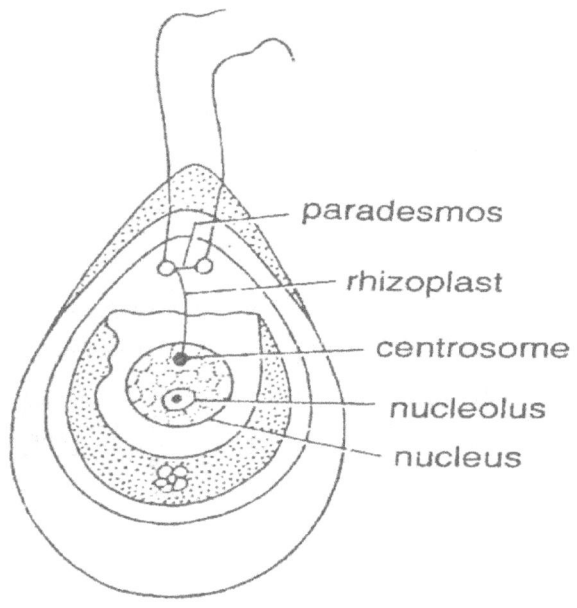

Chlamydomonas: Neuromotor apparatus.

Fig.2.

A tiny spot of an orange or reddish coloured spot known as stigma or eye spot (contains carotenoides) lies at the anterior end. It is a photoreceptive organ associated with the direction of the movement. The cell is bounded by the cell membrane and cell wall, which is thickened in the anterior portion. Each flagellum shows 9+2 pattern of structure. The chloroplast is surrounded by a unit membrane contain ing lamellae in pairs forming a disc. The number of discs occur in stacks ranging from 2-20. Each individual has a single nucleus lying in the colourless cytoplasm. Other structure include mitochondria, endoplasmic reticulum and dictyosome.

Reproduction

Chlamydomonas reproduce both sexually and asexually.

1) Asexual reproduction: It takes mainly by zoospores formation. But some species also reproduce by Aplanospores and palmella stage.

b) Zoospore formation: zoospore formation starts generally during night and under favorable conditions. At the time of zoospore formation the parent cell comes to rest, its flagella are withdrawn and becomes non motile. Then the protoplasmic content of the cell divide into 2, 4 or up to 16 daughter protoplast. Each daughter protoplast develops its cell wall and flagella. Thus, the structures are called zoospores. When zoospores are matured they are released out by rupturing parent cell wall. Thus formed zoospores are exactly similar with their parent cell but are similar in size. Each zoospore increases in its size and form individual.

b) **Palmella stage:** Under the unfavourable condition the cell becomes non motile by withdrawing its flagella then protoplasmic content divide into 2, 4 0r 8 daughter protoplast. But the daughter protoplast fails to develop their flagella. So they cannot escape out. Instead of releasing out, the protoplast divides into small daughter protoplast. As number of cells is increased the parent cell gelatinized and increases its circumference. As a result, certain no. of cells is embedded within parent cell wall and is called palmella stage. When the condition becomes favourable, each daughter protoplast is converted into zoospores and after releasing out it forms new indivi

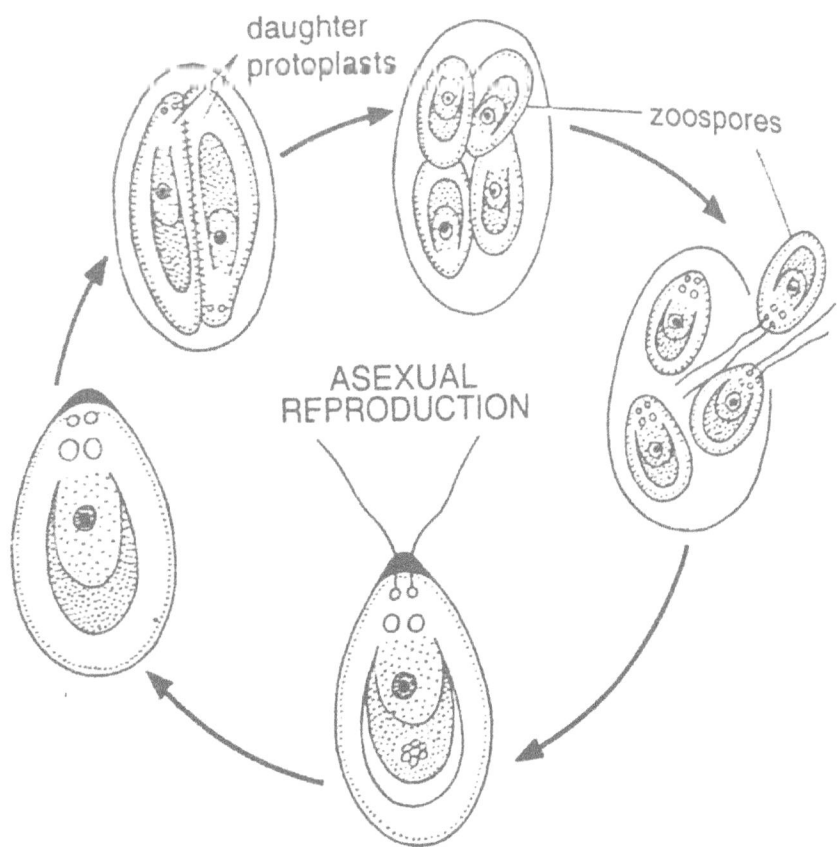

Fig. 3. Asexual reproduction: Zoospore formation

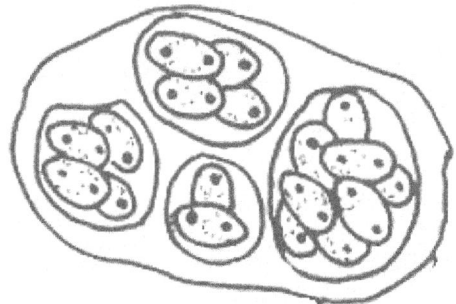

Fig.4. Palmella stage

c) **Aplanospores:** Chlamydomonas also reproduce asexually by the formation of Aplanospores/ hypnospore. Here, whole protoplasm rounds up to form a single Aplanospores. Each Aplanospores is capable of forming new plant under favorable condition.

2) Sexual reproduction

Sexual reproduction in Chlamydomonas takes place by isogamy, anisogamy and oogamy.

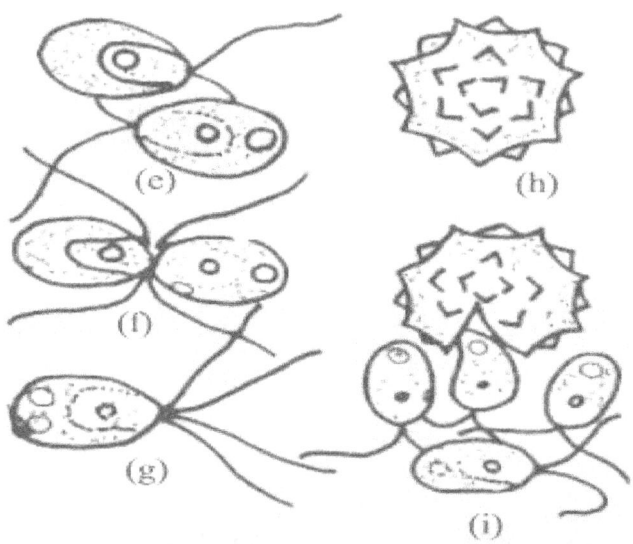

Fig.5. (e, f, g) Free swimming gametes and fusion of gametes (h) a resting zygote (i) 4 cells formed after meiosis of the zygote cell (zygospores)

a) Isogamy: it is most common and primitive type of sexual reproduction. During the process, any vegetative cell function as male gametes (+ strain) and next female gamete (-strain) which are similar in structure and size but physiologically different. When both of the gametes come to each other and attached to their anterior end. At the point of contact ell wall dissolves and is followed by fusion of protoplast. Here, cell appears as quadriflagellated since they swim in water with the help of flagella belonging to (+) strain. Then the protoplasmic content skip out form the gametes and is now is known as zygote. Then zygote comes from thick wall and undergo vegetative period.

b) Anisogamy: During this process, Microgamete swims towards macrogamete. Then one of the microgamete attached to the macrogamete with its anterior end. The cell wall between them gets dissolved and the content of micro gametes flows into macrogamete where fertilization takes place forming diploid zygote. Zygote after releasing out from female gametes undergo resting period.

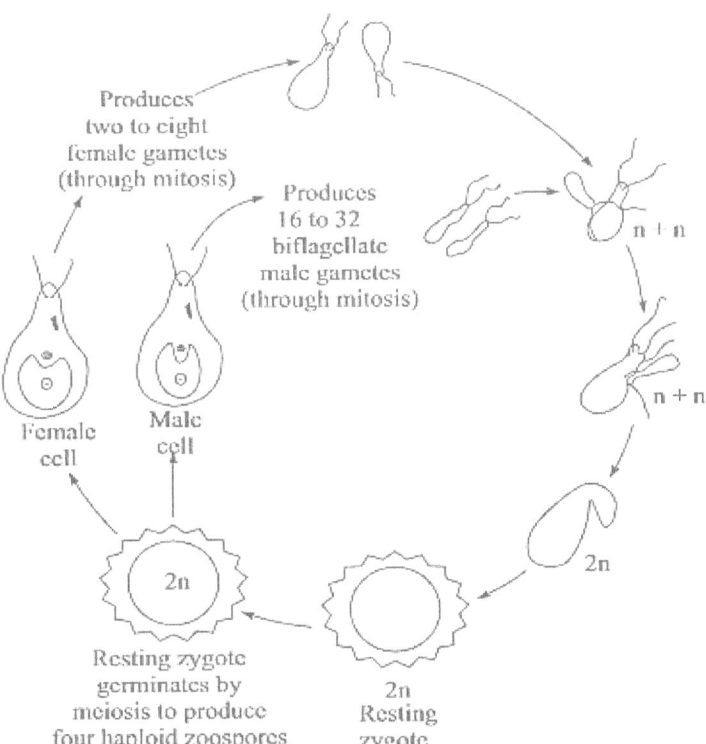

Fig.6. Sexual Reproduction by Anisogamy in *C.braunii*

c) Oogamous: Any vegetative cell withdraws its flagella then the protoplast content rounds off to form egg or ovum. The male cell produces 8-16 flagellated microgamete. The micro gametes move to female gamete and it is attached with the help of anterior end .cell wall between two gametes get dissolved and content of male gamete get migrated towards the female gamete resulting diploid zygote

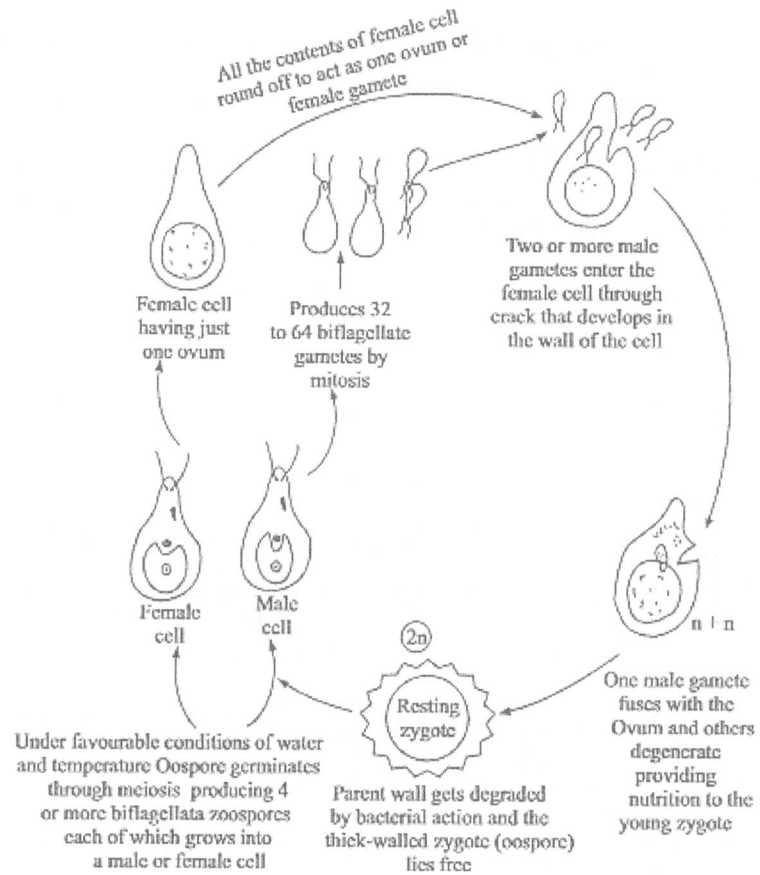

Fig. 7. Sexual Reproduction by Oogamy as in *C.oogamumand C. coccifera*

Germination of zygote:
After resting period, each zygote undergoes germination to produce new plant. Before germination, diploid nucleus of the zygote divides meotically forming 4 haploid nuclei. Each nucleus is surrounded cytoplasm forming 4 daughter protoplast. Each daughter protoplast is converted into zoospore after developing its cell wall and flagella.

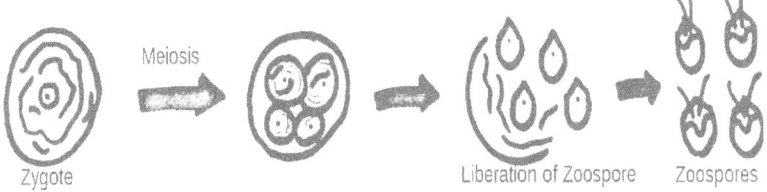

When zoospores are matured they are released out and each zoospore increases in its size to form new plant.

Fig.8. Germination of zygote

Family- Volvocaceae

The **Volvocaceae** are a family of unicellular or colonial biflagellates, including the typical genus *Volvox*. The family was named by Ehrenberg in 1834, and is known in older classifications as the Volvocidae. All species are colonial and inhabit freshwater environments.

Volvox

Systematic position
Class: Chlorophyceae
Order: Volvocales
Family: Volvocaceae
Genus: *Volvox*

Occurrence

Volvox is free floating fresh water green algae. *Volvox* grows as planktons on surface of water bodies like temporary and permanent ponds, lakes and water tanks. During rainy season due to its fast growth the surface of water bodies become green. The Volvox colonies appear as green rolling balls on surface of water.

Volvox is represented by about 20 species:
Some common Indian species are—*Volvox globator, V aureus, V. prolificus, V. africanus and V. rousseletii.*

Thallus Structure

Volvox occurs in the colony because it is a coenobial form (hollow ball) like a structure. In the young colony, the vegetative cells are similar in size and green in

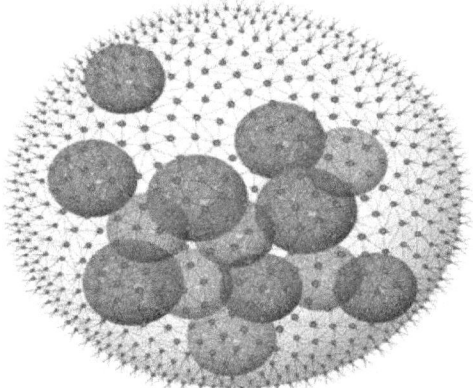

Fig.9. Volvox: **colony**

color. Each motile colony (coenobium) is free-swimming and appears as small pinhead like spherical to ovoid shape with hollow mucilaginous mass which consists of numerous small pear-shaped cells arranged in a single layer joined with one another by delicate strands of cytoplasm within the periphery of the gelatinous colonial matrix. The total number of cells in the colony varies from about 500 (*Volox aureus*) to about 2000 or more (*Volvox globate*). Each colony develops the following three types of cells:

Cell Structure of *Volvox*

The cell of the coenobium varies based on species and is mostly ovoid-shaped. Each cell measured about 16.25 μm in length. The cell wall encloses a mass of protoplast. The cell wall is thin and firm in nature composed of cellulose. Protoplast contains a basal cup-shaped chloroplast with several pyrenoids (*Volvox aureus*) or plate-shaped with a single pyrenoid (*Volovox globator*), a central nucleus, reddish-brown eyespot surrounded by a plasma membrane. The central cytoplasm possesses mitochondria, endoplasmic reticulum, ribosome, dictyosomes, etc. At the apical portion of the cell, two equal length whiplash types of flagella arise from the two basal granules, i.e. the blepharoplast. At the base of the flagella, 2-3 contractile vacuoles are present. They act as excretory organs.

Reproduction

Asexual Reproduction in *Volvox*

Asexual reproduction takes place during summer in a rapid manner under favorable conditions. In a young colony known, as coenobium, all the cells are the same but later, a few cells of the posterior half of the *Volvox* colony increase in size by storing up the food. These cells become enlarged in size and form asexual reproductive cells, called gonidia or parthenogonidia. The number of gonidia varies from 2-50 in each coenobium. They drop their flagella, become rounded in outline, contain dense cytoplasm and lie within the globose mucilaginous sac which projects towards the inside of the colony.

Each gonidium divides repeatedly and produces a spherical group of daughter cells. In this case, all cells are held together to form a new daughter colony. The divisions of the gonidial protoplast occurring in the formation of a daughter colony are always longitudinal and all cells of each cell generation divide at the same time. Continue longitudinal divisions of daughter cells occur simultaneously and produce several cell generations.

In the second generation, four cells are arranged quadrately while in the third cell generation, the 8 cells are crucially arranged, to form a curved plate, known as the plakea stage. Plakea takes the shape of a hollow sphere at the end of the 16-celled stage. Next, a pore called the phialopore is formed at the anterior pole of the daughter colony, when the cell division stops.

The young daughter colony turns itself out by inverting through the phialopore. After this, the cells develop flagella and the daughter colony escapes by moving through a pore-like opening at the free face of the sac. Finally, the daughter colony comes out due to the rupture or decay of the mother colony or coenobium.

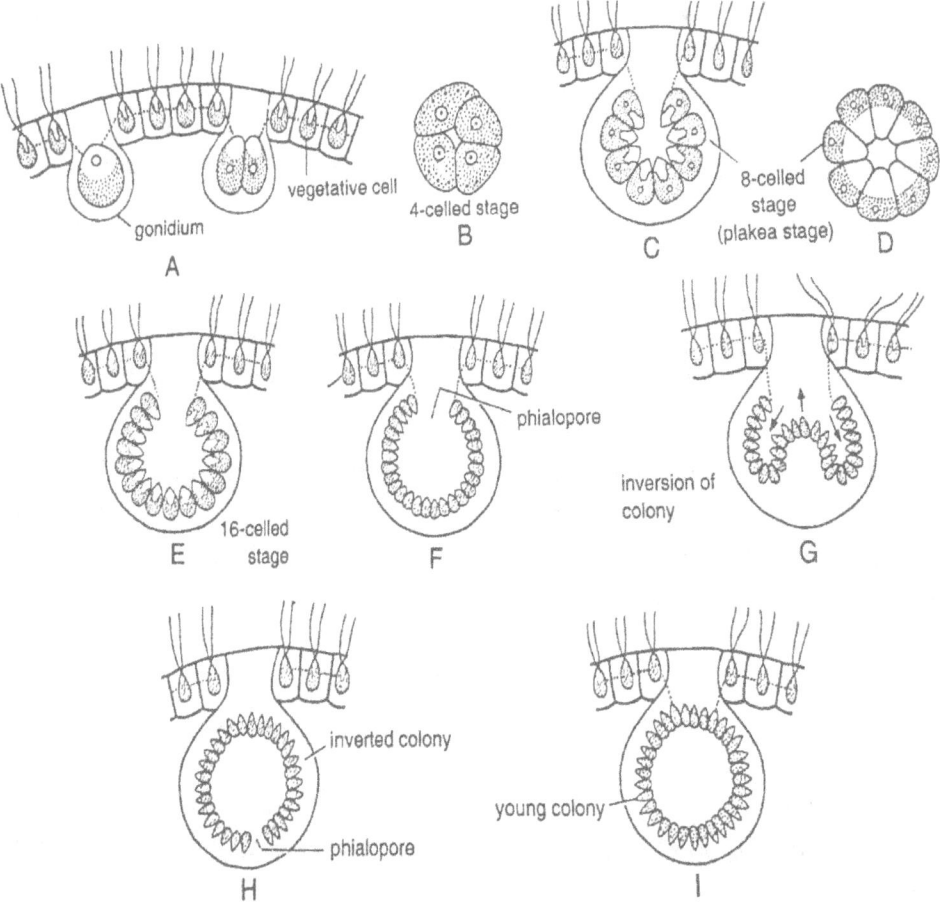

A-I, *Volvox*: **Asexual reproduction** (development of daughter colony); A-D. Formation or plakea stage, E. 16-celled stage, F-H. Stages of inversion, I. Daughter colony.

Fig.10.

Sexual Reproduction in *Volvox*

Sexual reproduction of *Volovx* is of oogamous type. The coenobium may be homothallic or heterothallic based on species. The posterior half of the coenobium forms some specialized enlarged cells or gametangia which may be either the female sex organs (oogonia) or the male sex organ (antheridia). They are produced fewer in number.

During the development of gametangia (oogonia or antheridia), the cell becomes rounded and enlarged and cast off flagella but they remain linked with other cells through fine protoplasmic threads. In this case, the male sex organ or gametangium is called antheridium and the female sex organ or gametangium is known as oogonium.

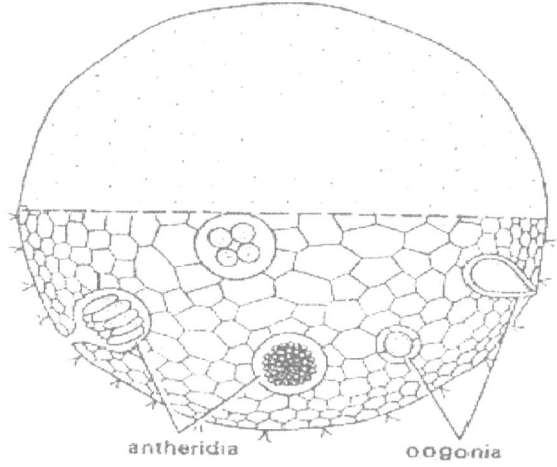

Fig.11. Sex organs in a coenobium.

In the monoecious species, such as *Volvox globator*, antheridia and oogonia are formed on the same coenobium but in the dioecious species such as in *Volvox aureus*, antheridia and oogonia are formed on different coenobium. In monoecious species, antheridia develop first and the fertilization occurs between the antherozoid and ovum of other plants.

Antheridium

Antheridium also possesses an enlarged structure similar to gonidia. The protoplast of an antheridium divides repeatedly to form 16, 32, 64, 128 or more small, spindle-shaped, yellowish, biflagellate antherozoids. Each antherozoid contains a single nucleus and a small pale green or yellow-green chloroplast.

Oogonium

Oogonium is a unicellular, enlarged, semi flask-shaped cell, with a gelatinous sheath-like wall. The protoplast of each oogonium forms a larger uni-nucleate spherical oosphere or egg with a beak-like protrusion towards one side. Antherozoid enters into the oogonium through this end.

The oosphere possesses a parietal chloroplast, pyrenoids, and a centrally placed large nucleus. Oogonium absorbs reserve substances from the neighboring cells through the protoplasmic strands.

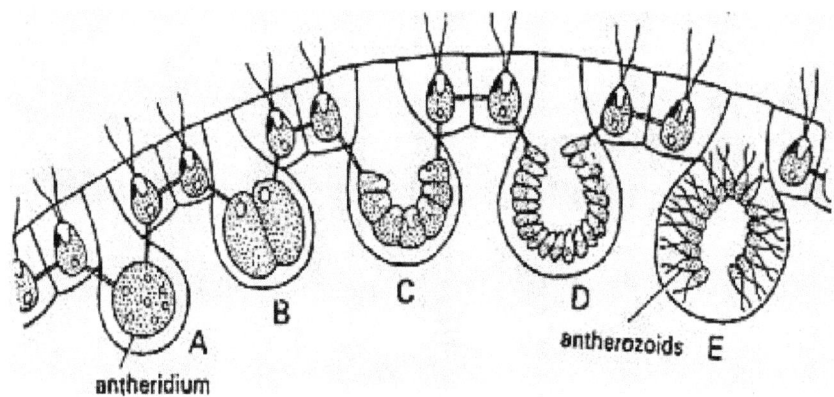

Fig.12. Development of Antherozoids

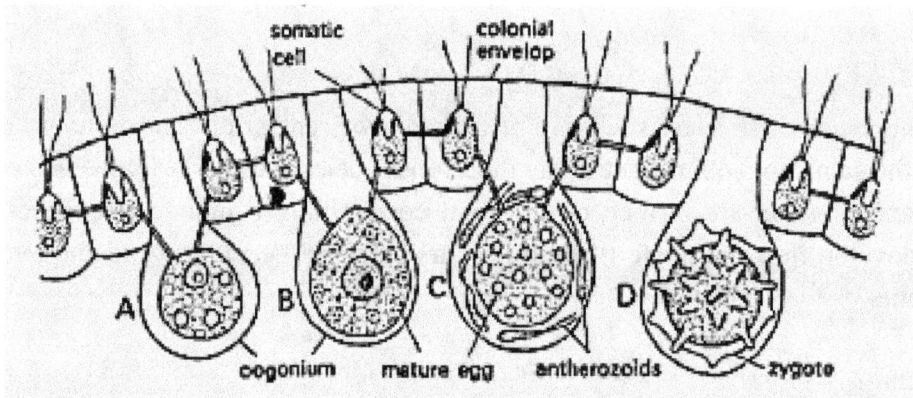

Fig.13. A.D. *Volvox* A-D. Development of ogonium, C. Fertilization, D. Zygotes

Fertilization

During the fertilization, antherozoids after liberation from the antheridium swim about as a group and remain intact until they reach the egg. Only one antherozoid fuses with the egg resulting in the formation of a zygote or oospore. After fertilization, the zygote develops a thick wall around it.

The protoplast of the zygote becomes orange-red in color. The zygote comes out of the parent coenobium by the disintegration of the gelatinous matrix of the coenobium and sinks to the bottom of the water and undergoes a period of rest.

Germination of Zygote

The zygote reserves enough food materials with other inclusions. The two outer layers of the zygote split and gelatinize. The outer layer is known as exospore which may be smooth in *Volvox globator* or spiny in *Volvox speematospaera*. The middle layer is known as mesopore while the inner layer is endospore. After releasing from the coenobium by disintegrating the gelatinous matrix, the zygote settles down at the bottom of the water body and may remain intact for several years.

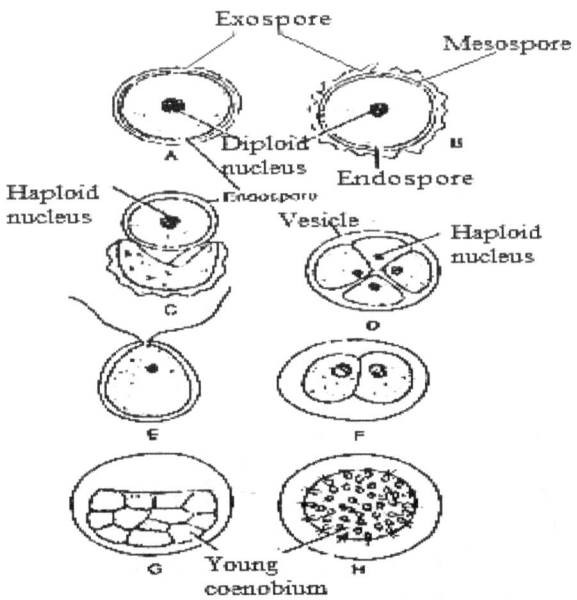

Image showing germination of zygote and formation of young coenobium

Fig.14.

Under favorable conditions, the inner wall layer extrudes out in the form of a vesicle and surrounds the protoplast of the zygote. Diploid zygote nucleus divides meiotically into four haploid nuclei; of these, 3 degenerates and the remaining one nucleus survive with cytoplasmic contents escapes from the vesicle. At this stage, it is known as a swarmer who swims freely and forms a zoospore and develops into a new coenobium (colony).

Each coenobium also contains a smaller number of cells which perform asexual reproduction for the next several generations. In the case of *Volvox rouseletti* and *Volvox minor*, the zygote's protoplasm is changed into a single zoospore and it divides again to form a new coenobium.

Pandorina

Systematic position

Class: Chlorophyceae
Order: Volvocales
Family: Volvocaceae
Genus: *Pandorina*

Occurrance

This alga is abundant in fresh-water ponds and ditches.

Thallus Structure

Typical chlamydomonas cells are closely packed together to form spherical to ellipsoidal coenobium (Fig.15A). The coenobium is composed of relatively definite number of cells ranging from 4 to 32, which are enclosed in a homogeneous gelatinous matrix with an outer watery sheath.

In a coenobium cells are so compact that they are often angular as a result of mutual pressure. The coenobia are hollow with the cells forming a single layer in the peripheral gelatinous matrix. The two flagella of each cell pass out through specially differentiated canals of the gelatinous matrix. The coenobia exhibit a marked polarity with the constituent cells showing a definite orientation.

Reproduction

Asexual reproduction is by the formation of daughter coenobia, prior to which the parent coenobium ceases to be motile and sinks to the bottom of the pool. Then each cell produces by several divisions, daughter coenobia (Fig. 15B to D). All the cells of the parent coenobium behave alike in reproduction. The daughter cells become inverted and assume a spherical shape and ultimately each cell produces two flagella.

The daughter coenobia are ultimately liberated by the disintegration of the parent envelope (Fig. 15E).

Sexual reproduction is anisogamous. Divisions leading to the formation of gametes are identical with those of asexual reproduction process. But the fusing gametes are always unequal in size with the larger one more sluggish than the smaller counterpart.

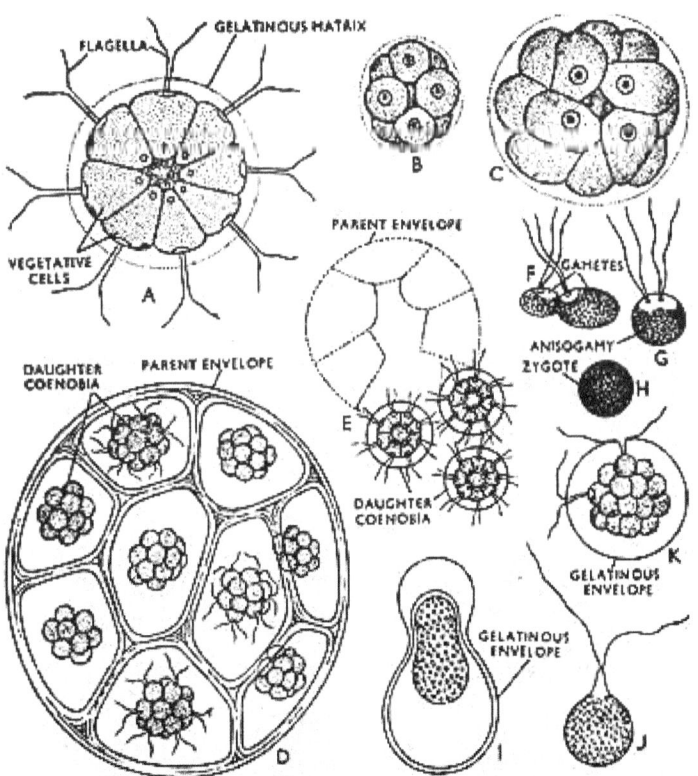

Pandorina sp. A. Mature coenobium. B-D. Stages in coenobia development. E. Liberation of daughter coenobia breaking parent envelope. F-H. Stages in anisogamy resulting in zygote formation. I. Germination of zygote. J. Swarmer. K. New coenobium developed from the swarmer.

Fig.15.

Gametes from different coenobia only fuse. Fusion may be terminal or lateral (Fig.15F). The quadriflagellate zygote remains motile for a while before it loses its flagella and secretes a wall (Fig. 15G & H).

Only one of the four resulting cells of the germinating zygote normally survives forming a swarmer (Fig. 15 I & J). After swarming for a short time the swarmer withdraws its flagella and secretes a gelatinous envelope, and divides and re-divides to form a new coenobium (Fig. 15K).

Some Indian species:

Pandorina morum(Mull.) Bory.; P. morumf. major Iyeng.

Special Features:

1. The coenobia are oblong to subspherical and are comparatively of small size.

2. Coenobium of typical Chlamydomonas-cells.

3. Anisogamous sexual reproduction.

Eudorina

Systematic position
- **Class:** Chlorophyceae
- **Order:** Volvocales
- **Family:** Volvocaceae
- **Genus:** *Eudorina*

Occurrance

Commonly found in fresh-water pools and ponds, particularly during the rainy season.

Thallus Structure

Spherical or ellipsoidal coenobia are distinguishable into anterior and posterior portions. There may be sixteen to sixty-four biflagellate, globose or somewhat pear- shaped cells present in the coenobium, which are not closely packed and are sometimes arranged in well-marked transverse series distributed in both anterior and posterior tiers (Fig. 16 A). Cells with usual Chlamydomonad-structure are embedded in a mucus investment in the coenobium (Fig. 16B). Almost all the cells of the coenobium give rise to new daughter coenobia with the exception of a few anterior cells which are much larger and do not reproduce. The process of daughter coenobium formation is same as that of Pandorina.

Reproduction

Sexual reproduction is advanced anisogamy indicating close approach to oogamy with the formation of typical antherozoids and non-motile female gamete—ovum (Fig.16E).

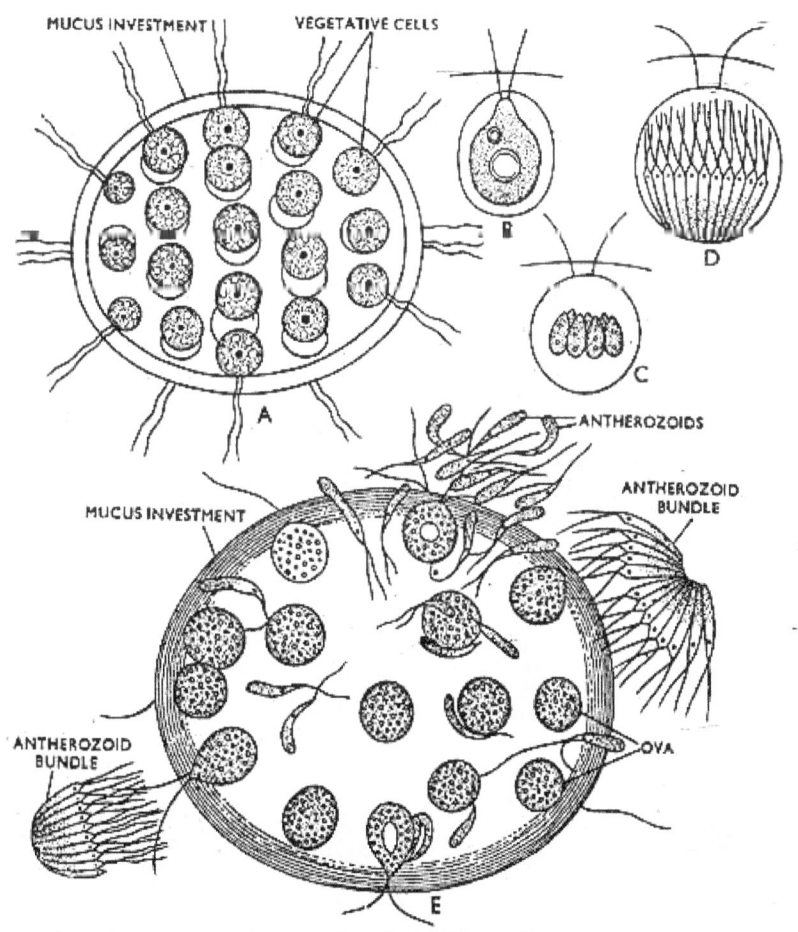

Eudorina elegans. A. Young coenobium. B. Individual cell of the coenobium. C–D. Formation of antherozoids. E. Coenobium showing stages in sexual reproduction.

Fig.16.

Coenobia are monoecious or dioecious. The anterior cells of a monoecious coenobium give rise to the antherozoids and the posterior ones form ova. But in heterothallic species there are distinct male and female coenobia which develop male and female gametes respectively.

The vegetative cells of the coenobium divide and revived forming anterozoids (Fig.16C & D). Whereas, during female gamete formation, the vegetative cells arc directly converted into female gametes—ova without undergoing any division. The zygote remains within the parent coenobium until it disintegrates.

Zygote germinates to give rise to a single swarmer which swims about for a time and then, in the same manner as a vegetative cell, divides and redivides to form a new coenobium.

Some Indian species:

Eudorina elegans Ehrb.; *E. indica* Iyeng.

Special features:

1. The coenobia are ellipsoidal and larger than in *Pandorina*.

2. Anisogamous sexual reproduction between an immobile large ovum and an elongated spindle-shaped biflagellate antherozoid exhibiting oogamous condition.

****

9.

Chlorococcales: Chlorella, *Chlorococcum, Hydrodictyon* and *Scenedesmus*

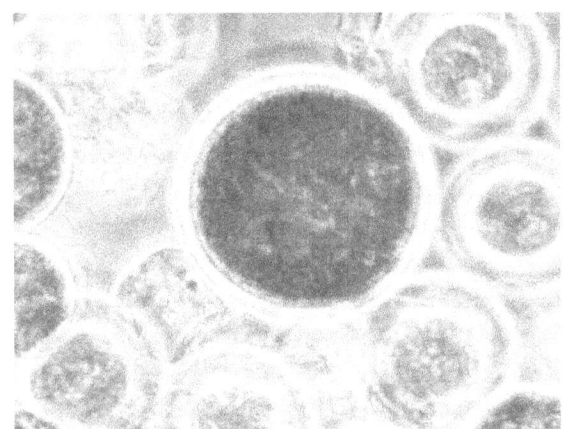

The members of the order Chlorococcales are exclusively freshwater. The vegetative thallus either consists of only a single cell or a coenobium composed of a definite number of cells arranged in a specific manner. The vegetative cell is uninucleate and non-motile. Organelles like eye spot and contractile vacuoles, though absent in the vegetative cell, may appear in motile reproductive cells. Vegetative reproduction occurs by zoospores or aplanospores which resemble the parent cell. Sexual reproduction is isogamous, anisogamous or oogamous.

Fritsch (1935) divided Chlorococcales into eight families:

(1) Chlorococcaceae

2) Eremosphaeraceae

(3) Chlorellaceae

(4) Oocystaceae

(5) Selenastraceae

(6) Dictyosphaeriaceae

(7) Hydrodictyaceae

(8) Coelastraceae.

He included the genus Scenede smus in the family Coelastraceae alongwith Coelastrum, Hofmania and Tetradesmus. Smith (1955) also recognised eight families in Chlorococcales but placed Scenedesmus in a separate family, Scenedesmaceae.

Family: Chlorellaceae

The family Chlorellaceae includes non-motile, solitary (e.g. *Chlorella*) or colonial (e.g., *Radiococcus*) forms. The colonial forms show indefinite orientation of cells. The uninucleate cell has a single chloroplast, usually without pyrenoid. Reproduction takes place by autospores.

Chlorella

Systematic Position

Class: Chlorophyceae

Order: Chlorococcales

Family: Chlorellaceae

Genus: *Chlorella*

Occurrence:

With its eight species, *Chlorella* has a very varied habitat. It may be found on damp soils, walls and exudation of trees. *C. parasitica* (called zoochlorella) is found as endosymbiont in the tissues of invertebrates such as Hydra and Paramecium. *C.lichina* is found as phycobiont in lichens. Common Indian Species: Chlorella vulgaris, *C. conductrix, C ellipsoidea* and *C. gonglomerata* have been recorded from India.

Thallus Structure

Chlorella is a unicellular, non-motile green alga (Fig. 1.A). The cell is solitary, small (2-12 gm), and spherical to ellipsoidal in shape. It is surrounded by a thin cellulose cell wall. The cell contains a centrally placed nucleus and a cup-shaped parietal chloroplast with a pyrenoid.

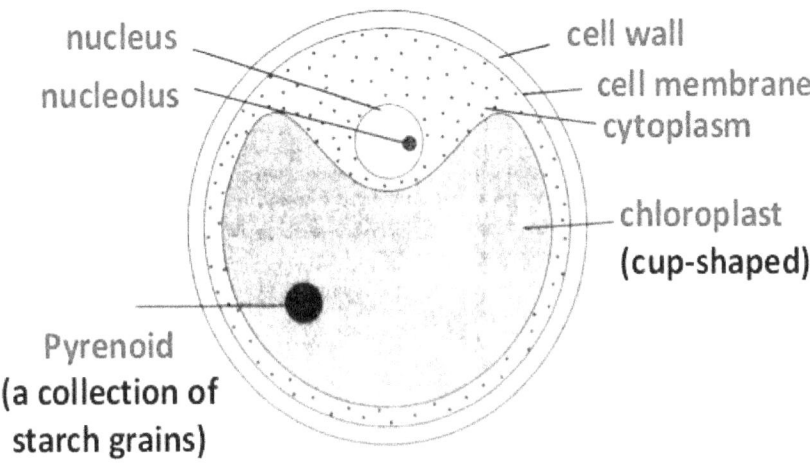

Fig. 1. *Chlorella*: Cell Structure: Under light microscope

Under electron-microscope the Chlorella cell seems to be enclosed by a two-unit membrane and its C-shaped chloroplast occupies the major portion of the cell. The nuclear membrane is bilayered and porus. Mitochondria, Golgi bodies, and a few vacuoles are also found in the cytoplasm.

Reproducion

Reproduction is effected by the production of autospores which are developed within a cell by the successive division of the protoplast (Fig. 3B).-Two or even sixteen autospores may be formed in each cell which are liberated out by the rupture of the parent cell wall (Fig. 3C). These immobile spores form new individuals.

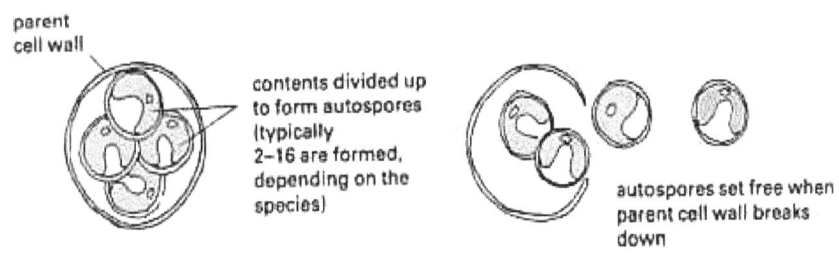

Fig. 2. Formation and release of autospores

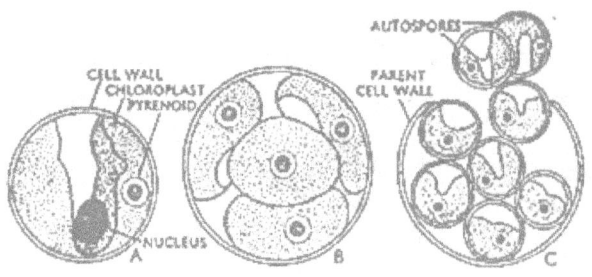

Chlorella vulgaris. A. Single cell showing structural details. B. Stage in division. C. Liberation of autospores.

Fig.3.

Chlorococcum

Systematic position:

Class: Chlorophyceae
Order: Chlorococcales
Family: Chlorococcaceae
Genus: *Chlorococcum*

Occurrence:

Chlorococcum is generally found in beneath the soil surface. It occurs in abundance on damp soil or brickwork.

Thallus Structure

Chlorococcum is unicellular with spherical or slightly oblong cells of varied size. The cells may be solitary or in irregular clumps, sometimes forming films on moist or submerged surfaces. The mucilage is thin and inconspicuous. Each cell has a single cup-shaped, parietal chloroplast with a single pyrenoid.

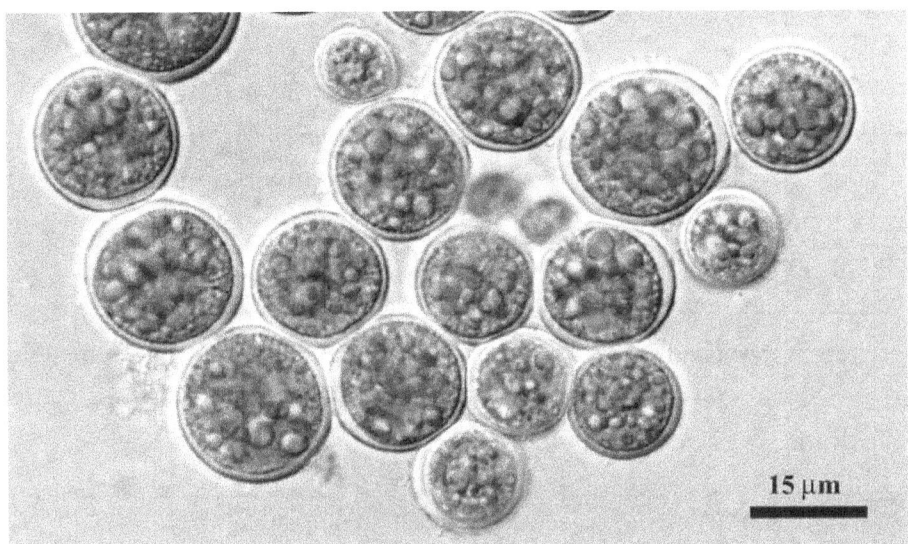

Fig.4. *Chlorococcum*

Reproduction:

Asexual reproduction:

- By zoospores,
- By aplanospores
- By Palmella stage.

Zoospores are formed under favourable conditions of growth. Mitotically nucleus divided many times followed by the division of the protoplast. Each uninucleated bit of protoplast converted a biflagellate zoospore. They are liberated by the dissolution of the parent cell wall. After a sort swarming period, withdraws the flagella in each zoospore, secretes a wall and develops directly into a vegetative cell.

Aplanospores are produced as zoospores under unfavorable conditions. On return of favorable conditions each aplanospore converted a new plant.

Sometimes, under desiccation, a **palmella stage** is produced due to the gelatinization of cell wall in a group of cells.

Sexual reproduction: It takes place by the formation of isogametes. Motile isogametes are formed as zoospores, but large in number. After Syngamy formed quadriflagellate zygospore. As the zygospore matures, the flagella are degenerated and nucleus divides meiotically once, followed by mitosis to produce several zoospores, which upon release lose the flagella and directly develop a new plant.

Family Hydrodictyaceae

The members of this family form non-motile coenobia which are more advanced than the coenobia of Coelastraceae. They reproduce asexually by zoospores. The zoospores become apposed to each other within the parent cell wall and as such a miniature colony is formed within the parent wall. The sexual reproduction is isogamous and the gametes are motile.

Hydrodictyon

Systematic Position

Class: Chlorophyceae

Order: Chlorococcales

Family: Hydrodictyaceae

Genus: *Hydrodictyon*

Occurrence

Hydrodictyon or 'water net is a free floating fresh water alga. It grows luxuriantly during late summer and spring season and sometimes almost completely covers the pond. The genus is represented by five species. *Hydrodictyon indicum* and *H. reticulatum* are common Indian species which occur in temporaryor permanent water reservoirs.

Thallus Structure

Hydrodictyon, known as the "water net", has large colonies composed of elongate cells linked in a reticulated, net-like pattern. Each cell is connected at its end walls to two other cells, forming meshes of five or six cells. The colonies can be as large 4-6 cm wide and 1 m long. The cells are coenocytic and multinucleate. Young cells have a single parietal chloroplast with one pyrenoid, while in older cells the chloroplasts become net-like with multiple pyrenoids. Large vacuoles take up most of the cellular space and push the cytoplasm around the periphery of the cell.

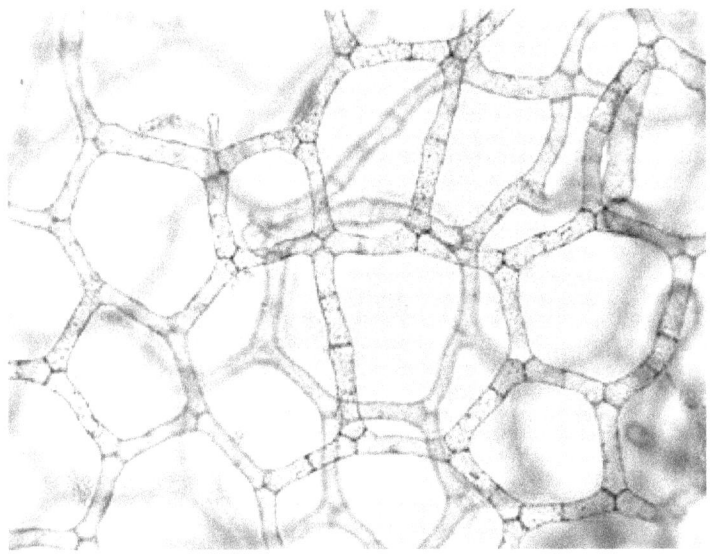

Fig.5. *Hydrodictyon*

Cell Structure of Hydrodictyon:

Each cell is long, cylindrical or ovoid in shape. Its internal structure can be differentiated into two parts: cell wall and protoplasm. Cell wall is two layered and is made up of cellulose. It encloses protoplasm. When young, the cells are uninucleate, but at maturity they become multinucleate (coenocytic).

Cells contain reticulate chloroplast with many pyrenoids (Fig. 6C). All the typical structures of green algae like ribosomes, mitochondria, dictyosomes are also present. As the cell matures, a central vacuole appears and the protoplasm becomes peripheral.

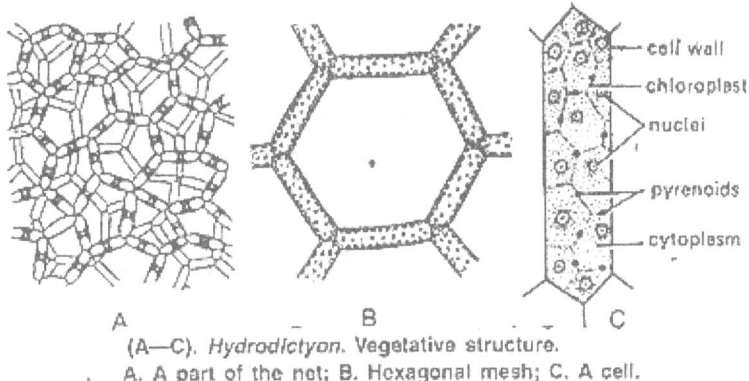

(A—C). *Hydrodictyon*. Vegetative structure.
A. A part of the net; B. Hexagonal mesh; C. A cell.

Fig.6.

Reproduction

It is of three types: Vegetative, asexual and sexual.

Vegetative Reproduction:

It takes place by fragmentation. Coenobium breaks up into small pieces called fragments. Which have capability to grow into new colonies. It may be due to water currents and movement of aquatic animals.

Asexual Reproduction:

It takes place by the formation of auto colonies or daughter colonies. These colonies are formed by the biflagellate, uninucleate zoospores. Under favourable conditions each coenocytic cell behaves as zoosporangium. Its nuclei undergo mitotic divisions to form a large number of nuclei (7000-20000). Protoplasm gets segmented into as many segments as there are nuclei. Each segment gets surrounded by small amount of cytoplasm, a limiting membrane and develops two whiplash type equal flagella and represents biflagellate zoospore. In Hydrodictyon a peculiar phenomenon is observed. The zoospores thus formed are never liberated outside the parent cell. They remain motile within the restricted region i.e., within the cell. After swimming inside the cell, they ultimately withdraw their flagella and get themselves arranged into characteristic hexagonal or pentagonal fashion to form a new net. This new net is called auto colony or daughter colony. The auto colonies are liberated by disintegration of the parent cell wall. The number of the cells in the daughter colony is fixed. Further growth of the coenobium is entirely due to increase in the cell size and not the number of the cells.

Sexual Reproduction:

It is isogamous. Any vegetative cell of the coenobium can function as gametangium. The biflagellate gametes are produced by the cleavage of the protoplasm of the gametangia like that of zoospores. Each segment metamorphoses into biflagellate gamete (smaller than 200 spores). Two isogametes from the same coenobium fuse laterally to form zygospore, which secretes a thick wall around them. At the time of spring season, the zygospore divides by meiosis to form four nuclei.

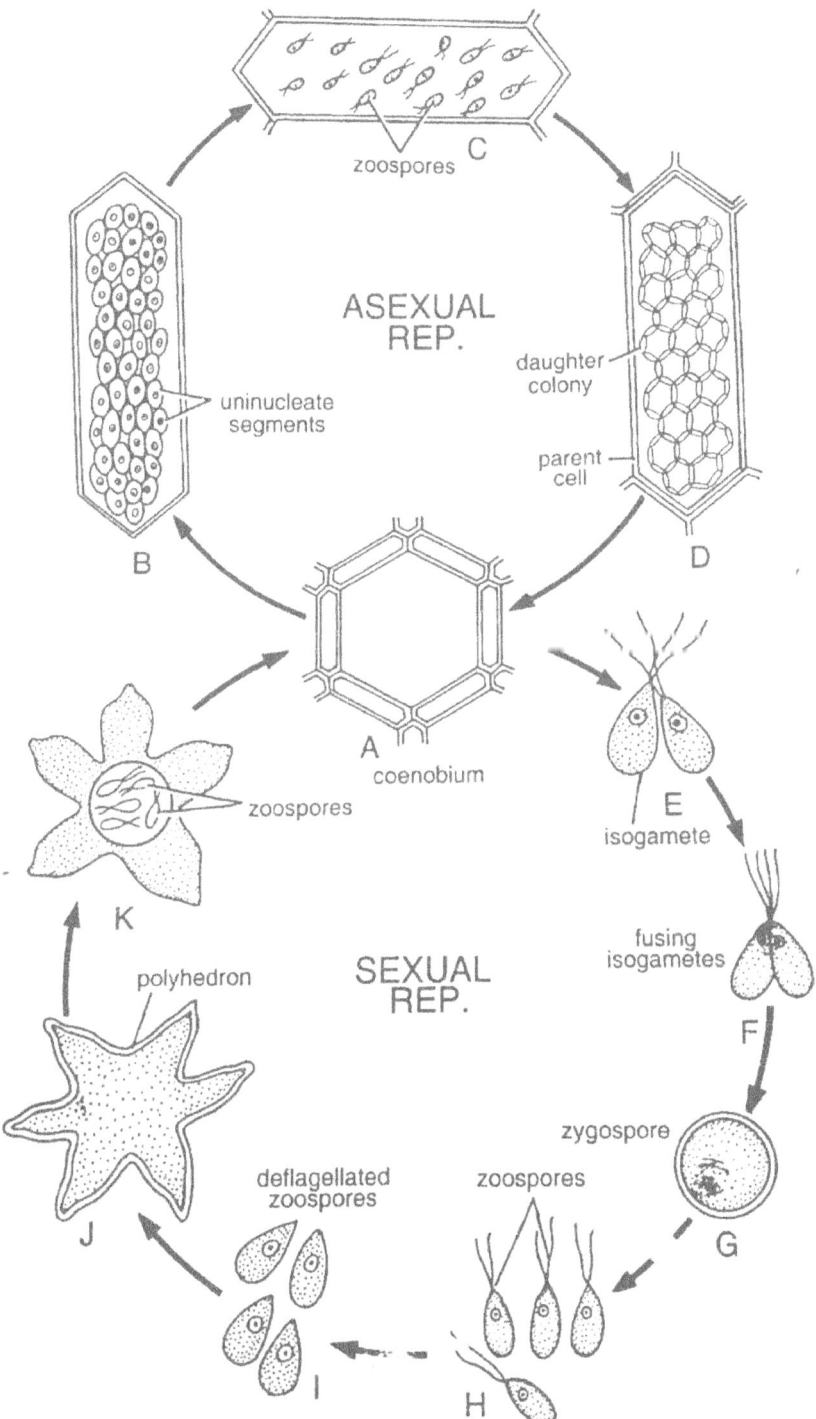

A-K. *Hydrodictyon*: **A-D.** Stages in asexual reproduction, **E-K.** Stages in sexual reproduction.

Fig.7.

Each uninucleate segment develops into biflagellate zoospore having discoid chloroplasts. After a short period of motility, the zoospore looses its flagella, comes to rest, secretes a thin wall and develops into polyhedron (polyhedral cell). Its haploid nucleus divides to form 200-300 zoospores eventually arranged in new daughter net within the polyhedron. The young net is liberated by rupture of this polyhedron wall. The net of *Hydrodictyon* is formed by joining together of large number of zoospores (each acts as individual cell).

Family: Coelastraceae:

In this family the vegetative cells of the coenobia are grouped in one plane with the long axes of the cells parallel to one another. Each cell possesses laminate chloroplast. Reproduction is by division of cells of the coenobium.

Scenedesmus

Systematic Position

Class: Chlorophyceae

Order: Chlorococcales

Family: Coelastraceae

Genus: *Scenedesmus*

Occurrence

Scenedesmus, a generally freshwater alga, has more than 100 species found in in ponds and lakes. It occurs as an almost pure culture in aquaria and in jars where water is left standing for some time. *Scenedesmus obliquus, S. dimorphus, S. quadricauda* and *S. denticulatus* are some common species of the genus.

Thallus Structure

Scenedesmus is a green alga that forms small colonies equipped with spines for flotation. The morphology varies with conditions and species, but this arrangement of 4 cells with 4 spines is quite common. Each cell is about 10 micrometres long. The spines are flotation devices, keeping the colony floating high in the water column near the source of light needed for photosynthesis. This may seem surprising, but at this microscopic scale water behaves much like treacle (low Reynold's number) and spines are efficient flotation devices.

The morphology of the coenobium can be modified considerably by varying the composition of the culture medium. In a medium with low phosphorus or low salt Concentration, Scenedesmus can be induced to grow as unicells. The shape of the cells varies from fusiformn (e.g., *S.obliquus* and *S. dimorphus*) to ellipsoidal (e.g., *S. quadricauda* and *S. denticulatus*). The cell wall is smooth (e.g., *S. bijuga*) or have processes that are elaborations of the cell envelope and are of taxonomic importance. For example, in *S. quadricauda* and *S. armatus* the terminal cell bears long spines in *S. denticulatus and S. hystrix* small spiny projections are present on the entire cell surface and in *S. acuminatus* the cell has a tuft of bristles at each end. The presence of these processes is probably related to the planktonic mode of life and they help in floatation of the colony. Each cell of the coenobium is uninucleate and possesses a single laminate chloroplast, containing

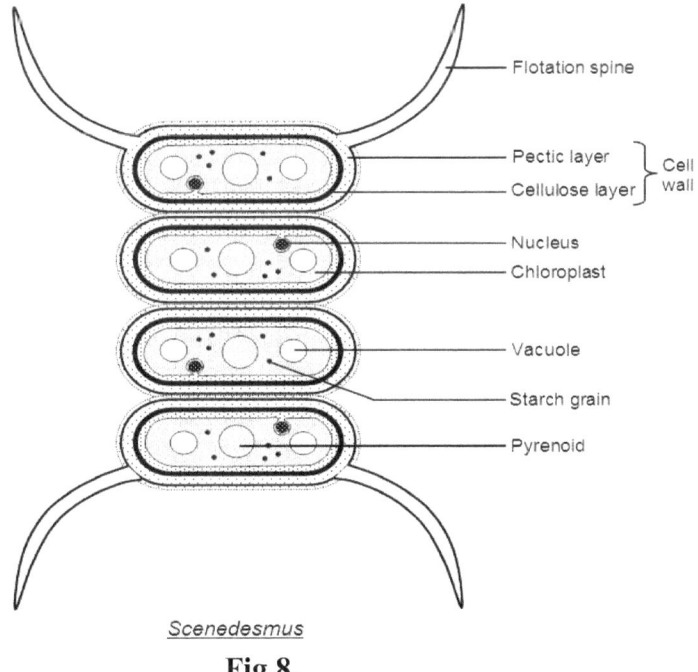

Scenedesmus
Fig.8.

Reproduction

During replication, the mother cell enlarges and becomes multinucleate after multiple divisions. The cytoplasm then is cleaved into uninucleate daughter cells, usually developing as non-motile autospores. These daughter cells typically link up with other daughter cells to form a colony within the parental cell wall to be later released. The cells progress through a typical mitotic cycle similar to other members of

Chlorophyceae, with the cytoplasm of the daughter cells becoming very dense. Eventually the mother cell wall breaks and releases the spores which adopt a normal cellular appearance. The cells at either end of the coenobium are different in morphology from those in the center. How the cells adhere to one another during development is still unclear, but it is known that a trilaminar sheath (TLS), composed of algaenan, is one of the first exterior structures to form, developing in patches before growing to connect into one continuous layer. The ornamented layer is the last component to develop.

****

10.

Ulotrichales: *Ulva, Enteromorpha, Ulothrix*

- Ulotrichales occur mostly in freshwater or on soil. A few species are marine.
- Plant body is filamentous, and the filaments are unbranched. Ulvaceae, however, show foliaceous and parenchymatous thalli.
- Cells are small, uninucleate, and contain a C-shaped, collar-shaped, parietal or axile stellate chloroplast. Some members have multinucleate cells.
- Chloroplast contains one or more pyrenoids.
- Many members show polarity by possessing a basal holdfast, which attaches the filament with the substratum.
- Asexual reproduction takes place by zoospores (bi or quadriflagellate), aplanospores or hypnospores.
- Sexual reproduction is isogamous but some forms also show anisogamy and rarely oogamy.

Fritsch (1935) divided the order into six families viz.

- Ulotrichaceae
- Microsporaceae
- Cylindrocapsaceae
- Ulvaceae
- Prasiolaceae
- Sphaeropleaceae

Family: Ulvaceae

In the members of the family Ulvaceae the plant body is an expanded sheet one or two cells in thickness, or a hollow cylinder with a wall one cell in thickness, or a ribbon two or more cells broad. The cells are uninucleate and with a single laminate chloroplast. Asexual reproduction takes place by means of quadriflagellate zoospores,

Sexual reproduction is isogamous or anisogamous. Common examples of the family are Enteromorpha, Monostroma and Ulva.

Ulva

Systematic Position

Class: Chlorophyceae

Order: Ulotrichales

Family: Ulvaceae

Genus: *Ulva*

Occurrence

Ulva is an essentially marine alga generally found in rocky shores where it occurs attached to stones, rocks etc. Some species of Ulva are also found in brackish water and polluted estuaries. The genus *Ulva* has around 30 described species, of which *U.lactuca* is the most common species.

Thallus Structure

Ulva is commonly known as 'sea lettuce' as the blades constituting the thallus are expanded leaf like structures which resemble garden lettuce in appearance. Thallus is conspicuous, macroscopic and consists of broad and flat blades. Blades are composed of two cell layers (distromatic) (Fig. 1). Each cell is isodiametric in shape, uninucleate with one parietal laminate to cup shaped chloroplast and a single pyrenoid. Blades of *Ulva* can be as long as 1 m. *Ulva* is attached to substrates in marine coastal waters by means of rhizoidal branches, and it also occurs in free-floating masses. The distromatic blades arise from zoospores. The blade is narrowed into a short basal stalk. Stalk is attached to the substratum with the help of an attaching disc. The attaching disc is formed by the rhizoidal outgrowths of the cells present on the lower side of the thallus.

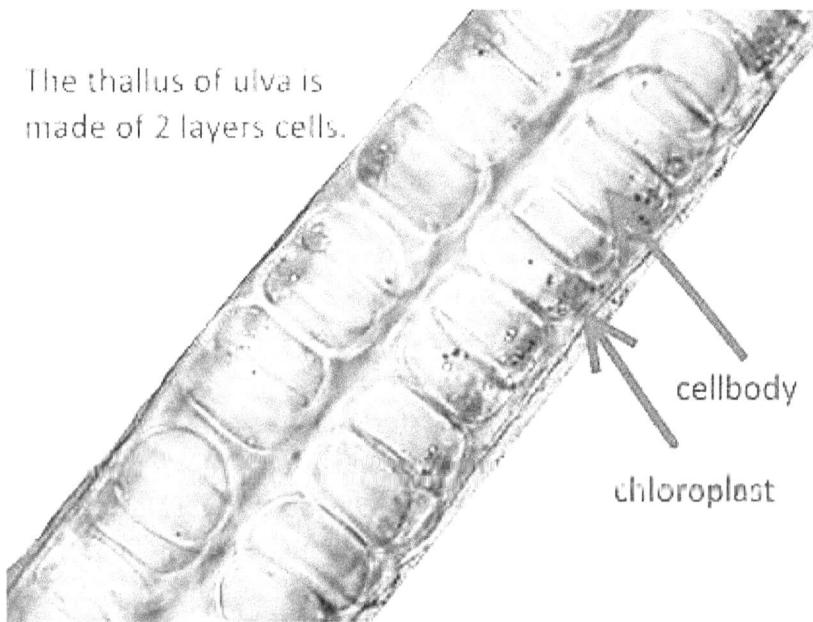

Fig. Ulva.1. Vertical section (V.S.) through a distromatic thallus (Image courtesy: Dr. Cindy Fernández)

Reproduction

Ulva reproduces vegetatively, asexually as well as sexually.

Vegetative reproduction: *Ulva* takes place by accidental fragmentation of the thallus growing usually in quiet estuarine waters.

Asexual reproduction takes place with the help of quadriflagellate zoospores, produced within the vegetative cells of the thallus. At first the cells which are near the margin of the thallus produce zoospores, and then the remoter ones. The production of zoospores goes on until, practically speaking; all the cells have behaved like zoosporangia. The zoospores come out through a pore in the cell wall, swim for a very short duration, come to rest, and then secrete a wall. A zoospore germinates to give rise to a new sexual plant.

Sexual Reproduction:

The majority of the species of *Ulva* are heterothallic. The gametes are generally isogametes, but *U. lobata* appears to be anisogamic. The protoplast of a vegetative cell undergoes repeated cleavages until sixteen or thirty-two daughter protoplasts are formed within the parent cell. Finally, each of these protoplasts becomes metamorphozed into a single biflagellate gamete. The gametes come out through a pore developed on the cell wall. The discharge of the gametes is sometimes so very copious that the water turns green in colour. Each gamete possesses a single chloroplast and a prominent eyespot, and is pyriform in shape. Gametes unite in pairs to form a zygote. The zygote is a first quadriflagellate; it swims for a short while, comes to rest by withdrawing the flagella, and secretes a wall around itself. The zygote undergoes germination within 24-48 hours after rest, and an equational division of the zygote nucleus takes place. As a result, ultimately new diploid plants are produced. In some cases haploid thalli are developed parthenogenetically from the gametes.

It is of interest to note that in the life-history cycle of *Ulva* an isomorphic alternation of generations can be traced. Morphologically the two types of plants, the sporophyte and the gametophyte, are identical. The sporophytic thallus produces the haploid zoospores, which develop into gametophytes. The zygotes, produced by the union of gametes developed on these gametophytes, give rise to new diploid thalli.

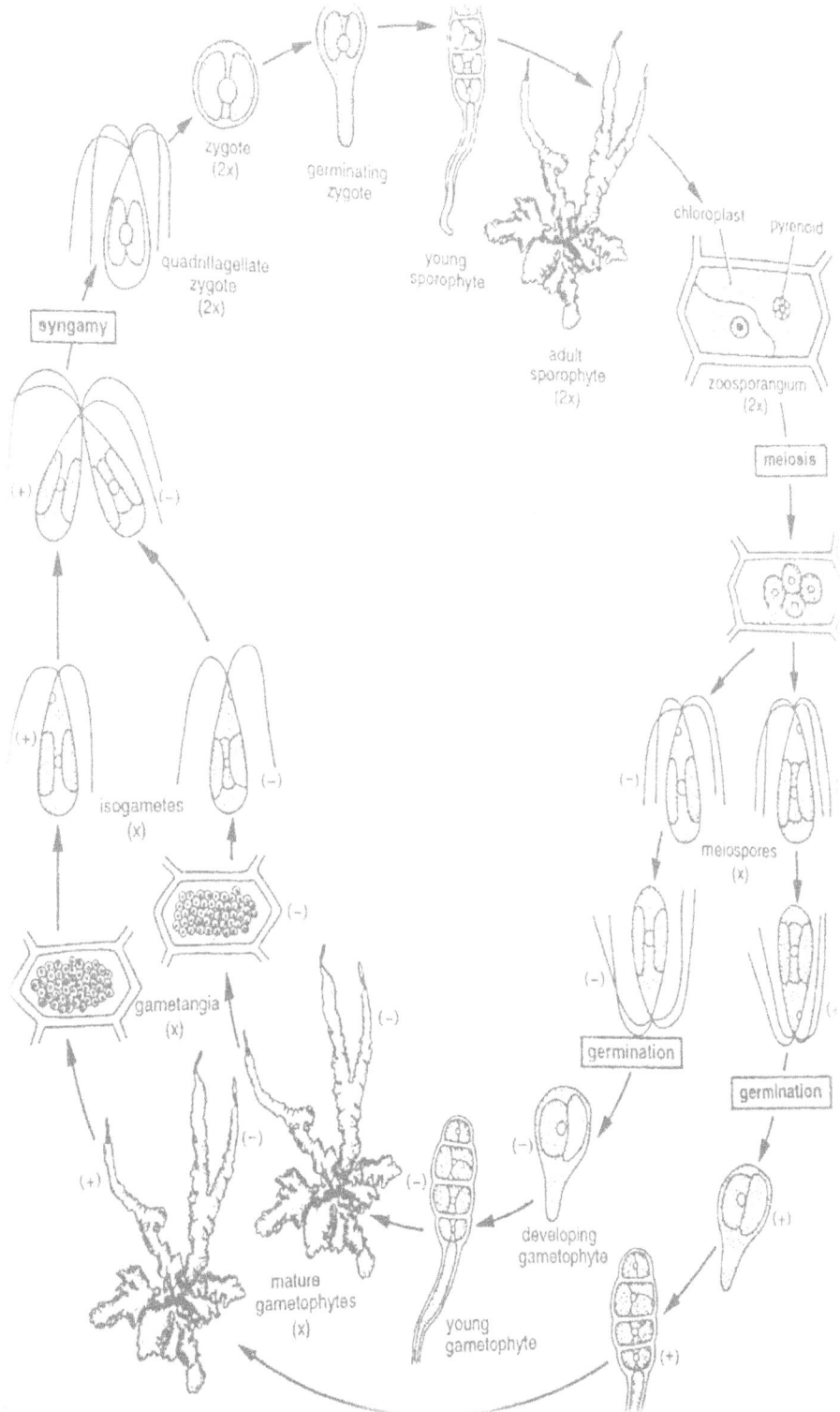

Fig.2. Ulva: life Cycle

Family Ulotrichaceae

Filaments are with or without an attaching base; typically unbranched, composed of cylindrical uniseriate cells with thin to gelatinous walls. Cells possess lateral plate-like or band-like chromatophores with or without pyrenoids.

Asexual reproduction is by bi- and quadriflagellate zoospores, although a few genera (e.g., *Stichococcus, Radiofilum,* and *Geminella*) fail to do so. Sexual reproduction is isogamous (Ulothrix), anisogamous, or oogamous (Cylindrocafisa).

Enteromorpha

Systematic Position

Class: Chlorophyceae

Order: Ulvales

Family: Ulvaceae

Genus: *Enteromorpha*

Occurrence. Speoies of Enteromorpha are essentially marine aid are widespread in the tropical waters. They occur chiefly in the upper littoral, intertidal zone and also near the low tide mark usually attached to rocks or muddy substrata, A few species occur in fresh-water.

Thallus Structure (Fig. 3):

At the very beginning the thallus is a uniseriate filament; very soon it becomes multiseriate and tube-like in appearance. The thallus is branched at intervals and exhibits some diversities of form. In a large number of species the thallus is attached to the substratum by means of rhizoids, while others, particularly those inhabiting salt marshes, remain free-floating, either throughout their life or at least a part of it; one species (E. nana) possesses a basal prostrate region.

The branching depends upon two major external factors, — the degree of salinity in the substratum and temperature. It has further beer recorded that the thallus has got some power of regeneration. If any injury takes place near about the apex, papillate outgrowths appear from the injured cells, whereas if the basal region is subjected to an injury rhizoids are developed. The cells of the thallus are somewhat tubular, much longer than broad, and contain a number of chromatophores, with pyrenoids.

Fig. 3.Enteromorpha: Plant Bidy

Reproduction in Enteromorpha:

Enteromorpha reproduces both asexually and sexually. The asexual reproduction takes place by means of zoospores, as in *Ulva*, Each zoospore on germination gives rise to a new haploid plant. In the species (*E. nana* and *E. procera*) diploid zoospores are produced, and these are termed as neutrospores, or neutral spores.

Enteromorpha is heterothallic and the mode of sexual reproduction is usually isogamous. An anisogamous type of reproduction is, however, found in *E. intestinalis*, where the smaller gamete with a rudimentary pyrenoid is regarded as a result of fusion between two motile gametes a zygote is formed, which also remains motile for about an hour. After this period the flagella are withdrawn and the motility ceases, but the division of the zygote does not commence before several days.

In some species like *E. clatharta* there is no union of gametes. On the other hand, some of the cells of the gametangial region, which fail to divide and thereby give rise to gametes (rest cells) germinate afterwards developing into new haploid plants. These ultimately get detached from the parent plant and live as independent ones.

Enteromorpha exhibits an isomorphic alternation of generations.

Ulothrix

Systematic Position

Class: Chlorophyceae

Order: Ulotrichales

Family: Ulotrichaceae

Genus: *Ulothrix*

Occurrence

Ulothrix (Gr. *oulos,* wooly; *thrix.* hair) is represented by about 23 species, the majority of which are found attached on rocks, stones or some other substratum in cold, calm and quite or slow-flowing freshwater reservoirs, such as ponds, tanks, pools, lakes, etc. (Ramanathan, 1964). Species such *as U. implexa. U.flacca* and *U. pseudoflacca* are marine. The most common freshwater species is *U. zonata*. Primarily it is a cold-water alga which grows during early spring, disappears during summer and reappears in rains.

Thallus Structure

Plant body is multicellular and filamentous, and the filaments are long, unbranched and uniseriate (cells arranged in a single row). When young, the filaments are bright green and remain attached to some substratum but later on they may become free-floating.

Three types of the cells (Fig. 24.47) can be distinguished in each filament:

(i) Apical cell, which lies at the tip of the filament. It is green and cylindrical or dome-shaped with hemispherical outer surface.

(ii) (ii) Basal rhizoidal cell, which is the lowermost cell of the filament. It is long, narrow, colorless, with its basal part modified *into a* disc-shaped structure. Because it holds the filament tightly with the substratum, it is also called *holdfast* or *hapteron*.

(iii) (iii) Middle cells, present in between the apical cell and rhizoidal cell, are green, cylindrical, more in breadth than in height in many species and also called intercalary cells. The cells may be 2-6 times more in height than in breadth in *U. subconstricta*.

Cell Structure: The cells are surrounded by a cylindrical cell wall, which is either smooth or stratified. It is generally bilayered. The outer layer consists of pectic substances and the inner layer or cellulose the chloroplast is single, Band or girdle-shaped, or ring-shaped, or collar-shaped or C-shaped, and parietal in position (Fig. 3). It contains one to many pyrenoids. The ring-shaped chloroplast may be closed or open at one end (Fig4). Internal to the chloroplast is present a single nucleus. Shyam and Saxena (1980) reported the chromosome number *(n = 10) in U. zonata*.

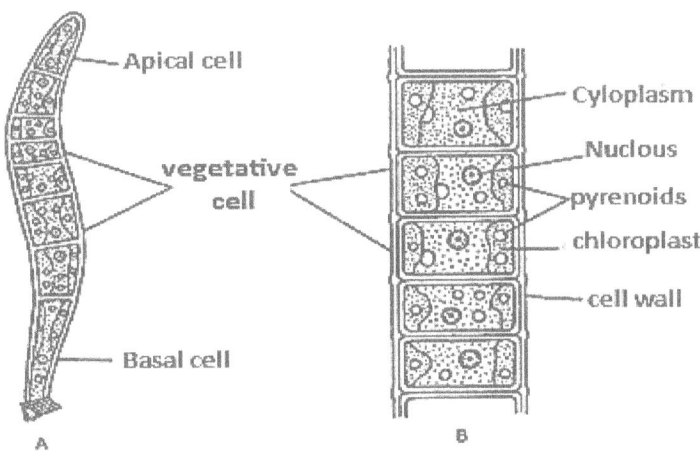

Fig.4. *Ulothrix*

Reproduction

1. Vegetative Reproduction:

It takes place by fragmentation.

Fragmentation:

In this process the filament breaks up into a number of parts. Each part is capable of developing a new plant like its parent.

2. Asexual Reproduction:

Asexual reproduction takes place in winter, during its active growth. It takes place by the formation of zoospores, akinetes and palmella stage.

a. Zoospore Formation:

Zoospores are formed during favourable condition with proper growth of the plant. Any cell except holdfast is capable of producing zoospores (Fig. 5A). During zoospore formation the protoplast becomes slightly contracted from the cell wall.

The protoplast either develops into single zoospore (U. fimbriata) or undergoes division and form 2, 4, 8, 16 or 32 units. Each unit contains single nucleus and cytoplasm. These small units form zoospores.

The zoospores are of different types:

Types of Zoospores:

Species of Ulothrix with narrow cells produce quadriflagellate zoospores of one kind. But species with broader cells like U. zonata produces 3 types of zoospores (Fig.5.B, C and D).

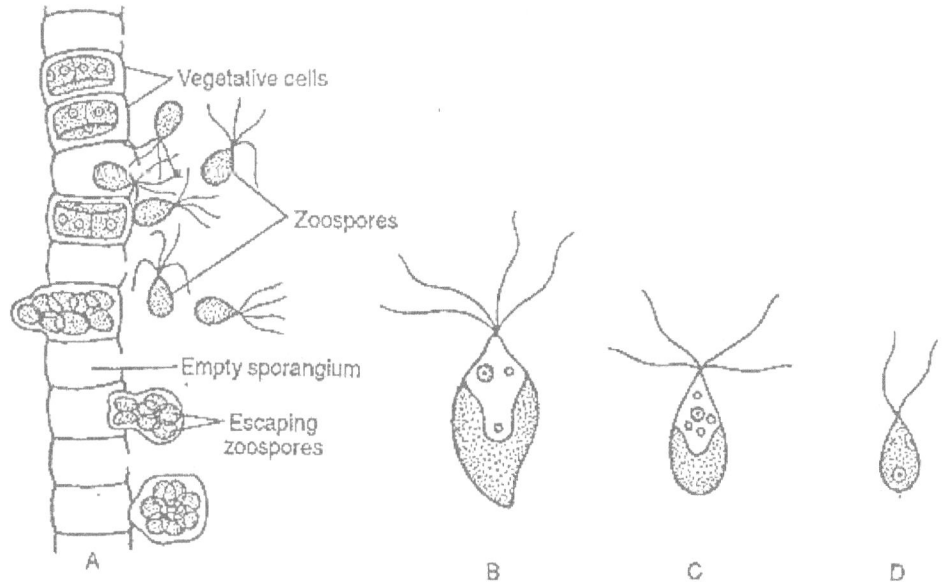

Ulothrix sp. ; A. Portion of filament with vegetative cells, sporangium and escaping of quadriflagellate zoospores, B-D. Different types of zoospore (B. Quadriflagellate macrozoospore, C. Quadriflagellate micro- zoospore and D. Biflagellate microzoospore)

Fig.5.

These are:

i. Quadriflagellate macrozoospores- usually 4 per cell,

ii. Quadriflagellate microzoospores- usually 8 per cell, and

iii. Biflagellate micro- zoospores- usually 16-32 per cell.

The zoospores are morphologically almost alike but differ in size, number of flagella, and position of the eye-spot. They also vary in their swimming period. The macrozoospores can swim for few to 24 hours, but the micro- zoospores (both types) can swim for 2-6 days. The macrozoospores are pear-shaped with tapering at posterior end, while the microzoospores are narrowly ovoid with rounded posterior end.

Germination of Zoospores:

After maturation the zoospores are liberated through a pore developed on the side wall and initially remain in a delicate mucilage vesicle (Fig.6.A). The vesicle soon dissolves and zoospores are liberated. The zoospores can swim for different period of time (few hours to even 6 days).

Coming in contact with suitable substratum they lose their flagella and get attached. The cell divides horizontally. The lower attached cell develops into holdfast and the upper cell forms the filament through repeated divisions.

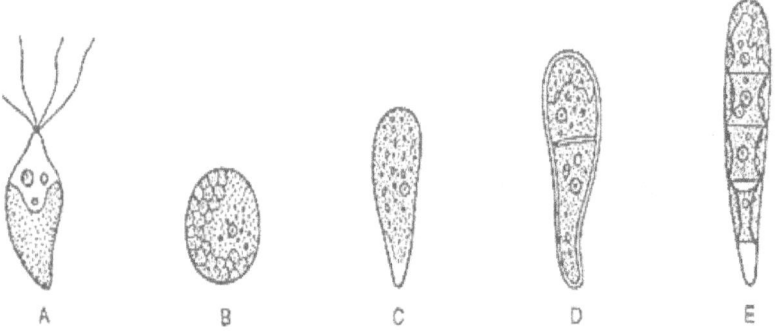

: *Ulothrix* sp. : A-E. Germination of zoospore and stages of development of new filament (A. Quadriflagellate zoospore, B. Encysted stage, and C-E. Development of new filament)

Fig.6.

b. Aplanospores:

With the sudden change of environment towards unfavourable condition during zoospore formation the protoplast units do not form flagella and remain inside the mother cell as non-motile units. These unicellular; uninucleate, thin walled non-motile units are called aplanospores. During favourable condition they germinate after or before liberation from parent cell.

c. Hypnospores:

During drought, sometimes the entire protoplast of a cell may round up and forms a single thick walled structure, the hypnospore. During favourable condition it germinates and develops into a new filament.

d. Akinetes:

Akinetes are formed during extreme unfavourable condition. During this process the cell becomes enlarged, protoplast accumulates food material and then forms thick wall around itself. This thick walled resting vegetative cell is called akinetes (Fig.7.B). It is found in *U. zonata, U.acrorhiza, U. oscillarina* etc.

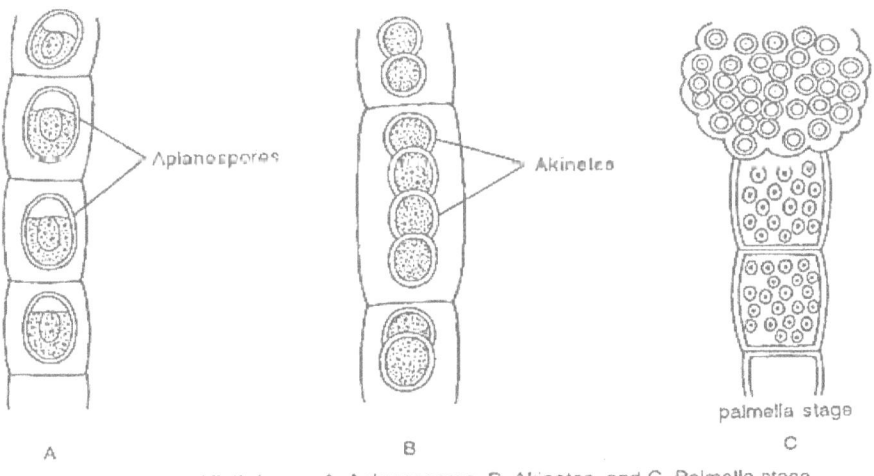

Ulothrix sp. : A. Aplanospores, B. Akinetes, and C. Palmella stage

Fig.7.

e. Palmella Stage:

Sometimes the wall of the aplanospore mother cell becomes mucilaginous. Consequently the wall of aplanospores also gets enveloped by mucilaginous substance. These coverings protect the aplanospores against desiccation.

In this way -many green round bodies become enclosed in a mucilaginous mass, called palmella stage (Fig. 7C). During favourable condition these green bodies come out by the dissolution of the mucilage covering and germinate into new plants.

3. Sexual Reproduction:

Towards the end of growing season they reproduce by sexual means. The sexual reproduction is isogamous i.e., the union between similar gametes (Fig. 8).

The fusion takes place between the gametes developed in different filaments i.e., they are heterothallic or dioecious (*U. rorida* is monoecious). Morphologically gametes are similar to the microzoospores. They are produced in the gametangium (Fig. 8.A) similar to zoosporangium. Depending on species the number of gametes may be 8, 16, 32 or 64. Like microzoospores they are uninucleate and biflagellate but smaller in size. Though the gametes are morphologically similar, they are physiologically different and designated as + and – strains.

Liberating from the gametangium, the gametes swim for some time. The gametes of opposite strain (+ and -) fuse and form a quadriflagellate spindle shaped zygote (Fig.8B, C). Initially the zygote keeps on swimming but later it settles down on some substratum, withdraws its flagella and becomes round off.

The zygote takes rest for about 5-9 months. After rest the nucleus of zygote (2n) undergoes meiosis and forms 4 haploid nuclei of different strains (2+ and 2- type). The mitosis may follow the meiosis and forms 8 (Fig. 8.E) to 16 haploid nuclei.

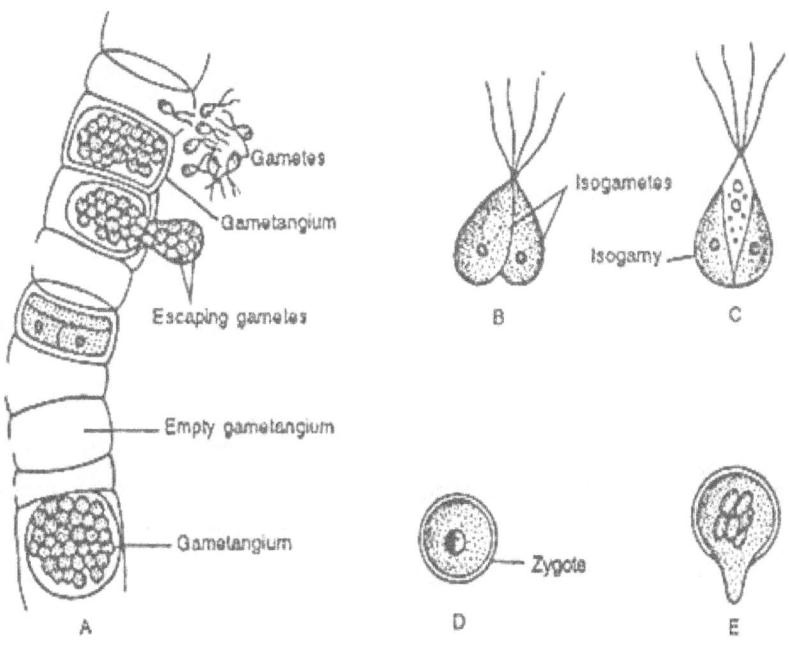

: *Ulothrix* sp. : A. Portion of filament with vegetative cells, gametangia and escaping gametes, B. Uniting gametes, C. Isogamy, D. Zygote, and E. Germination of zygote showing zoospore primordial inside

Fig.8.

The nuclei along with some cytoplasm form spores, called meiospores. The meiospores are haploid and quadriflagellate. On germination they develop into haploid *Ulothrix* filaments either + or – type.

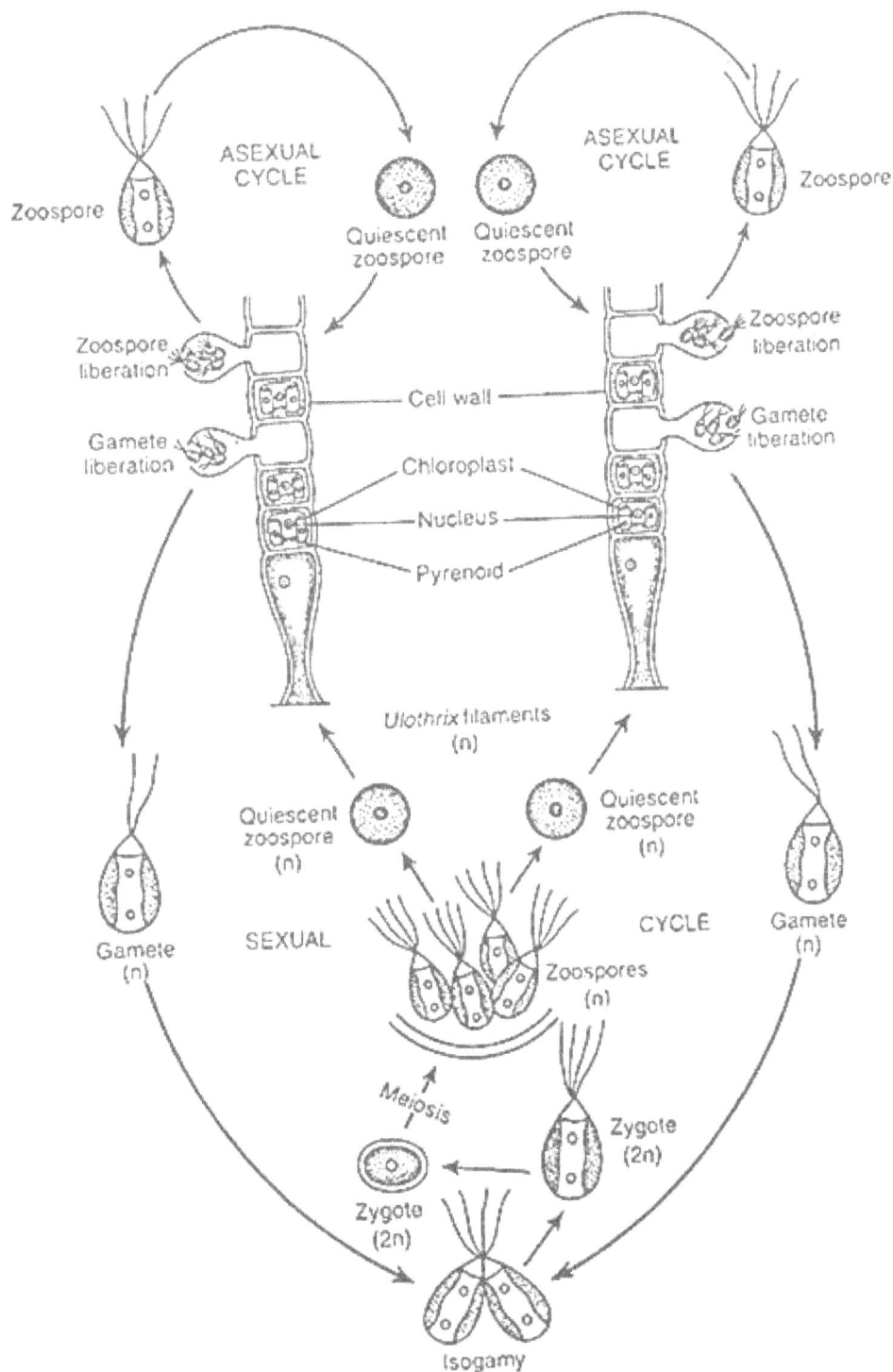

Fig. 9. *Ulothrix*: Life Cycle

Parthenogenesis:

Sometimes, the gametes fail to fuse, lose their flagella, and secrete a thick wall around them and are called parthenospores or azygospores. After rest, each parthenospore germinates directly into a new plant.

Indian Species:

Ulothrix zonata, U. tenerrima, U. aequalis, U. variabilis, U. oscillarina etc.

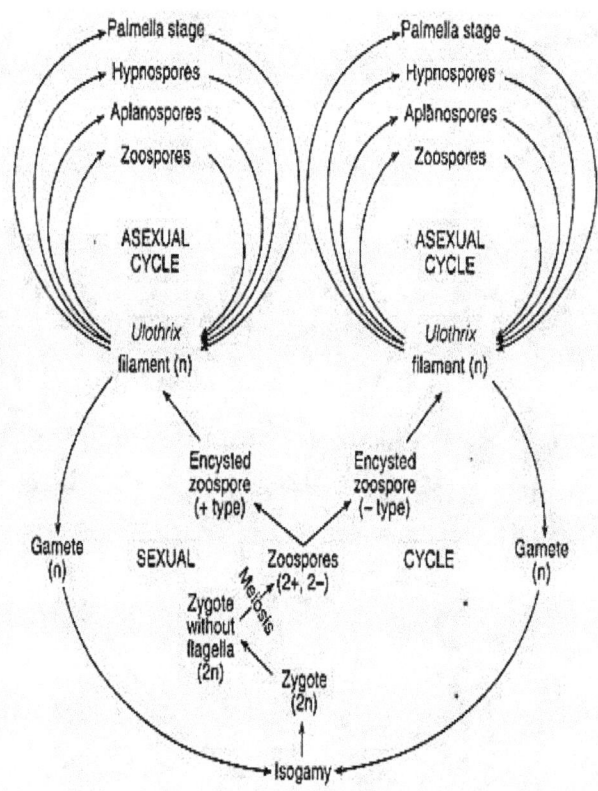

Fig.10. Graphic: Life Cycle

11.

Cladophorales: *Cladophora*

This order includes 12 genera and about 350 species. The forms are both freshwater and marine having branched filamentous thallus consisting of multi-nucleate cylindrical cells. Asexual reproduction takes place by means of quadriflagellate zoospores, aplanospores or akinetes. Sexual reproduction is either isogamous or anisogamous.

Family. Cladophoraceae:

This family contains many fresh-water forms. Some of the large genera, such as Cladophora, have both fresh-water and marine representatives, but there are several strictly marine genera. The plants are composed of multinucleate cells arranged in uniseriate fashion in simple or branched filaments.

The chloroplast is parietal and reticulate. The filaments are usually attached by rhizoids. Generic separations have traditionally been made on vegetative characters such as the presence or absence of branches, the kind of branches, and the form of the basal cell and its attachment rhizoids.

Cladophora

Systematic position:
Class: Chlorophyceae
Order: Cladophorales
Family: Cladophoraceae
Genus: *Cladophora*

Occurrence:

Cladophora has about 160 species. It occurs in both freshwater and marine water. Freshwater species are found in ponds, lakes, and rivers and remain attached to rocks, stones (i.e., **lithophytic**) and submerged aquatic plants (i.e., **epiphytic**). *Cladophora crispata* grows on shells of molluscs (i.e., **epizoic).** *Cladophora sauteri, C holozotica* and *C protunda remain* in the form of balls showing **aegagropilous** habit.

Common Indian Species: The common Indian species are *Cladophora crispata, C. fracta, C bengalensis,* and *C. glomerata.*

Thallus Structure

1. Thallus is rough to touch due to the presence of chitinous material.

2. If seen with the help of a magnifying lens, the fine small branches of the thallus look like the branches of a small herbaceous plant (Fig. 1. A)

3. Plant body is multicellular, filamentous and branched. It remains in the form of bush-like masses.

4. Thallus remains attached to the substratum by a holdfast formed by the lowermost rhizoidal cell.

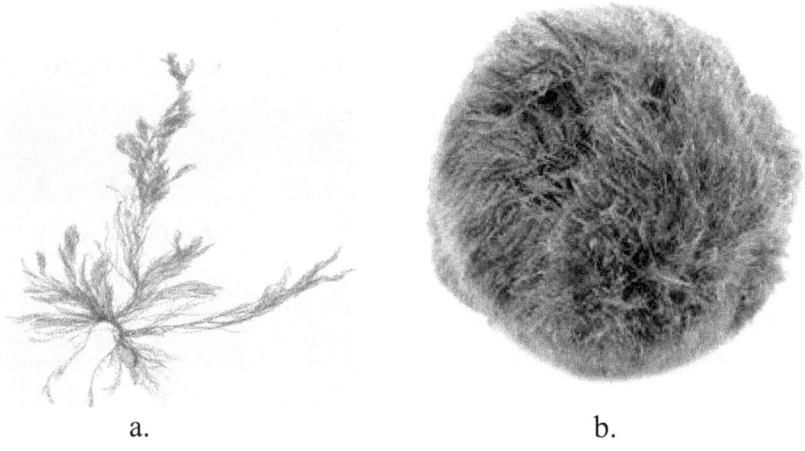

a.　　　　　　　　　　b.

Fig.1. a. *Cladophora* Thallus structure, b. Aegagropilous habit

5. A branch always arises just below or just above the septum.

6. Branching in *Cladophora* appears like a dichotomy but actually it is of lateral type. It appears so due to the process of erection, i.e., pushing of the original axis on one side.

7. Each cell (Fig. 2A, B) is much broader in length than breadth.

8. Cell is cylindrical and remains surrounded by a lamellated cell wall, the outermost layer of which consists of chitin, middle of pectose and the innermost of cellulose material.

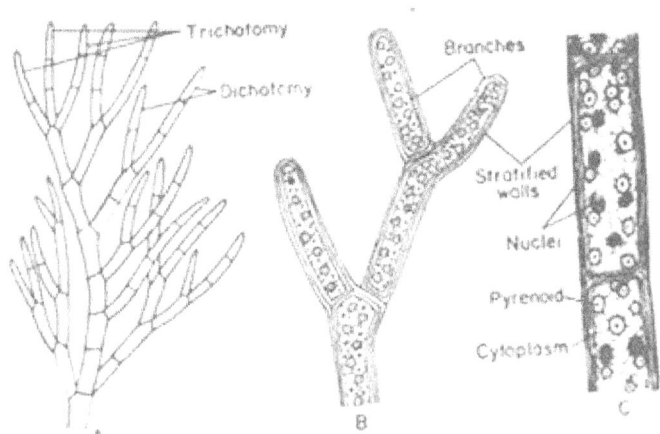

Fig.2. *Cladophora spp*. A. Filament, B. Enlarged view of filament, C. Cell contents

9. A large central vacuole is present in each cell.

10. A cell possesses either a single reticulate chloroplast or many discoid chloroplasts (Fig. 2C).

11. Many pyrenoids are present in each cell.

Reproduction

Cladophora reproduces vegetatively, asexually and also sexually.

Vegetative Reproduction

1. Fragmentation: Aegagropilous species mostly reproduce by fragmentation. Filaments may break into small fragments, which are capable of developing into new individuals. Some non-aegagropilout species also reproduce by fragmentation.

2. Storage cells: In some species *(C. glomerata* and *C. ophiophila)* the rhizoidal cells near the substratum divide and redivide to form some short branched filaments. The cells of these filaments become filled with reserve food material. Sometimes these cells are also impregnated with calcium carbonate. During the unfavourable conditions all the cells of the thallus disorganize and die, except the cells which are filled with reserve food material. On return of the favourable conditions, generally in the next season, these storage cells germinate into new filaments of *Cladophora.*

3. Stolons: **In** *C. glonterata* and some other species the rhizoidal branches sometime **develop into stolon-like** outgrowth, from which originate new erect **thalli during favourable conditions.**

4. Akinetes: In **some freshwater** species of *Cladophora,* some cells of the filaments become laden with reserve food material. Their walls become very thick and get swollen, assuming a pear-shaped form. These thick-walled cells having reserve food material are called akinetes. Sometimes akinetes are formed in many cells of the filament. Akinetes germinate into new filaments of *Cladophora.*

Asexual Reproduction

The zoospores are small pyriform bodies with a small anterior beak and two apical flagella, or in some species four. These flagella are differentiated before the emergence of the spores. There are two granules at the point where the flagella are inserted and the chloroplast appears as a ring in the posterior part of the cell and shows a slow streaming movement. Those zoospores nearest the orifice escape first, squeezing their way through the opening with their flagella behind them, and they are followed by a steady stream of others. After about twenty minutes they settle down by their anterior ends and elongate. A septum appears and gradually the coenocyte structure of the thallus is re-established.

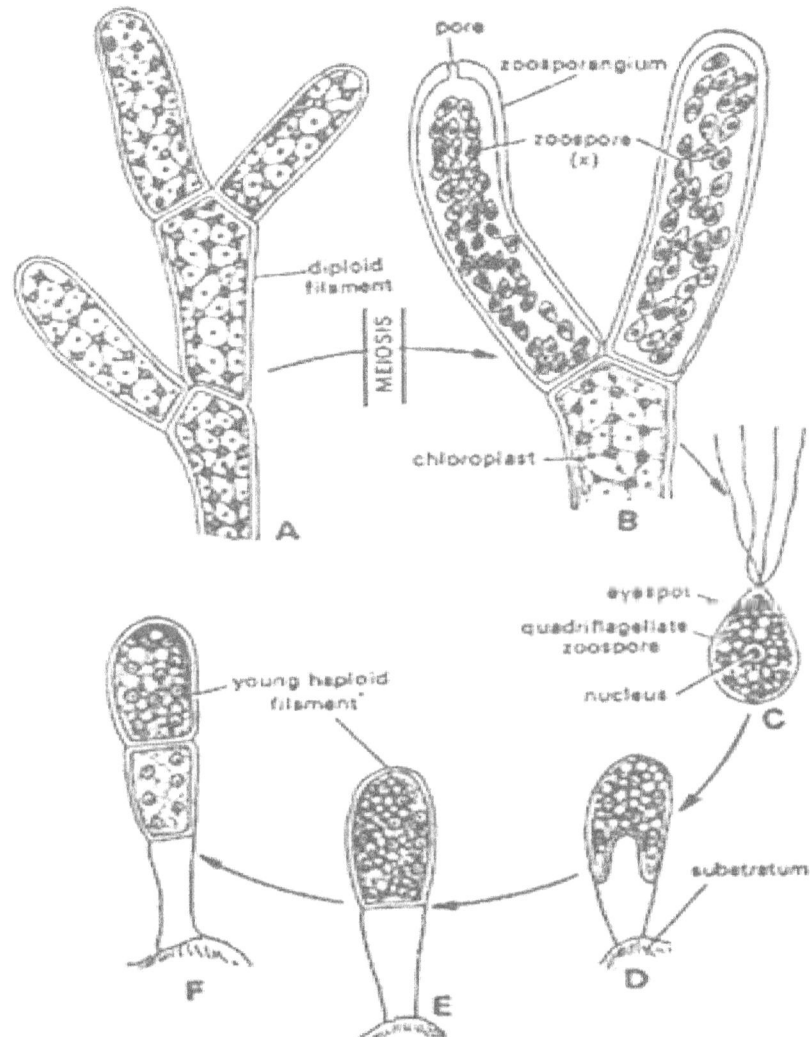

Fig. 3. A-F. Asexual Reproduction in *Cladophora*

Sexual Reproduction

The formation of gametes occurs in most species and probably in all. In some species it has been definitely established that the thallii are dioecious. Any segment of the thallus is potentially capable of functioning as a gametangium, and its development is similar to that of the zoosporangium. The gametes are isogamous and differ in no marked way from the zoospores, except that they are invariably biflagellate. They are liberated in a similar manner to the zoospores, and fuse externally to give rise to a zygote which germinates immediately to form a fresh Cladophora filament.

Alteration of Generation

It has now been definitely established that in almost all species there is a strict alternation of sexual and asexual generations. These generations are morphologically alike. Meiosis occurs prior to the formation of the zoospores which therefore give rise to male and female monoploid plants, the diploid phase being established after gametic fusion, resulting in the formation of a sporophytic plant, on which the zoospores are produced. A similar type of alternation has been found in a small number of related genera.

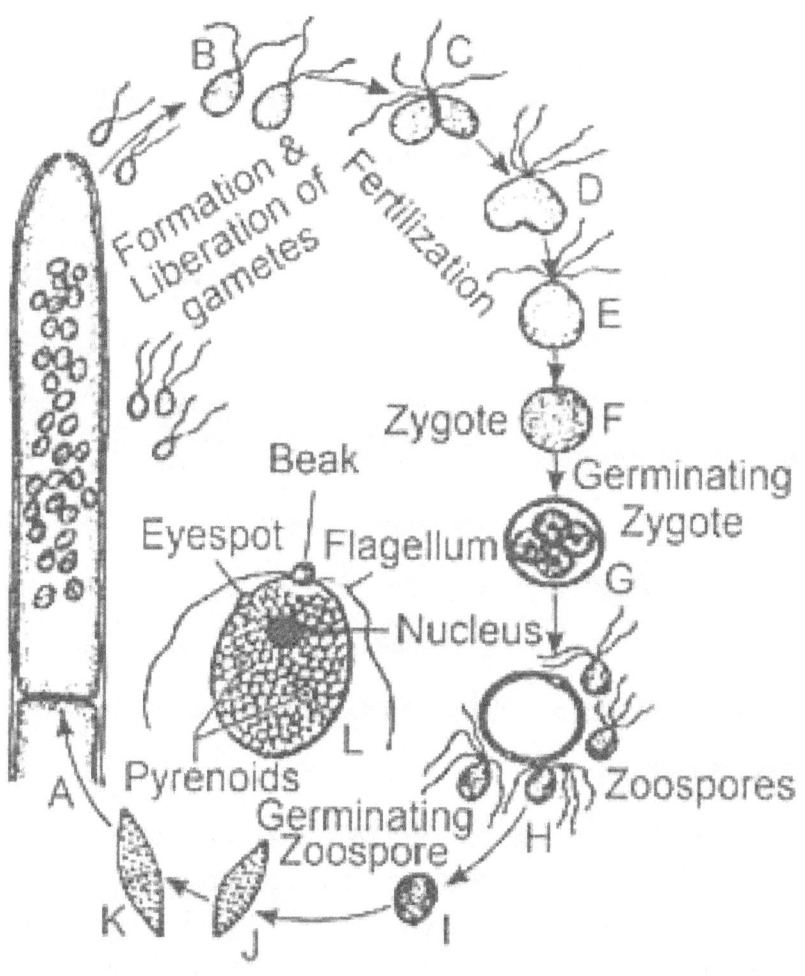

Fig. 4. A-K. Sexual Reproduction in *Cladophora*

In *Cladophora glomerata* it has been shown recently that zoospore formation occurs at intervals all through the year, while gamete formation happens only in the spring and at the end of a long series of zoospore discharges. Gametes and zoospores are developed in distinct plants, but there is no reduction division in the formation of the zoospores. It appears that this species can produce a succession of diploid zoospore generations, after which meiosis occurs during the formation of the gametes. Conjugation restores the diploid condition.

****

12.

Chaetophorales: *Fritschiella, Chaetophora, Draparnaldia, Drapamaldiopsis, Coleochaete, Trentepohlia,* and *Pleurococcus*

The members of Chaetophorales are characterised by the presence of a heterotrichous thallus, that is, the thallus possesses a prostrate and an erect system. In some genera one of the two systems is suppressed; for example, in *Drapernaldia, Drapermaldiopsis* and *Microthamnion* only erect system is present, whereas in Coleochaete and Ulvella there is only prostrate system. Certain members of the order have setae (e.g., *Coleochaete*) or hairs (e.g., *Stigeocloniunm*) and hence the name Chaetophorales (Chaetos = hair, phorales =plant: plants with hair or setae). Most of the menmbers of Chaetophorales occur in fresh water, but a few have a marked terrestrial tendency (e.g., Fritschiella tuberosa). The asexual reproduction takes place by quadri- or biflagellate zoospores, aplanospores or akinetes. The sexual reproduction is primarily isogamous (e.g., Fritschiella, Stigeoclonium), occasionally anisogamous (e.g., *Aphanochaete*) or oogamous (e.g., *Coleochaete*).

Fritsch (1935) divided the order Chaetophorales into the following five families:

(1) Chaetophoraceae

(2) Trentepohliaceae

(3) Coleochaetaceae

(4) Chaetos-phaeridiaceae

(5) Pleurococeaceae.

Family. Chaetophoraceae:

Chaetophoralean algae are heterotrichous in habit frequently bearing terminal or lateral hair-like projections. In their heterotrichy both the erect and prostrate portions of the plant may be well developed (e.g., Stigeoclonium) or with an elaborate erect portion and feebly developed prostrate portion (e.g., *Chaetophora, Draparnaldia, Draparnaldiopsis, Fritschiella*).

Cells are with thin to gelatinous walls, usually uninucleate having a single more or less dissected plate-like or ring-like chromatophore with pyrenoids.

Asexual reproduction is by biflagellate or qudriflagellate zoospores and sexual reproduction isogamous.

Fritschiella

Systematic Position

Class: Chlorophyceae

Order: Chaetophorales

Family: *Chaetophoraceae*

Genus: *Fritschiella*

Occurrence

Fritschiella a terrestrial alga which grows usually on moist alkaline soils was for the first time reported by **M.O.P. Iyenger** (1932) from South India and was named after his teacher and a great phycologist **Prof. F.E. Fritsch**. Later on it was also collected from other parts of India as well as from Nepal, Bangladesh, Burma, Sudan and Japan. *Fritschiella tuberosa* is the only reported species from India.

Thallus Structure

The thallus of *Fritschiella* shows heterotrichous organisation and is differentiated into four systems (a) Rhizoial system, (b) Prostrate system, (c) Primary projecting system and (d) Secondary projecting system. Out of these, rhizoidal and prostrate systems are underground and buried in the soil whereas primary projecting system is subaerial. Only secondary projecting system is aerial and emerges out of the soil surface. (Fig. 1).

The rhizoidal system consists of one or more septate rhizoid like elongated structure. It grows downwardly and does have colourless cells due to the absence of chloroplasts. However, it is absent in *Fritschiella* simplex, a species reported from Bangladesh. The prostrate system is made up of rounded or irregularly swoller clustur of cells. This is branched filamentous, tuberous or parenchmatous and in mature thallus comprises of short congested branches which are differentiated into nodes and internodes. Certain nodal cells of prostrate system give rise to rhizoid s towards lower side (which penetrate into the soil) and primary project-ing system towards upper side.

The prostrate system in young plants is not well developed but gradually it increases in size with the growth of the thallus and at maturity the thalli have very well developed prostrate system.

The primary projecting system Develops from prostrate system and is erect. It consists of uniseriate or biseriate filaments which may be simple or branched, sub-aerial and green. The cells are small and round in shape and resembles the cells of prostrate system.

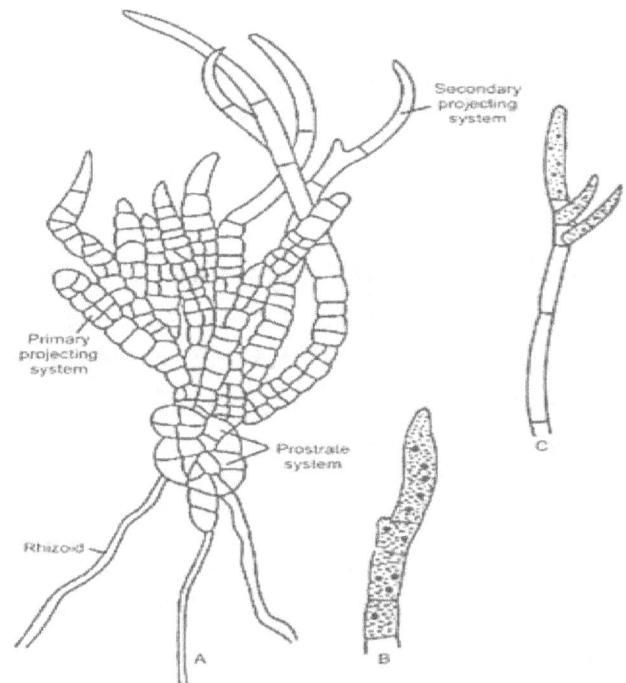

Fig. 1. *(A-C). Fritschiella:* (A), showing habit of plant; (B) and (C), segment of thallus showing secondary projecting system.

The secondary projecting system is aerial and consists of freely branched uniseriate filaments. It is given out from the primary projecting system and has elongated cells. It is less developed in the thalli growing in exposed areas and is absent in F. simplex.

Cells Structure The cells are uninucleate and thin walled. Except the cells of rhizoidal system all the cells are green due to the presence of chloroplasts. The cells of secondary projecting system possess curved plate like chloroplast which have 2-8 pyrenoids. The cells of primary projecting sys-tem and prostrate system have usually less developed chloroplasts with 2-4 pyrenoids and dense cell contents. However, the cells of rhizoidal system do not have chloroplasts and are colourless.

Methods of Perennation During unfavourable conditions sub-aerial primary projecting sys-tem and aerial secondary projecting system alongwith the rhizoidals system usually degenerates and remaining prostrate system serves as preventing structure. The cells of the prostrate system perennate during dry conditions without undergoing any remarkable change in structure and on return of favourable conditions germinate directly to give rise to new plants. However, the nodal cells of prostrate system sometimes get detached from the parent thallus due to death and decay of internodal cells and serve as perennating bodies. These cells on return of favourable conditions germ inate directly to give rise to new plants and thus help in vegetative propagation of alga as well. Another method of perennation in **Fritschiella** is by means of formation of small (**about 1 mm** in length) funnel shaped or club shaped tuber like bodies which are dark in colour and are en-sheathed with the thick layer of cuticle. These structures germinate or return of favourable conditions to give rise to new plants.

Reproduction

Fritschiella reproduces by asexual and sexual methods. The methods of perennation described above also help in vegetative propagation of the thallus.

Asexual reproduction

The asexual reproduction takes place by biflagellate and quadriflagellate z00spores. The zoospores are formed in the diploid filaments (2n-8) on the prostrate system (F tuberosa) or the primary projecting system (F. simplex).

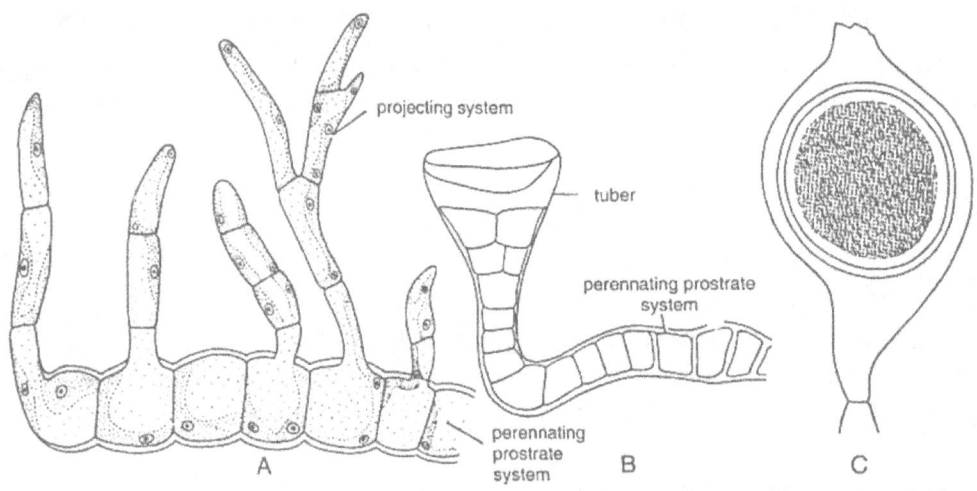

Fig. 2 A-C. *Fritschiella:* Asexual reproduction; A. Direct germination of the perennating prostrate system. B. Part of the prostrate system with club-shaped endings. C. A tuber like structure.

The filaments which form zoospores develop club-shaped endings that are covered with a thick pellicle of dark-brown colour. The cells of these club-shaped structures are similar to the other cells of the filament in their shape and size, but their contents are more homogeneous and brick-red in colour. When conditions are favourable and sufficient moisture is available, the nucleus of each of these cells undergoes meiosis and the contents form four haploid (n=4) zoospores (Fig. 3 A, B). These zoospores are at first colourless or dull brick-red in colour but after liberation they assume usual green colour. There are 2-4 biflagellate zoospores per cell, or only one quadriflagellate zoospore is formed in a cell. Both types of zoospores are produced simultaneously on the sporophyte plant. The zoospores are liberated by the rupture of the sporangial wall and on germination they give rise to new plants. The plants produced by quadriflagellate zoospores are diploid sporophytic plants, while those produced by biflagellate zoosporesare haploid gametophytic plants. Both haploid and diploid plants are however, morphologically similar (Fig. 4 A-F).

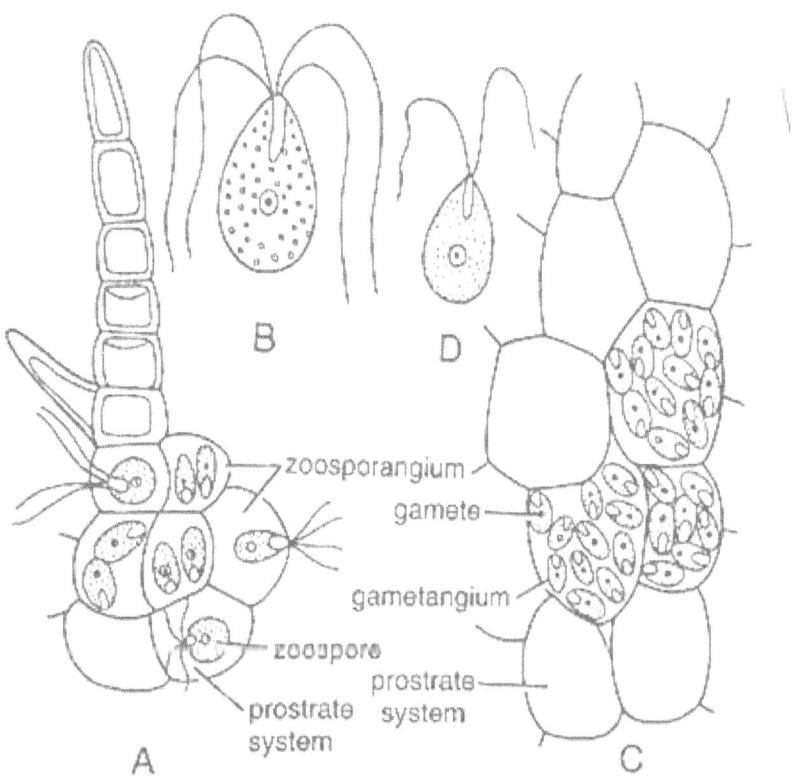

Fig.3. A-D. *Fritschiella*: Reproduction: A. Part of the thallus showing quadriflagellate zoospores, B. a quadriflagellate zoopore, C. Part of the prostrate system showing gametangia, D. A biflagellate gamete.

Sexual reproduction

The sexual reproduction is isogamous. The gametes develop in the cells of the prostrate system of the haploid plants. The protoplast of the gametangium divides repeatedly and forms numerous biflagellate gametes which are structurally similar to zoospores but are smaller in size (Fig. 3 C, D). The fusion of gametes results in the formation of a diploid zygote (2 x). The zygote germinates without undergoing any resting period and gives rise to a diploid sporophytie plant which is identical to the haploid gametophytic plant.

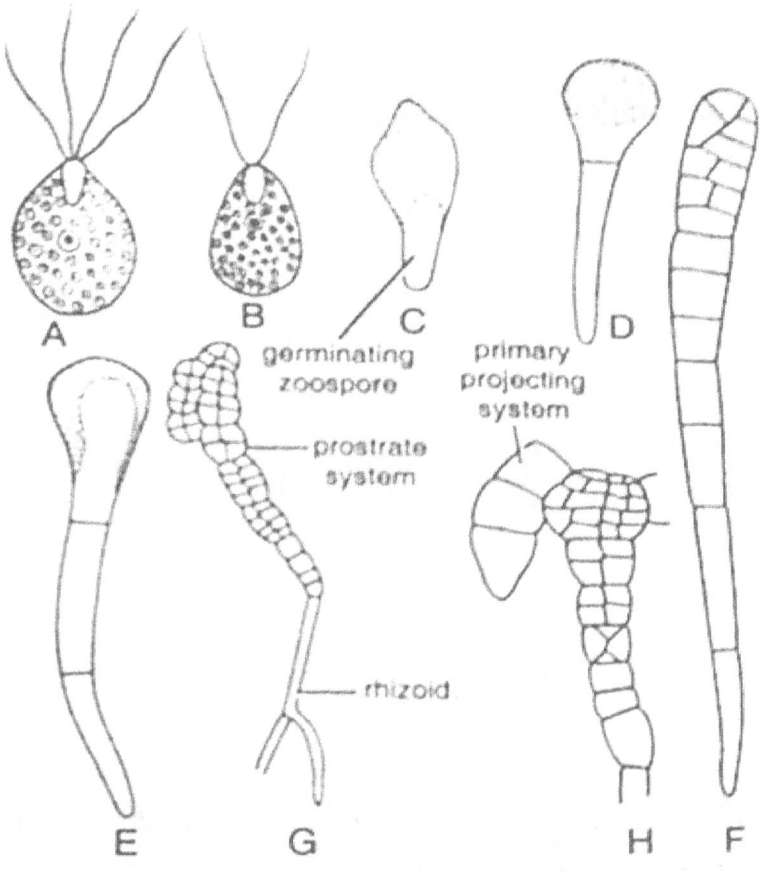

Fig. 4 .A-H. Successive stages in the development of thallus from a zoospore

Alternation of Generations

Fritschiella shows alternation of two morphologically generations (isomorphic alternation of similar generation). The haploid and diploid filaments are alike morphologically. The diploid filaments produce haploid zoospores which give rise to haploid gametophytic filaments on germination. The cells of the prostrate system produce gametes. Similar gametes fuse to form zygote (2 x) and the latter gives rise to diploid filaments on germination.

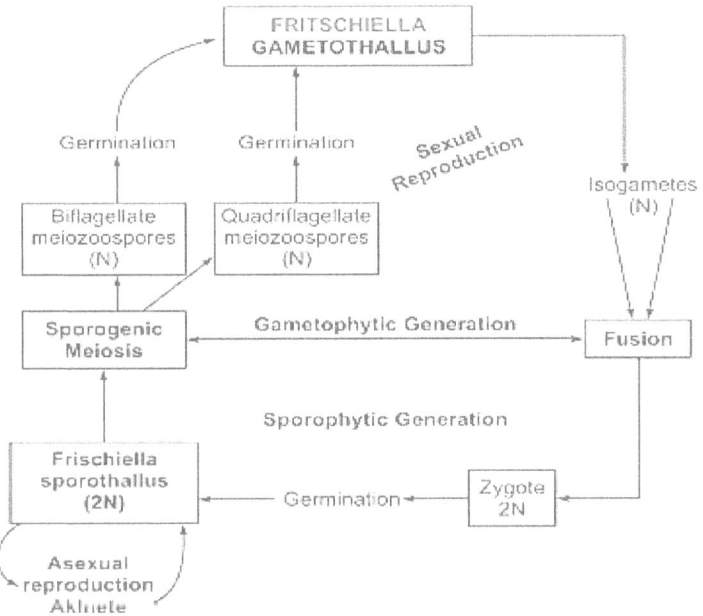

Fig.5. *Fritschiella* **Life Cycle**

Chaetophora

Systematic Position

Class: Chlorophyceae

Order: Chaetophorales

Family: Chaetophoraceae

Genus: *Chaetophora*

Occurrence

This alga occurs abundantly in fresh-water being attached to some substratum preferably to submerged plant parts exhibiting a gelatinous macroscopic growth.

Thallus Structure

Outstanding characteristics lie in the form of the vegetative body known as the heterotrichous filament being differentiated into feebly developed prostrate system often represented by loosely connected rounded cells, and a projecting system of branched filaments, apices of which are terminated by long hyaline hair-like structures known as setae (Fig. 6A & B).

The aerial branches are held together within a compact mucilage forming a tough envelope. The prostrate system may be impregnated with calcium carbonate. Cell structure and the mode of reproduction being very similar to Ulothrix may suggest common ancestry and development being along two divergent lines. Each cell contains a single banded chloroplast with several pyrenoids (Fig. 6C).

Reproduction

Asexual reproduction is by quadriflagellate zoospores.

Sexual reproduction is isogamous by the formation of biflagellate gametes.

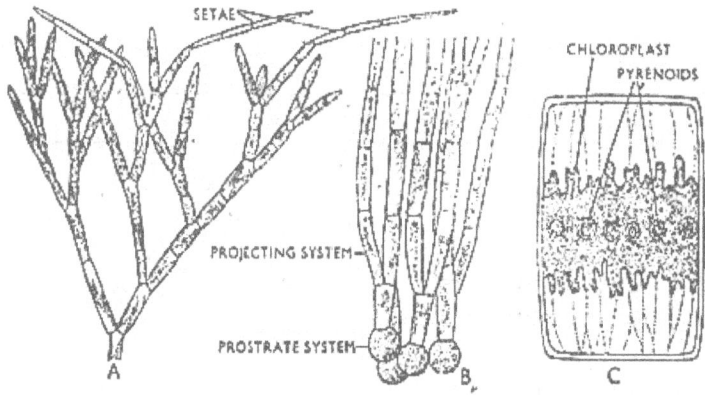

Fig. 6. *Chaetophora elegans*. A. Projecting system of branched filament. B. Prostrate system of rounded cells along with branched aerial filaments. C. A single cell showing chloroplast and pyrenoids.

Draparnaldia

Systematic Position

Class: Chlorophyceae

Order: Chaetophorales

Family: Chaetophoraceae

Genus: *Draparnaldia*

Occurrence

Draparnaldia is a fresh-water alga. Which occurs in flowing cold water attached to sand are sticks and also grows epiphytically on other aquatic plant.

Thallus Structure

Its vegetative body indicates evolutionary tendency towards the development of elaborate aerial system as against a very insignificantly developed prostrate portion. The plant body is macroscopic and is pale-green in colour (Fig. 7A). The projecting part of the alga is composed of a main axis consisting of large, more or less barrel-shaped cells with barrel-shaped chloroplast having toothed edges forming a girdle in the equatorial region (Fig. 7B).

The chloroplasts have several pyrenoids (Fig. 7D). The cells of the main axis are all alike and are uninucleate. From these cells arise short lateral branches in clusters, alternate, opposite, or verticillate, richly branched whose apices are drawn out into long hyaline setae.

These branches may be borne on any cell of the main axis except the lower region from which arise numerous multicellular rhizoidal branches serving the function of anchorage of the alga to its substratum (Fig. 7C).

Cells of the lateral, branches are cylindrical, much smaller than those of the main axis, and are uninucleate with a more or less parietal chloroplast with a single pyrenoid. The main axis is completely covered by lateral branches which grow very profusely. Formation of elements connected with reproduction is confined to the short branches.

Asexual reproduction is by means of quadriflagellate macro- and micro-zoospores and by aplanospores (Fig. 7E & F). The zoospores are pear-shaped.

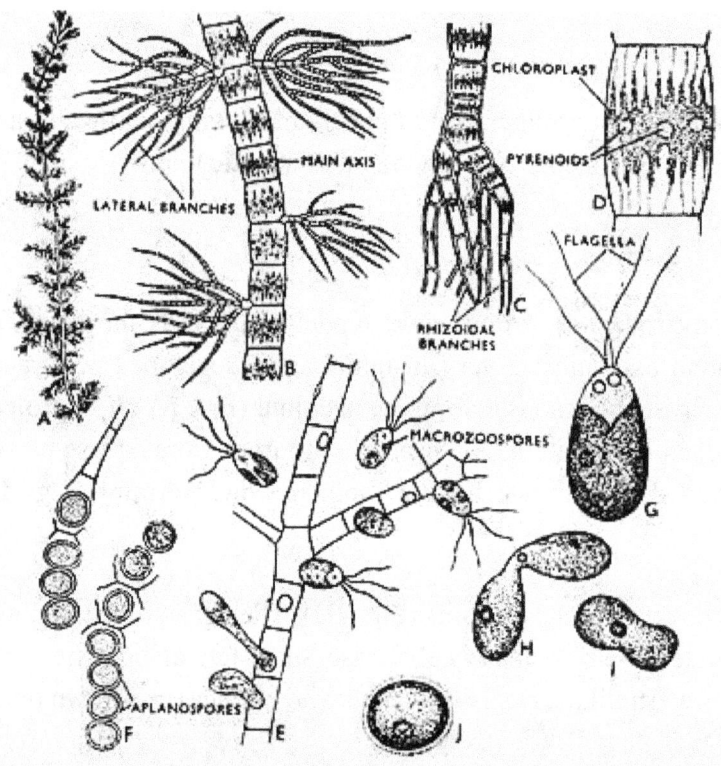

Fig. 7. *Draparnoldia* sp. A. General appearance of the plant body. B. Portion of plant body showing main axis and lateral branches. C. Prostrate portion of the plant body. D. A single cell showing barrel-shaped chloroplast with toothed edges and having several pyrenoids. E. Liberation of macrozoospores. F. Aplanospores. G. Gamete. H-I. Stages in isogamy. J. Zygote.

These branches may be borne on any cell of the main axis except the lower region from which arise numerous multicellular rhizoidal branches serving the function of anchorage of the alga to its substratum (Fig. 7C).

Cells of the lateral, branches are cylindrical, much smaller than those of the main axis, and are uninucleate with a more or less parietal chloroplast with a single pyrenoid. The main axis is completely covered by lateral branches which grow very profusely.

Reproduction

Formation of elements connected with reproduction is confined to the short branches. **Asexual reproduction** is by means of quadriflagellate macro- and micro-zoospores and by aplanospores (Fig. 7E & F). The zoospores are pear-shaped.

They are produced one to four in each cell. The zoospores are formed simultaneously in all the cells of the lateral branches and are liberated through an aperture in the lateral wall (Fig. 7E). The zoospores swim around in water, lose their flagella and develop into new plants. During unfavourable, conditions cell contents of the lateral branches may develop into thick-walled aplanospores instead of producing zoospores. The contents of each cell recede from the cell wall and the whole structure comes to rest by developing a thick wall. On the return of favourable conditions the old wall is cast off and the new walls are developed. It then directly gives rise to a new individual.

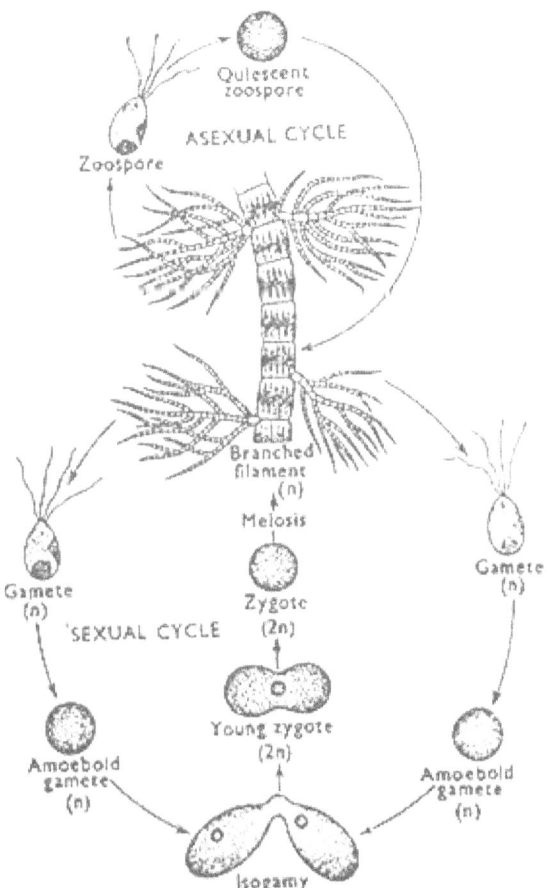

Fig.8. Life cycle of *Draparnaldia*

Sexual reproduction is isogamous. The isogametes are quadriflagellate (Fig. 7G) like the zoospores, they are also produced in the cells of the lateral branches and are liberated out through an aperture of the lateral wall. Prior to union, the gametes lose their flagella and become amoeboid (Fig. 8). The amoeboid gametes fuse in pairs resulting in the formation of a zygote (Fig. 7H & I).

The zygote develops a wall (Fig. 7 J) and germinates into a new plant. Behaviour of the zygotic nucleus has not yet been confirmatively worked out. Gases of parthenospore development have been reported.

Some Indian species of Draparnaldia:

Draparnaldia distans Kg. var. elongata-, D. glomerata Agardh; D. groenlandica Bachmann; D. mutabilis (Roth.) Bory.; D. plumosa (Vauch.) Ag.

Special Features of Draparnaldia:

1. Heterotrichous macroscopic plant body having an elaborate aerial system and a very insignificantly developed prostrate system.

2. Barrel-shaped chloroplast with toothed edges.

3. Asexual reproduction by quadriflagellate macro- and microzoospores and by aplanospores.

4. Isogamous sexual reproduction by quadriflagellate gametes. Prior to union the gametes lose their flagella and become amoeboid and then fuse in pairs.

Draparnaldiopsis

Systematic Position

Class: Chlorophyceae

Order: Chaetophorales

Family: Chaetophoraceae

Genus: *Draparnaldiopsis*

Occurrance

Draparnaldiopsis is a fresh-water alga usually found in shallow pools and ponds. .It grows affixed to aquatic angiosperms. The heterotrichous plant body has a remarkably elaborate aerial portion and an inconspicuous prostrate portion represented by rhizoidal branches with which the alga attaches itself to its substratum (Fig. 9A & E).

Thallus Structure

The main axis of the erect portion exhibits advanced characteristics of being composed of two types of cells: the short cells bearing branches of limited and unlimited growth developed in whorls, alternate with long cells in regular succession resembling nodes and internodes of higher plants (Fig. 9C, D & F).

In some species the branches have the tendency to coil around the main axis (Fig. 51 A) giving a somewhat corticated appearance. The entire plant body is enveloped by a thick gelatinous sheath. Cells are uninucleate having zonate chloroplasts with several pyrenoids (Fig. 9IB).

Reproduction

The alga **reproduces asexually** by three types of zoospores: quadriflagellate ellipsoidal to oval macrozoospores; quadriflagellate spherical microzoospores; and biflagellate spherical, ovoid to obovoid microzoospores (Fig. 9G to J). All the three types of zoospores are formed in the same plant and only from the cells of the lateral branches and never from the main axis. All the cells of the lateral branches except the basal one are capable of producing zoospores. Each cell produces a single zoospore (Fig. 9E & F). Both quadriflagellate macro- and micro-zoospores are developed out of the cells of the lateral branches of unlimited growth (Fig. 9A to F). They are diploid in nature and produce diploid plants—the sporophytes. Whereas, the biflagellate micro-zoospores are formed from the cells of the lateral branches of limited growth, during which meiosis takes place. These zoospores after being liberated swim around in water, lose their flagella, round up and produce new individuals resembling parent plant the sporophyte. But these are haploid gamete producing plants—the gametophytes.

Sexual reproduction is usually isogamous, but cases of anisogamy have also been reported. Similar to the zoospores, biflagellate spherical to ovoid gametes are produced only from the cells of the lateral branches of limited growth borne on the gametophytic plants. Prior to the development of the gametes the cells divide into a row of daughter cells. The contents of each daughter cell metamorphose into one to several gametes which on being liberated out by the gelatinization of the cell wall of the parent plant swim in water.

The gametes from two different gametophytes pair together at the anterior end and finally fuse with each other resulting in the formation of a zygote

The zygote germinates immediately into new plant without going to rest and produces a diploid plant —the sporophyte which resembles the gametophytic plant morphologically. The alga Drapamaldiopsis thus exhibits isomorphic alternation of generations.

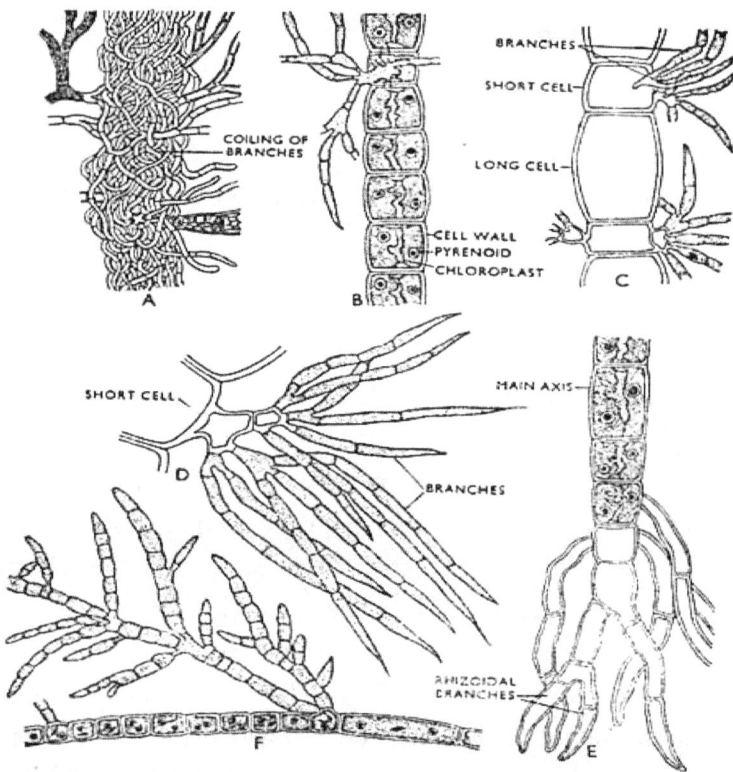

Fig. 9. *Draparnaldiopsis indica*. A. Portion of the plant body showing coiling of branches. B. Portion of the plant body showing cell structure. C. Main axis with long and short cells, short cells bearing branches. D. Short cell bearing branches. E. Main axis showing rhizoidal branches. F. Nature of branches.

Draparnaldiopsis has four main features which suggest affinity with the Charophyceae:

(i) It is differentiated into nodes and internodes,

(ii) It has a cortex-like covering derived from the basal cells of the laterals,

(iii) The laterals have limited, growth and,

Some Indian species:

Drapanaldiopsis indica Bharadwaja; *D. indica* var. *robusta* Islam & Ahia.

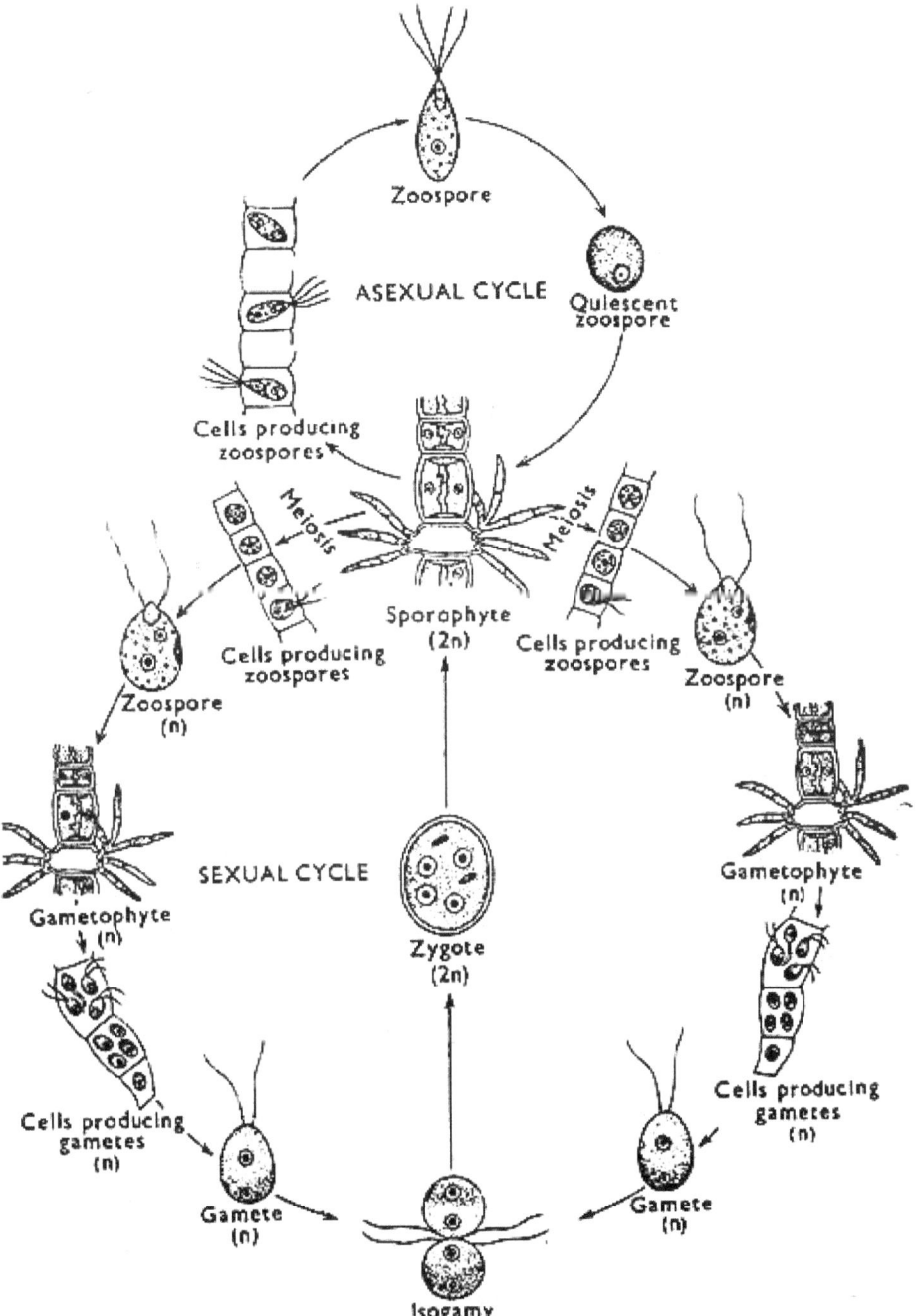

Fig.10. Life Cycle of *Drapanaldiopsis Indica*

Special features:

1. Heterotrichous plant body composed of upright thread and poorly developed rhizoidal branch performing the function of anchorage.

2. The upright thread is differentiated into main axis and lateral branches. The main axis again, is composed of long and short cells arranged alternately, and. the lateral branches are arranged in whorls originating from the short cells.

3. The alternate arrangement of short and long cells resembles nodes and internodes of higher plants.

4. Zonate chloroplast.

5. Asexual reproduction by quadriflagellate macro- and micro-zoospores and by biflagellate micro-zoospores.

6. All the zoospores are developed from the cells of lateral branches.

7. Sexual reproduction is isogamous, cases of anisogamy have also been reported.

8. Gametes are formed from the cells of the lateral branches.

9. Isomorphic alternation of generations.

Family. Coleochaetaceae:

The family Coleochaetaceae is represented by a single genus. Coleochaete. This alga usually occurs as an epiphyte on fresh-water plants such as, Hydrilla and Potamogeton, and also occurs attached to the fronds of Lemna. Its plant body exhibits different degrees of heterotrichy like—a definite prostrate system giving rise to the projecting system, and only prostrate system present and the projecting system has disappeared.

The projecting system grows by an apical cell and the prostrate system by means of marginal meristem. The vegetative cells are characterized by a laminate chloroplast and by the presence of sheathed bristles or hairs. Reproduction is both by asexual and sexual means. Asexual reproduction takes place by biflagellate zoospores. Sexual reproduction is oogamous. The ovum is borne in an oogonium having a long neck, the trichogyne. It is fertilized by a biflagellate antherozoid produced in an antheridium. The zygote is retained within the enlarged basal portion of the oogonium which is surrounded by pseudoparenchymatous investment.

The first division of the zygotic nucleus is meiotic and by subsequent mitotic divisions 32 biflagellate zoospores are produced which give rise to new thalli.

Coleochaete

Systematic Position

Class: Chlorophyceae

Order: Chaetophorales

Family: Coleochataceae

Genus: *Coleochaete*

Occurrence

The genus *Coleochaete* (Gr. Keleon, sheath; chaetos, hair) is represented by about 10 species, out of which 3 species are found in India. They grow in fresh water either as epiphytes on different angiosperms (*Trapa, Hydrilla, Ipomea, Sagittaria, Typha etc.*) and algae (*Oedogonium, Vaucheria*) or as endophytes (*C. nitellarum*) within the cells of *Chara and Nitella* (Chlorophyceae).

Thallus Structure

The thallus is multicellular, heterotrichous and branched. Heterotrichous habit is best represented by *C. pulvinata* which has well-developed prostrate and erect systems (Fig. 6 B). In the other species and one of the two systems (prostrate or erect) is either absent or reduced, e.g., in *C. nitellarum* and *C. divergens* there is a well-developed

erect system but the prostrate system is filamentous or reduced and in C. scutata and C. orbicularis erect system is absent and the lateral branches of the prostrate system form discoid thalli which may be irregular (e.g., *C. scutata*) or regular (*C. orbicularis*) in (e.g.outline (Fig. 11).

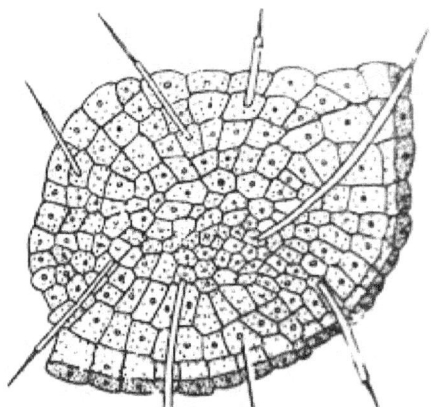

Fig. 11. Thallus Structure

Cell structure.

The cells of the thallus may be quadrangular, polygonal or cylindrical. They are uninucleate and each cell has a large laminate parietal chloroplast with a single pyrenoid. Some of the cells of the thallus have a long unbranched cytoplasmic thread, called seta. The base of the seta is covered with a sheath of gelatinous material.

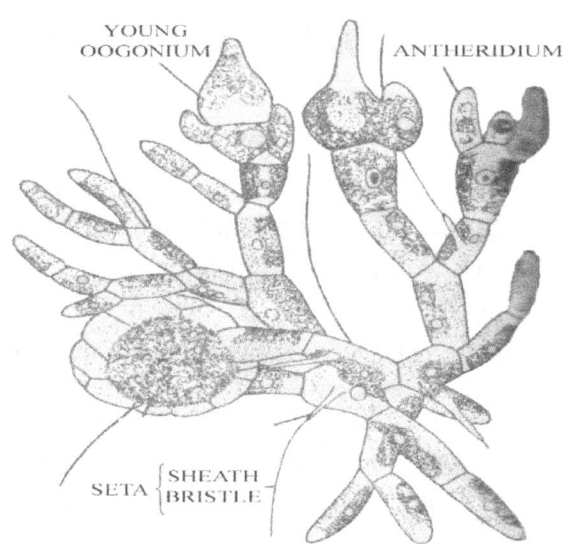

Fig : Vegetative structure of *Coleochaete pulvinata*.

Fig. 12.

Reproduction in *Coleochaete*:

The reproduction in *Coleochaete* takes place by asexual and sexual methods:

1. **Asexual Reproduction:-**

It takes place by means of motile asexual spores called zoospores. Zoospores are developed from the terminal or sub-terminal vegetative cells of the thallus. Each zoospore is biflagellate, ovoid or rounded structure. It possesses a single chloroplast and is without eye spot. On maturity the zoospores escape through the opening formed at the apex of a short papilla-like outgrowth. After liberation each zoospores takes a short period of rest and secretes a wall, restore flagella and producea one-celled germling. The germling soon begins to germinate. During germination it divides transversely into two daughter cells. The upper one forms the seta and the lower one divides and re divides, finally producing a discoid thallus. In heterotrichous form the germinating zoospore first produces a branched prostrate system from which the branches of erect system arise later.

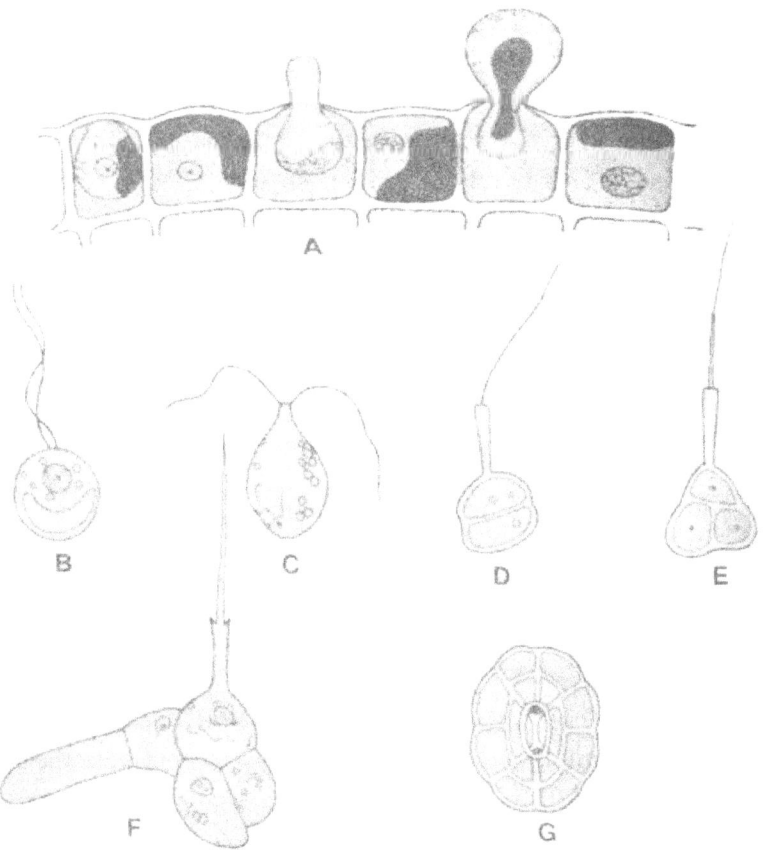

Fig : Asexual Reproduction of *Coleochaete spp.*

Fig.13.

2. Sexual Reproduction:-

Sexual reproduction is of advanced oogamous type followed by the fruit formation. The plants may be monoecious (*C. pulvinata*) or dioecious (*C. scutata*). The male sex organs are called antheridia and the female oogonia or carpogonia.

(a) Oogonia:-

The female sex organs i.e., oogonia are formed terminally on the marginal cells of the shortlateral branch of the projecting filaments. Each oogonium is a cone shaped or flask shaped structure with a swollen basal portion and a long neck-like extension called trichogyne. Hence oogonia are called carpogonia. The basal portion contains a single nucleus and chloroplast with a pyrenoid. All these combined to form the egg. At maturity the trichogyne disintegrates.

(b) Antheridia:

The antheridia are usually borne terminally in groups on the branches of erect system. The antheridia arise as colourless, bluntly conical outgrowths from the terminal cells of the lateral branches. The protoplast of each antheridium metamorphoses into a single colourless and biflagellate antherozoid or sperm.

(c) Fertilization:

The liberated antherodoids swim towards the opened neck of the oogonium. On reaching the neck, they are attached to the disintegrated trichogyne. Finally, one antherozoid enters the oogonium and fuses with the egg to complete the fertilization. As a result a diploid zygote (2n) is formed. After fertilization the zygote is retained within the oogonium, secretes a thick wall and increases in sizegreatly. Under favourable condition, the zygote nucleus (2n) divides meiotically until 16 – 32 daughter nuclei are formed. Simultaneously, the wall formation between the nuclei takes place, which results in the formation of daughter cells.Each daughter cell is metamorphosed into a single biflagellate zoospore or swarmer. All the swarmers are liberated by breaking of the zygote wall, and after a little period of rest, it develops into a new individual.

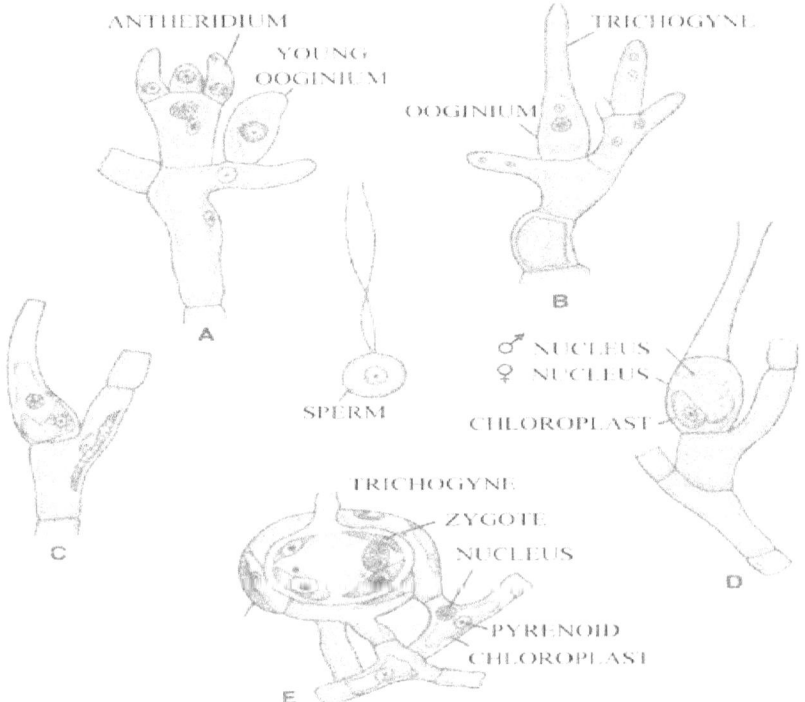

Fig : Sexual Reproduction in *Coleochaete spp.*

Fig.14

Affinities of Coleochaete:

Coleochaete possesses advanced characters like terrestrial tendency, heterotrichous thallus differentiated into prostrate and erect system. The male and female sex organs are distinct and the sexual reproduction is advanced oogamous type.

The post fertilization changes in Coleochaete are parallel to Rhodophyceae and certain characters like presence of setae are comparable to Phaeophyceae. According to Fritsch (1935) Chaetophorales may represent the possible ancestors from which the terrestrial higher plants originate.

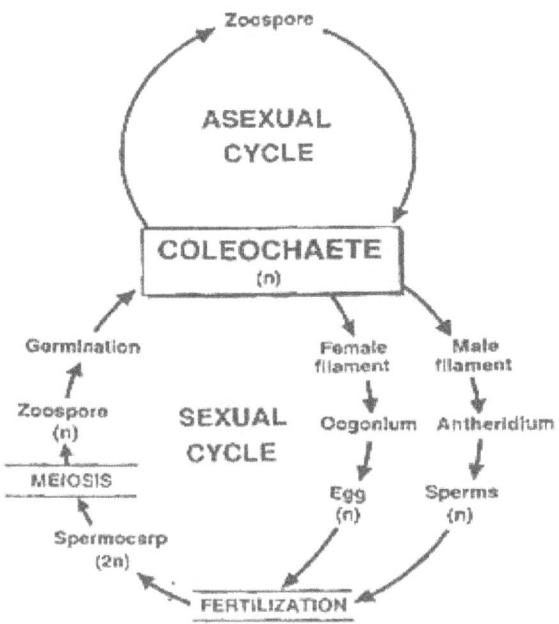

Fig.15. Graphic Life Cycle of Coleochaete

Family. Trentepohliaceae:

Members of this family grow as epiphytes and parasites on and in the leaves of various angiospermic plants. Some constitute the algal component of a number of lichens, Cephaleuros virescens causes the red rust of tea plant. The widely distributed genus Trentepohlia occurs attached to rocks, tree-trunks, etc. The plant body is composed of branched threads comprising of heterotrichous conditions.

In the majority of cases the Trentepohlias appear as orange-yellow cushions due to the presence in the cells orange-red haematochrome dissolved in fat. Besides vegetative, reproduction is usually by the development of sessile, stalked, and funnel-shaped sporangia.

The sessile sporangia often behave as gametangia and produce isogametes, the stalked sporangia bear bi- or quadri-flagellate swarmer's, and the behaviour of funnel-shaped sporangia is still obscure.

Trentepohlia

Systematic Position

Class: Chlorophyceae

Order: Chaetophorales

Family: Trentepohliaceae

Genus: *Trentepohlia*

Occurrance

It is an aerial alga growing epiphytically as orange-yellow cushions or felt-like layer on bark of trees and other damp situations. Plants can withstand prolonged desiccation without any appreciable change. Some species also constitute the algal component of a number of Lichens.

Trentepohlia aurea var. polycarpa at Point Lobos State Reserve. *The orange color is from one of the pigments,* beta carotene.

Thallus Structure

The plant body is composed of heterotrichous filament. In the majority of the species, the thallus exhibits a differentiation of aerial as well as prostrate systems represented by richly branched threads. But there are some species where both prostrate and aerial systems are not equally well developed (Fig. 16D, E & L).

The branches of the aerial system are opposite, alternate, or unilateral arising either from the tip of the parent cell or sub-terminally or even from the middle and are generally slightly attenuated (Fig. 16A to C). The cells are cylindrical or barrel-shaped and often length being twice the breadth (Fig. 16L). The cells have thick cellulose walls which are stratified with parallel or divergent layers (Fig. 16C & F).

The free ends of the apical cells are often covered by pectose caps (Fig. 16F to I). In some species, during the growth of the apical cells, the pectose caps get pushed aside and remain attached to the lateral walls of the older cells (Fig. 16J & K). The cells have numerous band-shaped or discoid chloroplasts without pyrenoids.

The characteristic orange-yellow colour of the cells is due to the presence of orange-red haematochrome dissolved in fat occurring as globules around the chloroplasts.

The function of this special pigment is not yet clearly known. According to some, the haematochrome is a kind of reserve food, whereas, others consider it as a light screen. Cells when young, are uninucleate and with maturity become multinucleate.

Bi- or quadriflagellate swarmer's are developed in three different kinds of sporangia:

(i) Sessile sporangia, these may be terminal, lateral, intercalary, or rarely axillary in position (Fig. 16D and Fig. 17A to G). They are formed by enlargement of cells and liberate biflagellate swarmer's without being detached from the mother plant. These swarmer's behave as isogametes (Fig. 17N to R). Sessile sporangia are quite commonly produced on the prostrate portion of the plant. They behave as gametangia.

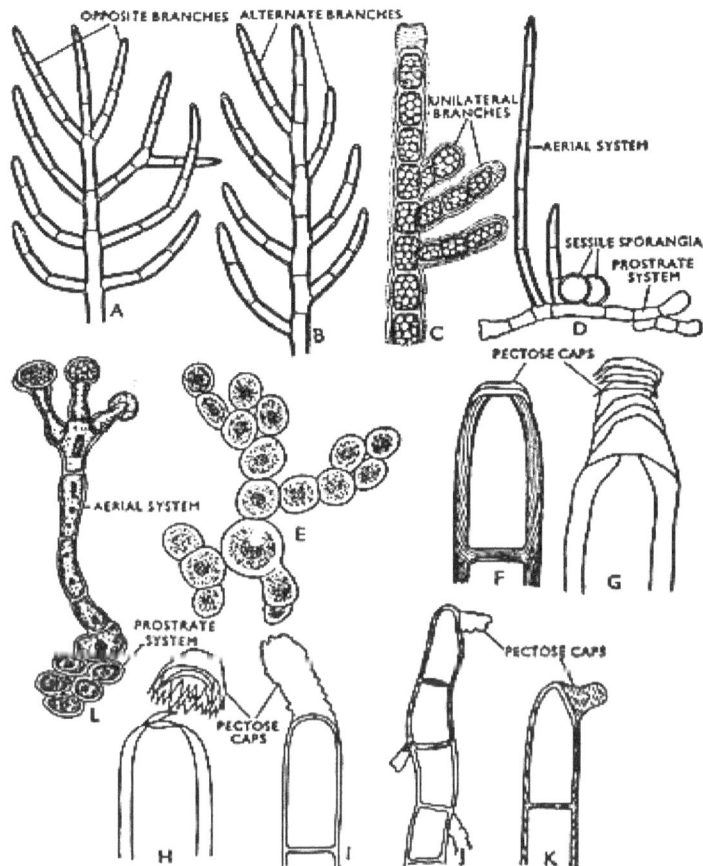

Fig.16. *Trentepohlia* sp. A–C. Different forms of branches in the aerial system. D. Prostrate and aerial systems with sessile sporangia. E. Prostrate system. F–I. Structure and mode of formation of pectose caps on free ends of apical cells with stratified wall. J–K. Pectose caps pushed aside and remaining attached to the lateral walls with the growth of the apical cells. L. Pigmented cells of prostrate and aerial systems, aerial system cells having length greater than breadth.

(ii) Stalked sporangia are formed terminally or laterally by cutting off from a tubular outgrowth developed from enlarged supporting cell. The apical portion of the outgrowth swells to form sporangium and separates from the stalk cell underneath, which is often bent to assume a knee-shaped appearance (Fig. 17D & E).

These sporangia produce bi- and quadriflagellate swarmer's which behave as zoospores. Stalked sporangia are usually developed on the projecting system of the plant (Fig. 17 F to H).

(iii) Funnel-shaped sporangia formed from the apices of the terminal cells, become detached at maturity (Fig. 17 I to M). Their function has not yet been established.

Both sessile and stalked sporangia may be developed on the same plant or on separate plants. The development of sessile and stalked sporangia in the life cycle of the alga is suggestive of the possibility of alternation of generations.

Indian species of Trentepohliaceae:

Trentepohlia monilia de Wilderman.

Special features of Trentepohliaceae:

1. Epiphytic alga forming orange-yellow cushions or felt-like layer on damp substrate.

2. Plant body with adaptations to withstand prolonged desiccation.

3. Heterotrichous plant body, aerial as well as prostrate systems with or without being equally well developed.

4. The branches of the aerial system are opposite, alternate, or unilateral.

5. Cells with stratified wall, special pigment—orange-red haematochrome and band-shaped or discoid chloroplasts.

6. Free ends of the apical cells are covered by pectose caps.

7. Bi- or quadriflagellate swarmers are developed in specialized structures which according to the behaviour of the swarmer's are gametangia or sporangia.

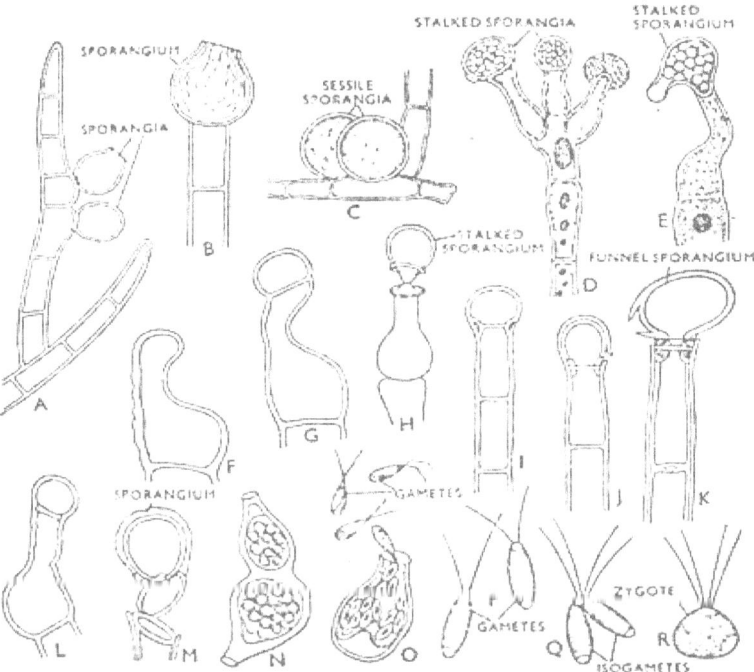

Fig.-17. *Trentepohlia* sp. A–C. Lateral, terminal and axillary sessile sporangia. D–E. Stalked sporangia F–H. Formation of stalked sporangium. I–M. Development of funnel sporangium. N–O. Sporangia behaving as gametangia, liberation of gametes. P–R. Gametes, gametic union and zygote formation.

Family. Pleurococcaceae:

This family is best known for its genus Pleurococcus, also called Protococcus. It includes terrestrial algae occurring in all kinds of damp situations as a thin incrustation of green coat on the windward side of stones walls, tree-trunks, etc. The cells can withstand prolonged desiccation without change.

The cells, which are globose in shape and occasionally branched, are single or else as many as four may be united into a group. Under certain cultural conditions branching may be copious. The sole method of reproduction is through vegetative division in three planes.

Pleurococcus (or Protococcus):

Systematic Position

Class: Chlorophyceae

Order: Chaetophorales

Family: Pleurococcaceae

Genus: *Pleurococcus*

Occurrance

This is a simple alga growing in moist situations on the bark of trees, old walls, flower pots, and similar other places. It may occur singly as isolated individuals or as small compact masses or groups of two, four or eight individuals. When adhering together, the sides in contact are rather flattened.

Thallus structure

This unicellular alga carries on all essential life processes within the single cell. From the physiological point of view it has the same characteristics as higher plants.

The cells can withstand prolonged desiccation. The resistance of the cells to desiccation is aided by a highly concentrated cell sap and a capacity to imbibe water directly from the air.

Reproduction

Occasionally, when submerged under water, cell division may continue with the production of a colony or the daughter cells may grow out into short filaments or threads (Fig. 18F to H).

For this character Pleurococcus is considered to be reduced and simplified form originating from a filamentous one in which both vegetative structures and reproductive functions have retrogressed.

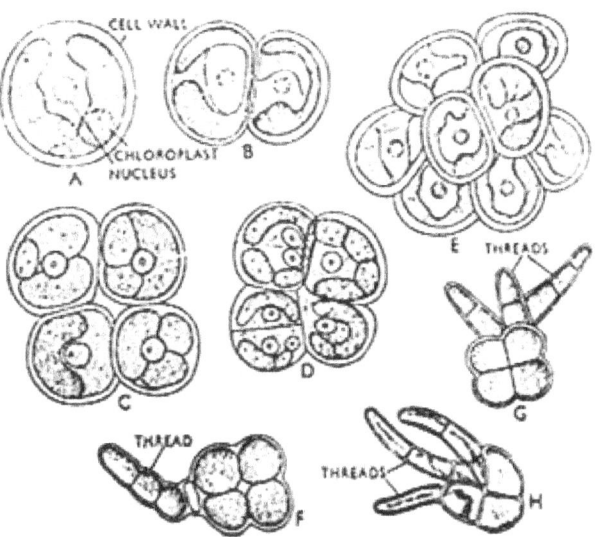

Pleurococcus sp. A. A single cell showing structural details. B–C. Cell division. D–E. Daughter cells adhered together. F–H. Thread formation.

Fig.18.

Each cell is spherical with a cellulose wall, and is densely filled with protoplasm containing a large lobed chloroplast occupying the peripheral portion of the cell, and a nucleus containing a nucleolus (Fig. 18A). The rest of the protoplasm is colourless. The only method of reproduction is by fission.

The cells divide freely into two, with successive divisions taking place in three directions and at right angles to each other (Fig. 18B & C). The daughter cells may either round themselves off or separate immediately after each division, or may remain adhered together (Fig. 18D & E) for a few generations and later they fall apart.

Some Indian species of Pleurococcaceae:

Pleurococctis cohaerens Kutz.; *P. viridis* Ag.

Special features Pleurococcaceae:

1. Unicellular alga occurring singly or in compact masses.

2. The cells can withstand prolonged desiccation.

3. Large lobed peripheral chloroplast.

4. The only method of reproduction is by fission. The daughter cells either may remain adhered together for a long time or may produce a short filamentous structure.

There is probably no other algal genus about which there has been so much confusion. Phycologists are not even in agreement as to whether the genus should be called Pleurococcus or Protococcus. This confusion is partly due to the fact that any unicellular aerial green alga has been considered a species of Pleurococcus.

Some authorities have even placed the genus Pleurococcus in the Chlorococcales, while others have treated it in a special group, the Pleurococcales.

In recent years it has been realized that the unicellular aerial green algae whose cells divide vegetatively and do not produce swarmer's, belong to the Pleurococcaceae. Besides this, as the alga Pleurococcus can occasionally develop branched filaments there would seem to be evidence for regarding it as a much reduced member of the Chaetophorales.

****

13.

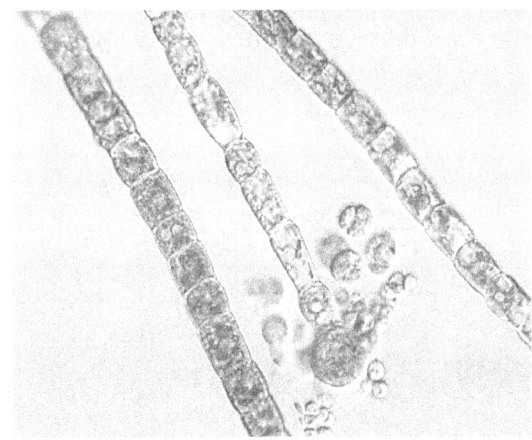

Oedogoniales: *Oedogonium*

(1) Oedogoniales are essentially a fresh water group, playing a very important role nearly all the world over in the algal vegetation of smaller water, the peculiar Oedocladium is, however, mainly terrestrial.

(2) The thalus is branched (Oedocladium and Bulbochacta) or unbranched filament (Oedogonium) made-up uninucleate, cylindrical cells.

(3) Cells show apical basal polarity.

(4) The choloroplast is a parietal network (reticulate) with pyrenoids

(5) Cell division and 'cap' formation is very characteristic and takes place by annular splitting of the lateral cell walls.

(6) There is a ring of numerous flagella around the anterior end of the zoospores,

androspores and male gemetes.

(7) Asexual reproduction takes place usually by zoopsores.

(8) Sexual reproduction is always oogamous.

Family Oedogoniaceae:

This family includes chiefly the fresh-water forms which prefer to grow in quieter situations and are often attached to submerged plant parts or any other object. The plant body may be of branched or un-branched filament. The distinguishing feature of the family is the presence of a large number of flagella at the anterior ends of the motile reproductive structures—zoospores, androspores, and the male gametes.

Oedogonium

Systematic Position

Class: Chlorophyceae

Order: Oedogoniales

Family: Oedogoniaceae

Genus: *Oedogonium*

Occurrence

The genus Oedogonium includes 285 species. It is a fresh water alga, found floating really in mature stage, but it is attached to stem and leaves of the submerged aquatic higher plants or other algae (Epiphytic) when young. However, in the species inhabiting running water, the filaments continue to remain attached to the substratum.

Thallus Structure

Each plant is an unbranched, deep green filament consisting of a series of long cylindrical cells, strictly differentiated into the base and apex (apical basal polarity). The cells are separated from one another by common septa. The plant body is called filament. The basal cell flattens into a disc like structure forming a hold fast, with which the filament attaches itself with the substratum, whereas the apical cell is rounded. Presence of caps is the characteristic feature of *Oedogonium* (Fig. 1).

Cell Structure

Cells are uninucleate and when mature have a central vacuole containing cell sap; and an elaborate, reticulate chloroplast containing many pyrenoids. A sheath of starch granules surrounds each pyrenoid. The cells of Oedogonium are similar to those of other green algae except that their chloroplasts contain microtubules each made up of two helically wound subunits. Such elements have also been observed in the chloroplasts of zoospores, zoospore germlings and eggs (Hoffman, 1967). The semicrotubules are considered to provide structural support to the massive reticulate chloroplast and perhaps also to facilitate its growth and development.

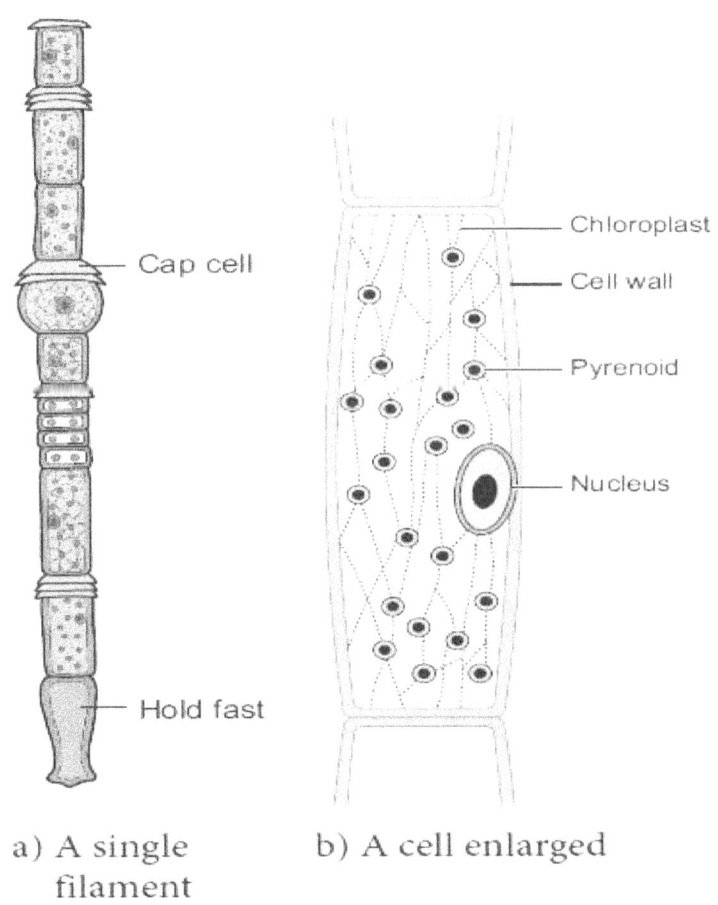

a) A single filament

b) A cell enlarged

Fig.1. *Oedogonium*, vegetative filament and cell structure.

Cell Division.

As the cell enters the division phase, the nucleus moves from the lateral position to the centre. So on a transverse ring of wall material appears on the inner face of the lateral wall just below the apical end of the cello the nuclear division, the growth of the ring in thickness and the formation of a groove enclosing the growing ring occur concomitantly. An unattached floating septum is formed between the two daughter nuclei. The middle and outer wall layers external to the groove then rupture, permitting free elongation of the ring which forms a new piece of cell wall lying between the cap and the sheath. Ultimately, the floating septum moves upward and becomes fixed near the terminus of the old cell wall. The new cell has the wall formed from the thickened ring and the newly synthesized piece. The membranous striation of the ruptured parental wall at the anterior region of the upper daughter cell is the cap and the cell bearing it is known as a cap cell. The number of caps on a cell indicates the number of cell divisions that have taken place (Fig.2.). A recent ultrastructural study (Fig. 2.) of the mode of cell division in *Oedogonium borisianum* (Hili and Michalis, 1968) has revealed that:

(1) The ring (Fig. 2. A) below the cap originates as a three-layered structure by a method which excludes the possibility of its origin from an infolding of the innermost wall layer.

(2) The ring grows in size with a gradual addition of vesicular material from the cytoplasm;

(3) The mother wall adjacent to the fully-formed ring splits off.

(4) The single, peripheral nucleus migrates to the centre and karyokinesis occurs; this is followed by the formation in the internuclear region of a row of microtubules (Fig. 2. B) in a plane parallel to that of the future cross wall;

(5) Concurrent with this development the ring expands into a cylinder (Fig. 2. C) which becomes the cuticle of the upper daughter cell. The septum, which for some time remains unattached to the lateral wall, later migrates upward to the base of the cylinder (Fig. 2. D).

(6) Soon ~ new lateral wall is formed between the cuticle and plasmalemma of the upper daughter cell.

(7) Finally, the upward migrated transverse septum grows into a mature cross wall which unites with the newly formed wall on either side (Fig. 2. E). the mode of cross wall formation rules out the possibility of cell division either by cytokinesis or by annular furrowing of the cytoplasm.

Reproduction

Oedogonium reproduces by vegetative, asexual and sexual methods.

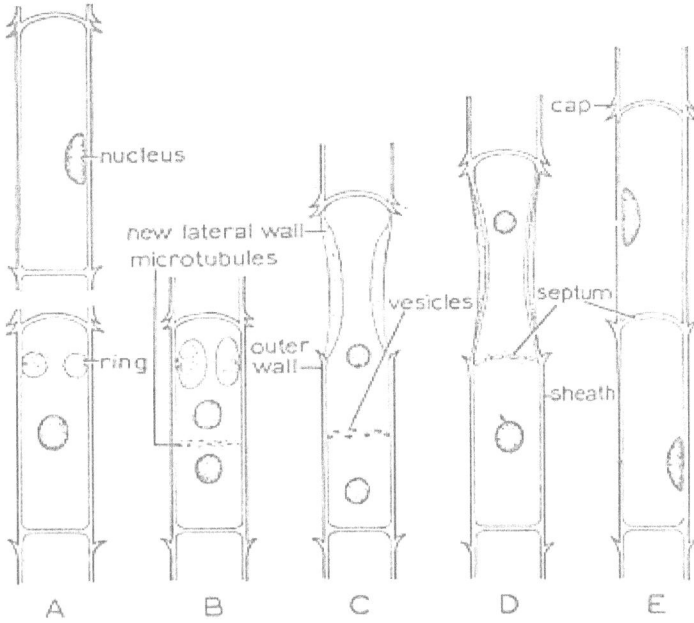

Fig.2. *Oedogonium borisianum*, stages in cell division.

Vegetative Reproduction

It occurs by fragmentation and akinete formation. Fragmentation takes place by dying out of some intercalary cells, due to accidental breaking of the fi lament, through the conversion of intercalary cells into zoosporangia or gametangia, or in adverse environmental conditions. In some species the vegetative akinetes are developed during unfavorable conditions for vegetative growth. Under favorable conditions akinetes germinate into new fi laments.

Asexual Reproduction:

Asexual reproduction takes place by means of zoospores (Fig. 3.A-C). Zoospores are formed singly within a cell. Comparatively younger cell i.e., the cell with cap behaves as sporangium mother cell. The zoospores are multiflagellate and ovoid, pyriform or spherical in shape. They are uninucleate with single chloroplast and occasionally with an eye-spot. During favourable condition, the zoospore formation begins in a cap cell of the filament. The entire protoplast of zoosporangium contracts from the wall and behave as a unit. The protoplast becomes round or oval in shape and its nucleus moves at one end.

Near the nucleus a semicircular hyaline area develops. Just below the hyaline area a ring of blepharoplast granules develops, connected with each other by fibrous strands (Ringo, 1967). Later on, from each blepharoplast granule, single flagellum develops. Thus a crown of flagella is present around the colourless semicircular area.

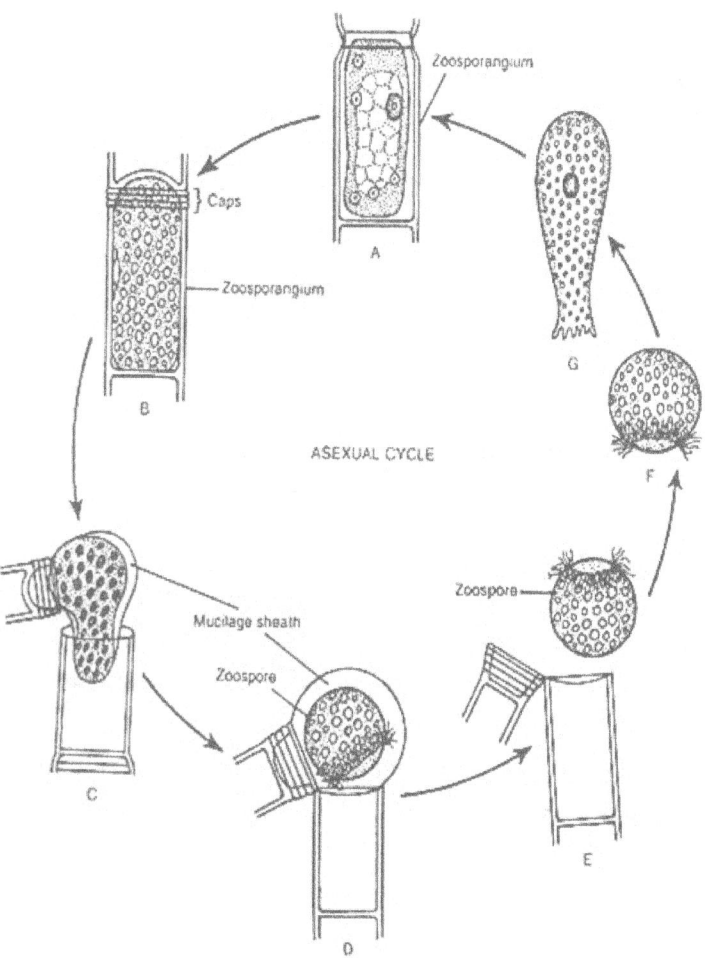

Fig. 3. *Oedogonium* Sp., Asexual reproduction: A-G Successive stages of Zoospore Formation and Germination

The fully developed zoospores are liberated by breaking the zoosporangium wall. The wall of the zoosporangium breaks near the cap region and the neighbouring cell bend on one side to make way for the liberation of zoospore. During liberation, the zoospore remains as a delicate mucilaginous vesicle for 3-10 minutes. After dissolution of vesicle the zoospore gets free and starts swimming in the surrounding water.

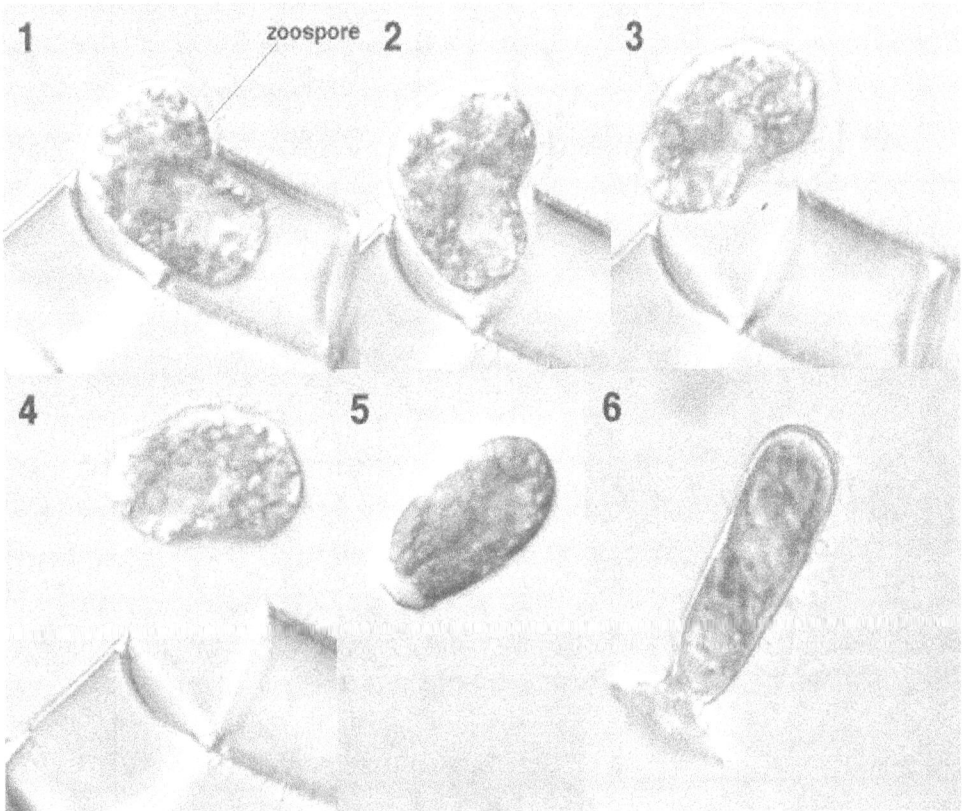

Fig. 4. Successive stages of Zoospore Formation (Mishima, Shizuoka Pref., Japan, April 2001 by Y. Tsukii)

Germination:

The zoospore can swim for about one hour or more. Coming in contact with substratum by the anterior end, it loses flagella and starts to elongate. The lower hyaline part becomes separated by cell wall, which forms the hold fast. Through the subsequent division and re-division in a single plane, new filament is formed.

Sexual reproduction. It is always oogamous. The species may be homothallic or heterothallic. Sexual reproduction may be of **rnaerandrous** or **nannandrous type. It frequently** occurs in still water and rarely in flowing waters.

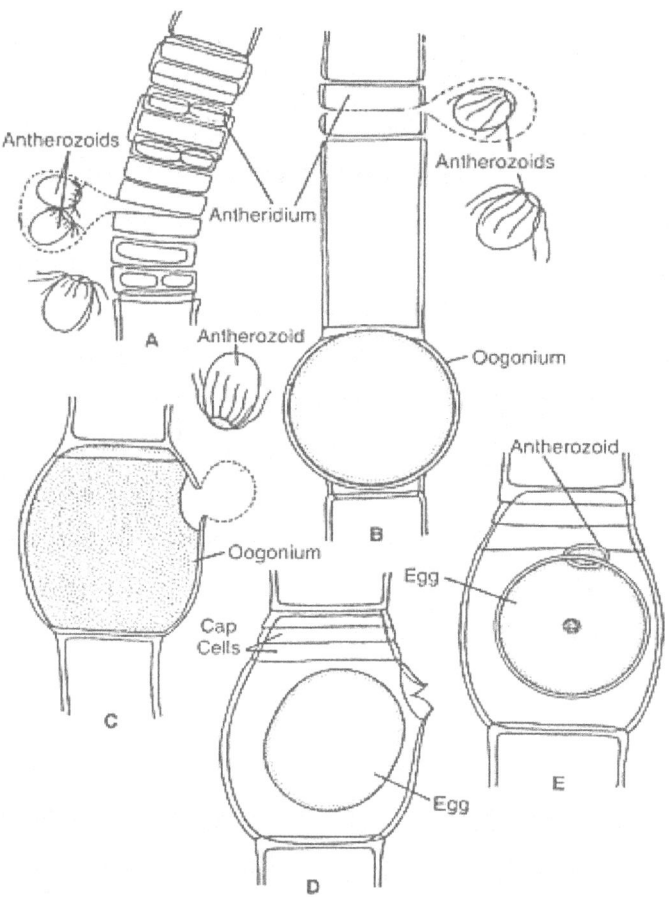

Fig.5. *Oeciogonium* sp. Sexual Reproduction-Macrandrous species A, antheridia and liberation of antherozoids; B, a monoecious species having antheridia and oogonium on the same thread; C, mature oogonium ready for fertilization; D, entrance of antherozoid in the oogonium through receptive spot; E, fertilization; F, single antherozoid.

Development of antheridia.

In macrandrous species, the oogoma and antheridia may be developed on the same filament (homothallic or monoecious species) or on two different filaments (heterothallic or dioecious species). The antheridia are either terminal or intercalary. They are produced by the division of antheridial mother cell. Any vegetative cell may act as antheridial mother cell. This cell dcvides into two unequal cells at its upper end forming an antheridium. The lower sister cell divides repeatedly and a chain of 2-40 antheridia is produced. The protoplast of each antheridium metamorphoses in a single antherozoid. Sometimes two antherozoids are produced in one antheridium. The antherozoids resemble zoospores in structure and shape only with the difference that they are somewhat smaller in size and with fewer number of flagella. The antherozoids liberate in the same way as zoospores. They come out in thin membranous vesicles by the split of antheridial wall. Very soon the vesicle disappears and the antherozoid moves freely in the water in any direction.

In nannandrous species, the antheridia are produced on special dwarf males or nannandria. Majority of the nannandrous species are heterothallic (dioecious), oogonia are produced on the ordinary filaments and antheridia on special filaments called 'dwarf males' or nannandria. These 'dwarf males' or nannandria originate by the germination of a special type of swarmers known as androspores produced singly in flat cells, i.e., androsporangia formed by repeated transverse divisions of the ordinary vegetative cells. The androsporangia may occur on the same filaments on which the oogonia are produced or on different filaments. The androspores resemble in shape and structure to zoospores. They are somewhat smaller than zoospores and larger than antherozoids. After their small swarming period, they settle down either on the oogonium or onone of the neighbouring cells. Here the androspore germinates and develops into a dwarf male. The nannandrium consists of one basal cell attached to the oogonium or adjacent cell, and one or more flat antheridia. In Oedogonium, the nannandrium consists of an attaching cell only producing two antherozoids develop in an antheridium. The structure of antherozoid is same as that of the antherozoids of macrandrous species. The antherozoids swim here and there in the water after liberation.

Development and structure of oogonia.

The development and structure of oogonia of macrandrous and nannandrous species is identical. The oogonia develop by terminal or intercalary oogonial mother cells. The oogonial mother cell divides transversely and the upper daughter cell always develops into an oogonium. Each oogonium always possesses one called the 'suffultory cell' which does or more caps at its upper end. The lower daughter cell is not develop into an oogonium and thus two oogonia may be seen in continuation on the filament. The oogonium is rounded or ovoid, in structure. On maturity, a hyaline spot develops at the upper end of the oogonium. Each mature oogonium contains a large spherical egg cell in it. The nucleus is situated in the centre of the egg. Prior to fertilization the centrally located nucleus shifts towards upward near the hyaline spot, the oogonial wall cracks at this spot and the antherozoid approaches the egg cell through it.

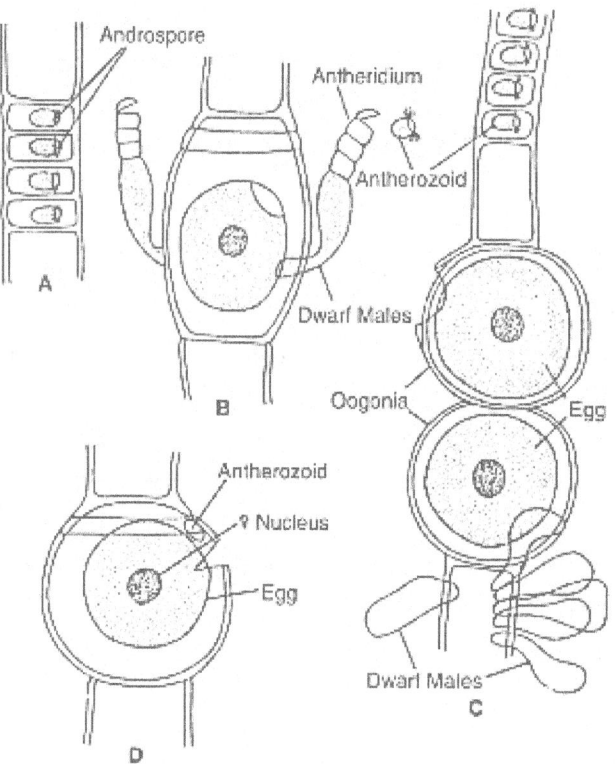

Oedogonium. sp. Sexual Reproduction. Nannandrous species; A. androspore fomation in androsporangia; B. oogonium with dwarf males and antheridia; C, two oogonia and unicellular dwarf males on supporting (suffultory) cell; D. fertilization.

Fig.6.

Fertilization. The swimming antherozoid approaches the oogonium at the receptive (hyaline) spot. The flagella are retracted and the plasmogamy takes place which is soon followed by karyogamy. The zygote or oospore which is somewhat retracted from the oogonial wall secretes a three layered-thick smooth wall around it. The colour of oospore changes from green to reddish brown.

Germination of oospore (Zygote)

The oospore germinates during favourable condition (Fig. 7). The nucleus undergoes meiosis and forms 4 haploid daughter nuclei. The nuclei accumulate some cytoplasm and form 4 daughter protoplasts. They liberate by rupturing the oospore wall. During liberation they develop flagella and are called meiospores or zoomeiospores.

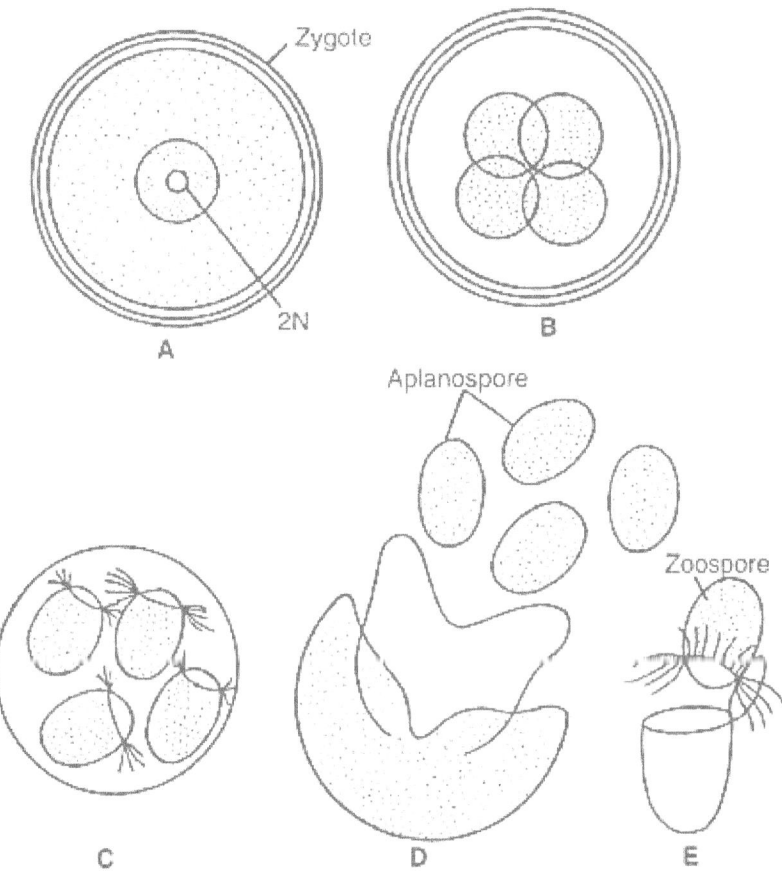

Oedogonium sp. A, a zygote; B-C, germination of zygote; D, liberation of aplanospores; E. zoospore comes out from an aplanospore spot, the oogonial wall cracks at this spot and the antherozoid approaches the egg cell through it.

Fig.7.

Initially they remain inside a delicate vesicle, which soon disintegrates and the zoospores get free into the environment. After swimming for some time in water they withdraw their flagella and germinate into new haoloid Oedogonium filament like zoospore in asexual reproduction. The nature of zoomeiospore development varies in monoecious and dioecious species.

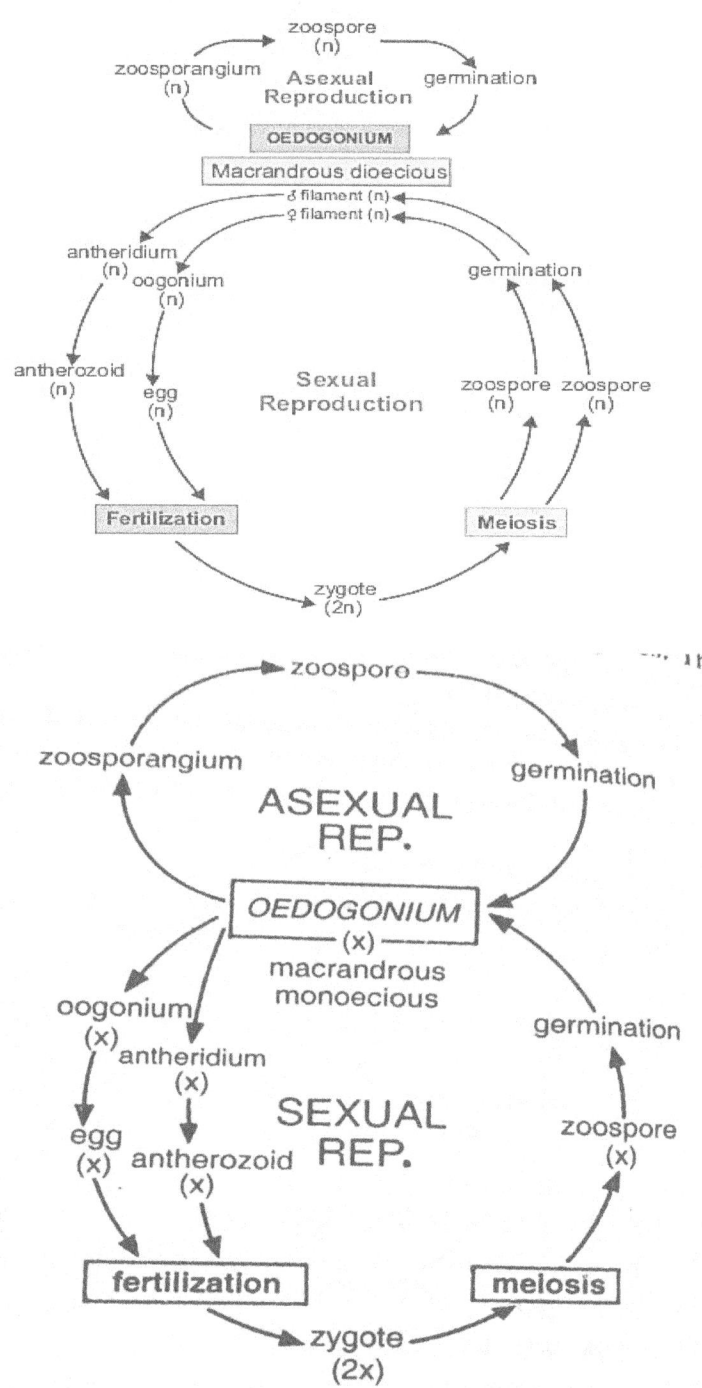

Fig.8. *Oedogonium*: Life cycle of macrandrous dioecious species

Fig. 9. *Oedogonium*: Life cycle of macrandrous monoecious species

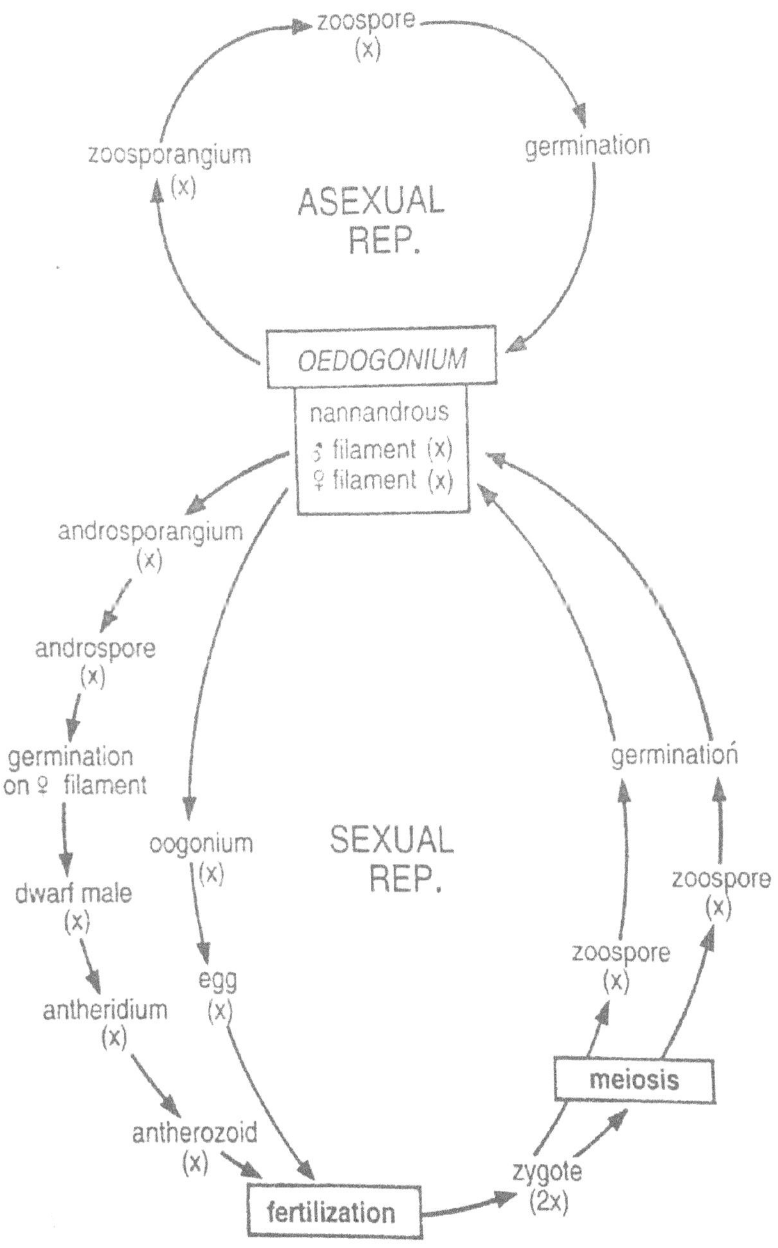

Fig.10. *Oedogonium*: Life cycle of nannandrous species

****

14.

Conjugales: *Spirogyra Zygnema, Cosmarium*

Algae belonging to the order Conjugales are commonly known as 'pond silks or water silks. They are generally found in fresh waters only. They are differentiated from other Chlorophyceae in the lack of flagellated reproductive structure and in the structure of chloroplast. These algae vary in their structure from unicellular to unbranched chloroplast. Filamentous forms. The unicellular forms are commonly called as desmids. The cell wall is composed of an outer pectie layer and inner cellulosie layer. The cells are uninucleate and the nucleus is centrally located. The chloroplasts are of three types : one or more peripheral spirally twisted ribbon-like bands as in Spirogyra: (i) an axial plate extending the length of the cell as in *Mesotaenium* and *Mougeotia*; and (ii) two stellate axial chloroplasts as in *Zygnema* and *Cylindrocystis*. Vegetative multiplication in unicellular forms takes place by simple cell division and in filamentous forms by fragmentation. Motile asexual spores are not fourd in these algae. They reproduce asexually with the help of aplanospores. Sexual reproduction takes place by the fusion of non-flagellate amoeboid gametes through conjugation. The zygospore formed by the fusion of two gametes secretes a thick wall with characteristic ornamentation.

Fritsch (1935) divided the order into five families:
Mesotaeniaceae,
Zygnemaceae
Mougeotiaceae,
Gonatozygaceae
Desmidiaceae.

Family: Zygnemaceae

Members of Zygnemaceae are among the mosty unbranched filamentous algae, favouring small stagnant bodies of water. They are especially abundant in spring months. The cells of the filament are cylindrical and more or less permanently united.

The cell wall lacks pores. Union of two aplanogametes usually takes place by the establishment of a conjugation tube. The family is represented by three genera,

- *Spirogyra*
- *Debarya*
- *Zygnema*

Spirogyra

Systematic Position

Class: Chlorophyceae
Order: Zygnematales
Family: Zygnemaceae
Genus: *Spirogyra*

Occurrence

Spirogyras are common free-floating freshwater algae that inhabit ponds, pools, tanks, lakes, ditches, etc. The word 'Spirogyra' is derived from the two Greek words, '*Speria*', meaning coil, and '*gyras*' meaning twisted. *Spirogyra* has many common names, including blanket weed, water silk, mermaid`s tresses, etc. It grows up to several centimeters in length and 10-100 μm in width. There are about 400 known species of *Spirogyra* worldwide. They are filamentous and slippery in natures due to the presence of external mucilaginous sheath; hence, they are called pond scum or pond silk. Some of the species of *Spirogyra* (*Spirogyra adnata, S. jogensis*) bear holdfast or haptera by which they remain attached to the substratum.

Thallus Structure

The plant body is un-branched filamentous, green-colored with cylindrical cells placed end to end. In free-floating species, there is no basal differentiation, but in some sedentary species, the basal cell is modified into a haptera or holdfast, which is the organ of attachment. The lateral cell wall is stratified and three-layered, the inner two layers are composed of cellulose, and the outermost layer is composed of pectose. The cross walls are also three-layered, the middle lamella is composed of pectose bounded on either side by layers of pectin, in some other species, there is an additional annular ingrowth of cellulose along with the cross wall, giving rise to the so-called replicate wall which again is of five types such as plane, replicate, semi-replicate, colligate (cross wall looks like a short H-like piece) and unduliseptate, respectively.

A vacuolated, granular protoplast is present in each cell in the form of a thin lining along the cell wall and is termed as the primordial utricle. The vacuole is separated from the surrounding cytoplasm by a semi-permeable membrane, the tonoplast. The eukaryotic nucleus with a distinct nucleolus remains embedded within the cytoplasm or resides at the center being supported by the cytoplasmic strands. Each cell bears a varying number of ribbon-shaped, spiral chloroplasts with either serrated or smooth edges. In the chloroplasts, dense, highly refractive, granular, protein body surrounded by starch called pyrenoids at short intervals. In some cases, pyrenoid like bodies without starch sheath is also present called protopyrenoids.

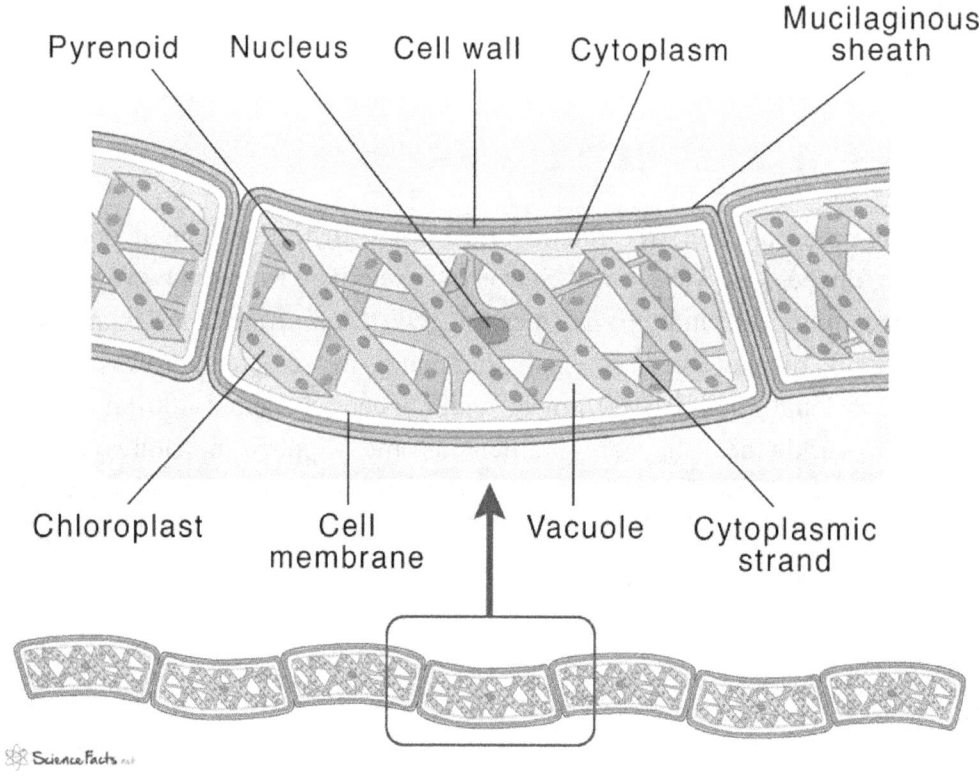

Fig.1. Vegetative structure of *Spirogyra*

They also contain other cellular organelles such as endoplasmic reticulum, Golgi bodies (dictyosomes), mitochondria, ribosomes, etc. which remains scattered within the cytoplasm.

Reproduction

A. Vegetative reproduction:

During the growing season, the vegetative filament of Spirogyra undergoes fragmentation, so that each fragment independently develops into a new filament by repeated cell division and elongation. In favourable conditions, fragmentation is a most common method of multiplication. The fragmentation is caused by (i) mechanical injury (ii) dissolution of middle lamella (Hi) development of H-shaped pieces (e.g. in *S. colligates*).

B. Asexual reproduction:

Asexual life cycle is less common and reported only in a few Spirogyra sp. Ascxual cycles involve the formation of aplanospores, akinetest, azygospores (Fig. 5.6).

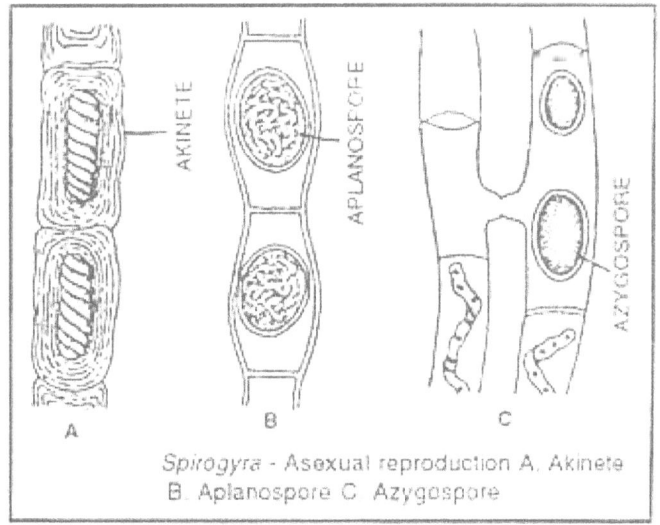

Fig.2.

(i)   **Formation of aplanospores:**

Under un-favourable conditions, the protoplast of each vegetative cell shrinks and develops a wall around it to form an aplanospore. Each non-motile aplanospore germinates to form a new filament. For example, S. articulate, S. mirabilis etc. In S. aplanospora the life cycle involves the formation of aplanospores only.

(ii)   **Formation of akinetes:**

Under unfavorable conditions, the vegetative cells of S. farlowii develop thick-walled cells called akinetes. On the return of favourable conditions each akinete germinate into a new filament.

(iii) Formation of azygospores or parthemspores:

In the life cycle of S. varians and S. groenlandica, sometimes the gametes fail to fuse and each get enclosed by thick cell wall to become azygospore or parthenospore. Each azygospore germinates in to a new filament.

Sexual Reproduction

The sexual reproduction occurs when the conditions are unfavorable for growth. It is primitive type of isogamous sexual reproduction called conjugation.

Mechanism of Conjugation

During the process two filaments come near each other and become glued together by their mucilage. Papillate (finger-like) outgrowths arise from the lateral walls of the cells. These papillae meet and their common walls dissolve to form a tube-like Structure, the conjugation tube. The protoplast in each cell rounds off due to loss of water to form a gamete. The gametes are isogametes; however, one gamete is active and move from one filament to the other. It is called male gamete.

The other gamete is passive and remains in the cell. It is called female gamete. Therefore, ***Spirogyra*** exhibit physiological anisogamy. All the gametes in a conjugating filament may be either male or female. The male gametes move towards the female gamete through conjugation tube. The protoplast and nuclei of both the gametes unite to form a diploid zygote which secretes a thick wall to become a zygospore. After a period of rest the zygospore germinates into a new filament by cell division, Low nitrogen supply and high light intensity usually stimulate sexual reproduction.

The conjugation may be scalariform conjugation or lateral conjugation:

Scalariform Conjugation: It is the most common type of conjugation and occurs between opposite cells of the neighboring filaments as described above.

Lateral Conjugation: In lateral conjugation, gametes are formed in a single filament. Two adjoining cells near the common transverse wall give out protuberances known as conjugation tubes, which further form the conjugation canal upon contact.

The male cytoplasm migrates through the conjugation canal, fusing with the female. The rest of the process proceeds as in scalariform conjugation.

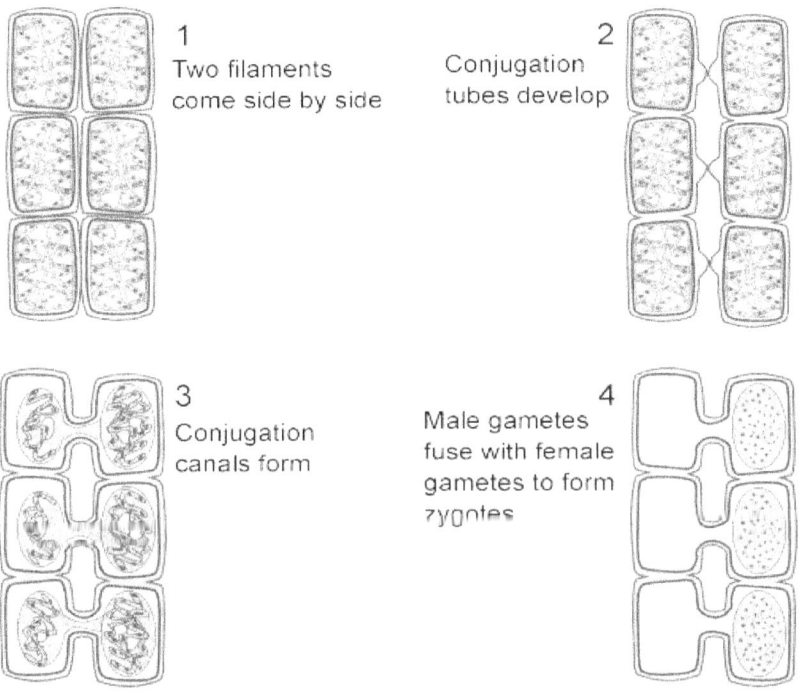

Fig.3. *Scalariform Conjugation*

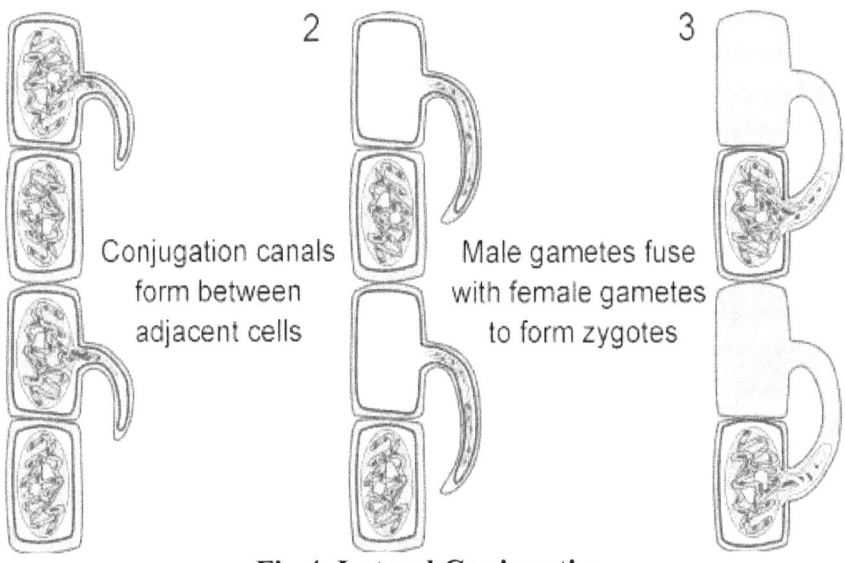

Fig.4. Lateral Conjugation

Structure and Development of Zygospore

The zygospore wall is 3-layered. The outermost layer is cuticularized whereas the innermost is thin and cellulosic. It is spherical or ellipsoidal and diploid. The zygospore becomes red in colour due to disintegration of chloroplast, accumulation of oil and development of haematochrome. The zygospore is liberated by the decay of the wall of the female gametangium. It settles down to the bottom of the water and undergoes a long period of rest.

Germination of Zygospore

At the return of favourable conditions, the zygospore germinates. The nucleus of the zygospore divides meiotically into four haploid nuclei. Three of these degenerates and the fourth one remains functional. The two outer layers burst and the innermost thin layer protrudes out in the form of a germ tube. This tube elongates and divides transversely into a colourless cell which functions as rhizoidal cell and attaches the germ tube to some support, and an upper green cell. The upper cell divides repeatedly to produce an adult filament which detaches itself from the support and floats freely in water.

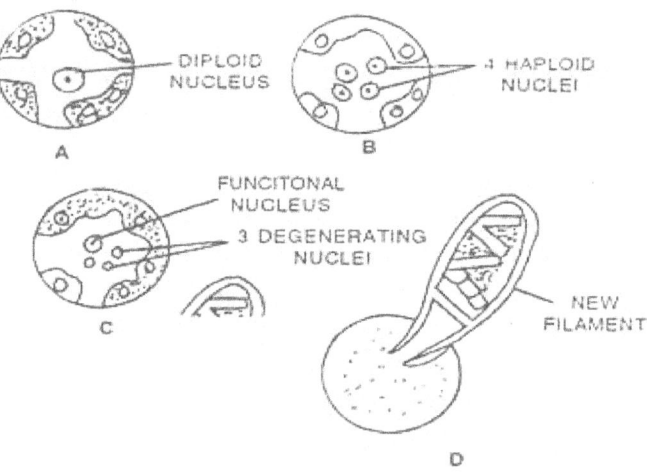

Fig.5. Germination of Zygospore

Zygnema

Systematic Position

Class: Chlorophyceae
Order: Conjugales
Family: Zygnemaceae
Genus: *Zygnema*

Occurrance

Zygnema is very common fresh water, unbranched, filamentous alga. It is commonly found in stagnant waters but can also be found in running water attached to the periphery by some attaching cells. Generally it also occurs in ponds, lakes, pools and other fresh water reservoirs along with some species of *Spirogyra*. The distinct identification of *Zygnema* is the presence of two star-shaped chloroplast. Each chloroplast has a pyrenoid. The filaments usually have soft mucilage sheath, are unbranched and are not very long.

Thallus Structure

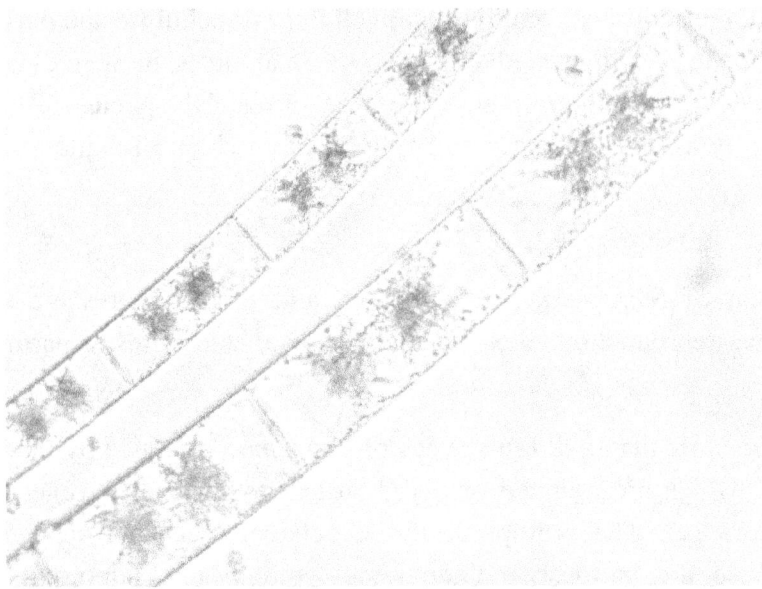

Fig.6. *Zygnema filaments*

Like Spirogyra, *Zygnema* is also filamentous. It differs from Spirogyra mainly in plastid morphology. Cells of Zygnema' contain a pair of axile stellate chloroplasts (Fig. 6), each harbouring a single central pyrenoid with radiating starch grains. The cells are uninucleate and the ' single nucleus lies embedded in the middle of the cytoplasm that separates the two chloroplasts. The nucleus contains one nucleolus with one nucleolar organizing chromosome.

Reproduction

Vegetative Reproduction: Zygnema reproduces vegetatively by two methods:

Fragmentation:

Filaments divide into short length fragments or sometimes even "into individual cells" (Randhawa, 1959), each of which is capable to develop into a new individual. Fragmentation occurs owing to some mechanical injury or accidents.

Akinete: Because of the accumulation of starch and fat, and also because of the development of some additional wall layers of cellulose or cellulose and pectose, some cells of the filaments get thickened in the form of akinetes in species such as *Z. gigantewn, Z. sterile, Z. terrestre,* etc. The one terrestrial species *(Z. terrestre)* reproduces only by akinetes whereas in *Z. giganteam* a chain of akinetes is formed.

Asexual Reproduction:

Similar to that of *Spirogyra,* no flagellated and motile spores are formed in *Zygnema.* Asexual reproduction takes place either by aplanospores or parthenospores (azygospores).

Aplanaspores: More than a dozen species of *Zygnema* reproduce by the formation of aplanospores, which are rounded or ovoid structures formed in vegetative cells, called aplanosporangia. The protoplast of the cell contracts and changes into an aplanospore. These are variously coloured and smooth or sculptured structures, resembling and often being confused with zygospores. Some of the aplanosporeproducing species are *Z. himalayense, Z. launaoense, Z. terrestre* (Fig. 6.B), etc. *Z. kumaoense* (Fig.7.C) reproduces only by the formation of aplanospores.

Parthenospores: In the absence of gametic union, parthenospores (azygospores) are formed in some species such as *Z. collinsianum*.

Sexual Reproduction

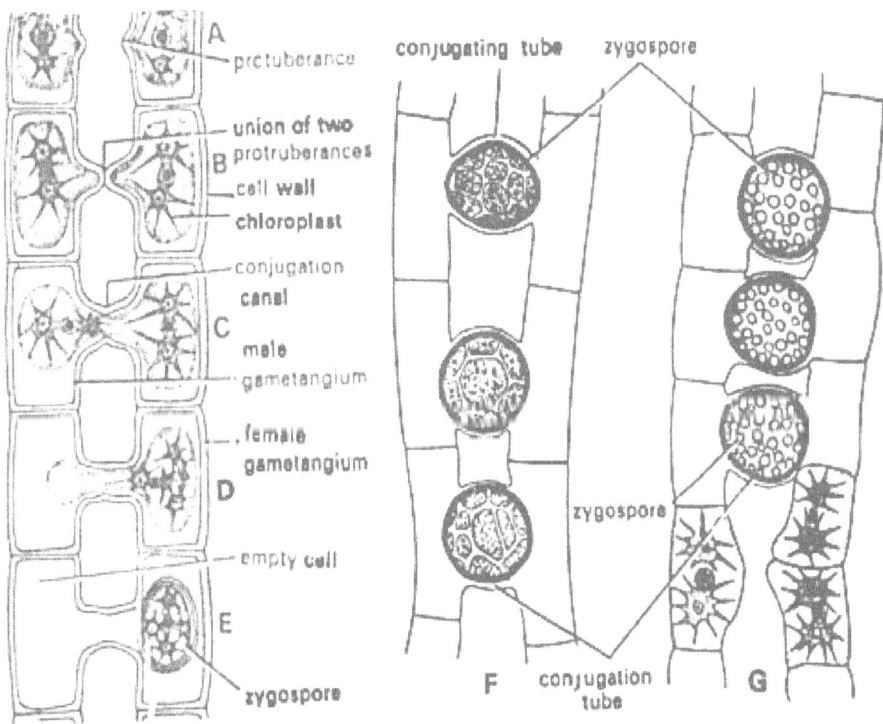

Zygnema. A-E, Scalariform conjugation and zygospore formation in anisogamous species; F, G, Zygospore formation in conjugation tubes in *Z. chungli* and *Z. pectinatum*. (F, after Taft; G, after Czurda)

Fig.7.

Sexual reproduction

Sexual reproduction is of two type.

1. Scalariform Conjugation : It take place between the cells of two different filaments of opposite strain. At the time of conjugation two opposite strain are attracted toward each other and lie parallel in close contact under common sheath of mucilage. The cell of opposite filament behaves as gamentangia and produce a tubular outgrowth called papillae. The papillae elongates and fused through anterior end with papillae of opposite filament forming conjugation tube. The conjugation tube is hollow structure formed as enzyme cytase dissolve the wall joining the two papillae. Several conjugation tube may appear along the length of filament giving ladder like appearance. So, this conjugation is known as scalariform conjugation. At the same time the protoplast of conjugating cell lose water, accumulate starch and function as gamete. Among two gamete, male gamete is motile and active while female gamete is stationary. Male gamete squeezes through conjugation tube and pass into another filament and get fused with female gamete forming zygote. The zygote produce thick wall around it forming Zygospore. One filament form series of zygospore while other filament become vacant. Zygospore under suitable condition germinate into new filament.

2. Lateral Conjugation:

It also follows the same pattern as in *Spirogyra*. It takes place between two cells of the same filament, of which one functions as male gametangium and the other as female gametangium. The longitudinal wall at the region of septum between two adjacent cells protrudes. Two outgrowths are than formed, one on either side of the septum (Fig.8.A). The contact walls of both the outgrowths dissolve and a passage is formed between two gametangia (Fig. 8.). This passage functions as a conjugation tube.

In the species showing isogamous lateral conjugation *(Z. hlmalayense, Z. terrestre* and *Z. csurdae)* both the gametes move and fuse in the conjugation tube (Fig.8.B, C). The zygospores are thus formed in the region of conjugation tubes (Fig.8. D). But in the species showing anisogamous lateral conjugation *(Z. stellinum)* the male gamete moves towards the female gametangium through the conjugation tube and fuses with the female gamete to form the aygospore. Thus the zygospores and empty cells in the anisogamous lateral conjugating filament are present alternately.

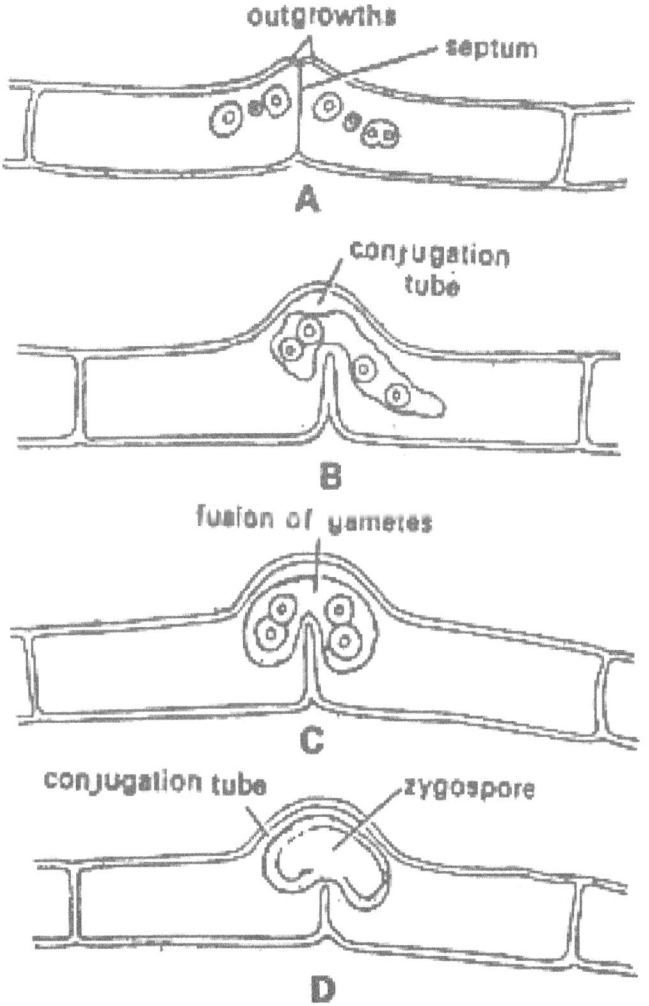

A-D, Isogamous lateral conjugation in *Zygnema czurdae*. (after Iyengar)

Fig.8.

Zygospore:

The sygospore represents the diploid (2n) phase in the life-cycle (Fig. 9.E). It contains *4* stellate chloroplasts (of two fusing gametes) and one diploid nucleus, for some time. But soon two chloroplasts belonging to male gamete degenerate (Fig. 9.F). The zygospore gets enveloped by a three-layered thick wall, of which outermost thin layer (exospore) is made up of cellulose or pectose, middle thick layer (mesospore) of cellulose which is chitinized and sculptured, and innermost thin layer (endospore) of cellulose (Randhawa, 1959).

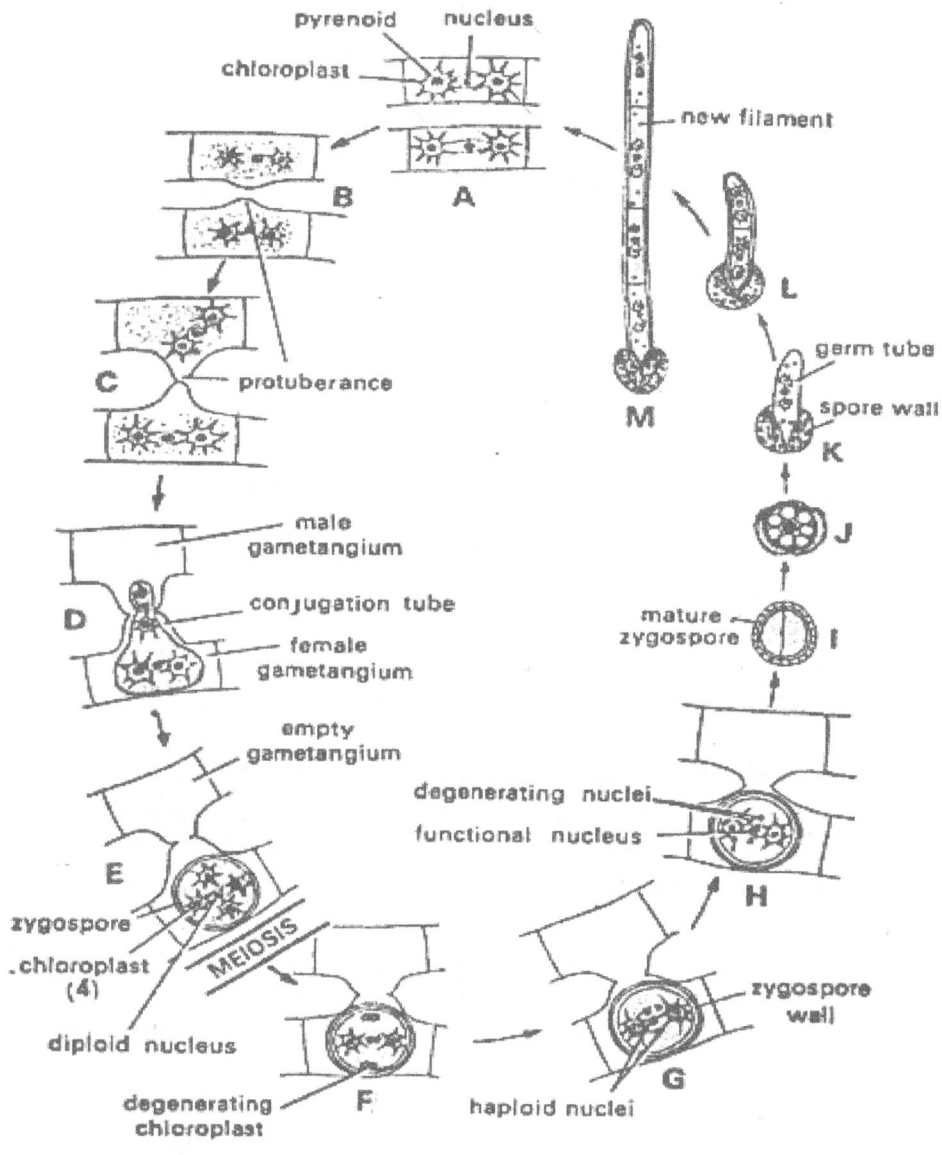

Fig.9. A.M. Life- cycle of *Zygnenea*

The colour of zygospore varies from pale yellow to browb in different species and develops in the middle layer. The gametangial wallgets decayed and the zygospores are liberated. They sink to the bottom of the pond and undergo a long resting period. The reduction division of the nucleus followed by an ordinary division takes place before the germination of the zygospore. It results into the formation of 4 haploid nuclei (Figs. 9.F), of which 3 degenerate and only one survives.

The zygospores germinate generally after rains, when water is available. The exospore ruptures and the protoplast remains surrounded by inner wall layers. 'The zygospore contents grow out into a long germ-tube (Fig. 9.K). The tube divides transversely into two cells (Fig. 9.L), both of which divide and redivide transversely to form the new filament of *Zygnema* (Fig. 9.M). *Zygnema* differs from *Spirogyra* also because there is no distinction of basal and apical cell at two-celled stage of germling.

Family: Desmidiaceae:

This family includes fresh-water unicellular forms of algae.

Cosmarium

Systematic Position
Class: Chlorophyceae
Order: Zygnematales
Family: Desmidiaceae
Genus: *Cosmarium*

Occurrence

Cosmarium is a free floating placoderm desmid. It occurs in fresh water ponds, pools and ditches, either singly or in amorphous colonies. Late spring and early summer seasons are most favourable for its growth. The genus includes over 800 species, of *C. bengalensis, C. connatun, C. distichum, C. granatum* and *C. tenne* are common in India.

Thallus Structure

Single-celled, non-motile plants of rare beauty with length rarely more than one and a half times the breadth, are characterized by the fact that the cells are usually divided into two symmetrical halves by a median constriction. The two halves are referred to as semi cells; the constriction is called the sinus 5 and the narrow connecting zone of the two halves is known as the isthmus. The cells may have smooth or ornamented wall which when mature is composed of two well-differentiated layers. The inner one of cellulose and the outer is firmer and thicker and is composed of cellulose impregnated with various substances including iron compounds. Numerous small pores occur in the cell wall except in the region of isthmus. Through these pores mucilage is secreted which sometimes resembles like a third layer of the cell wall.

Most of the species have two relatively large chloroplasts with radiating plates having centrally placed pyrenoids, one located in each semicell. The single nucleus, embedded in a small mass of cytoplasm, is in the isthmus.

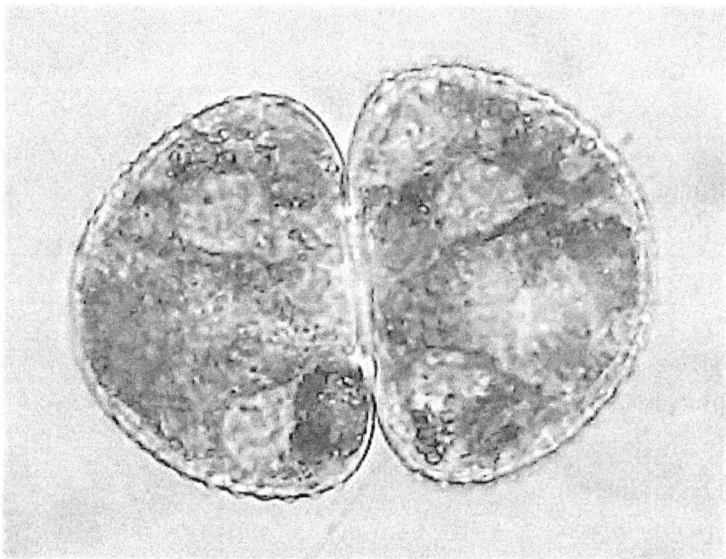

Fig.10. *Cosmarium:* **thallus structure**

Reproduction

Reproduction is by cell division and by the process of conjugation which resembles that in *Spirogyra* and *Zygnema*.

As soon as the cell division starts the nucleus divides mitotically into two daughter nuclei (Fig. 11E & F) and is followed by the elongation of the isthmus region. Then there appears a constriction in the isthmus in between the two daughter nuclei dividing the parent cell into two daughter cells by the formation of a wall (Fig. 11F & G). The portion of the isthmus of each daughter cell later grows into a new semi cell (Fig. 11H). A new chloroplast is formed in each of the new semi cells by the division of the existing chloroplast and pyrenoid of the old semi cell. Hence, the two semi cells that constitute the *Cosmarium* plant, one is always younger than the other (Fig. 11H). The younger semi cell is at first very small, but it gradually attains the same size as the other semi cell (Fig. 11 I).

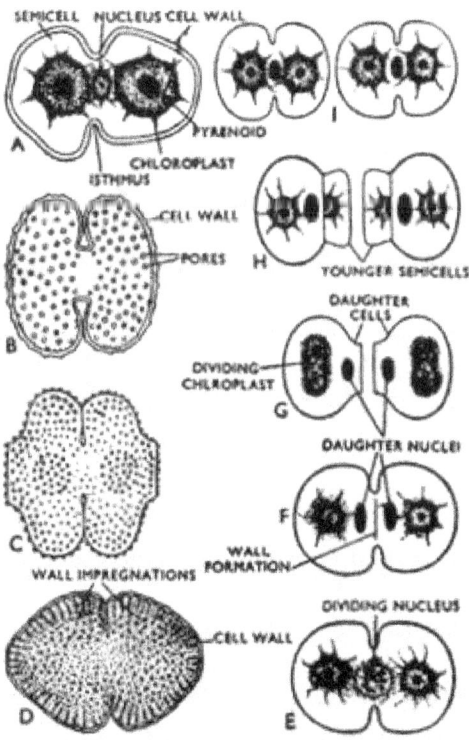

Cosmarium sp. A. Details of cell structure. B–D. Nature of cell wall. E–I. Stages in the cell division.

Fig.11.

Sexual reproduction is by conjugation, a process of aplanogametic isogamy 2 in which the contents of the two conjugating cells behave as gametes. To start with the process of conjugation, the two conjugating cells come together, lie side by side, and are embedded in a common mucilaginous sheath. The cells split open at the isthmus and the semi cells separate allowing the protoplasts to emerge. Each protoplast behaves as a gamete. The gametes move towards each other by slow amoeboid movement and finally fuse together forming a zygote—zygospore which surrounds itself with a three-layered heavy wall, the outermost layer of which may be smooth or spiny.

The zygospore sinks to the bottom of the pool and after a period of dormancy, the diploid zygote nucleus divides meiotically forming four haploid daughter nuclei, of which two are functional and the other two gradually disintegrate. The zygospore chloroplast then divides into two daughter chloroplasts.

Each daughter chloroplast with one functional nucleus forms a new plant. The formation of parthenospore has also been reported in some species of *Cosmarium*.

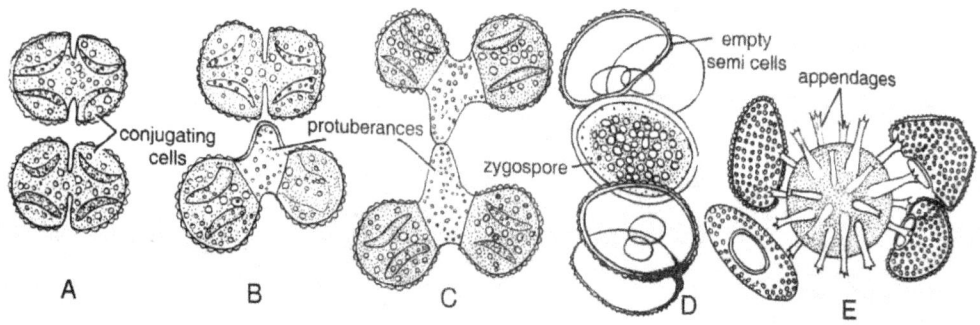

Fig. 12 A-E. *Cosmarium*: Sexual reproduction; A. Two cells of opposite strains, B-C. Stages in conjugation, D. Smooth walled zygospore, E. ZygoSpore with ornamentations in the wall.

****

15.

Siphonales: *Bryopsi, Caulerpa, Codium and Vaucheria*

The order Siphonales includes 50 genera and 400 species, which are mostly marine. The disgintuishing characters of order are the following.

(1) The plant body is siphonous and has no cross walls or septa, being one continuous multinucleate (coenocytic) protoplast, enclosed by the peripheral wal, thus resembling a siphon or a tube.

(2) In the cytoplasm surrounding the central vacuole are present numerous discoidal chloroplasts.

(3) Pyrenoids are commonly present.

(4) The numerous nuclei lie internal to the chloroplasts. The arrangement is reversed in growing apices.

(5) The cross walls remain absent and confined to the formation of reproductive structures.

(6) The chloroplast contain two distinctive xanthophylls siphonein and siphoxanthin in addition, to the pigments characteristic of the green algae.

(7) Asexual reproduction takes place by zoospores, synz0ospores or aplanospores.

(8) Sexual reproduction ranges from isogamy to anisogamy.

Fritsch (1935) included the following nine families in the order Siphonales.

(1) Protosiphonaceae, example *Protosiphon*.

(2) Caulerpaceae, examples Caulerpa, Bryopsis.

(3) Derbesiaceae, example *Derbesia*.

(4) Dasycladaceae, examples *Dasycladus, Acetabularia*.

(5) Codiaceae, example *Codium*.

(6) Valoniaceae, examples *Valonia, Siphonocladus*

(7) Chaetosiphonaceae, example- *Chaetosiphon*.

(8) Phyllosiphonaceae, example *Phyllosiphon*.

(9) Vaucheriaceae, example *Vaucheria*.

Family. Caulerpaceae:

Plants are coenocytic, branched, slenderly filamentous, or quite large and then differentiated into rhizoidal, stoloniferous and erect portions, the latter assuming a great variety of forms. The coenocyte has firm wall braced internally by a system of trabeculae.

Reproduction is by segregation of portions of the contents of a branch to form dimorphic swarmer's, probably gametes, which are discharged through elevated papillae.

Bryopsis

Systematic Position

Class: Chlorophyceae

Order: Siphonales

Family: Caulerpaceae

Genus: *Bryopsi*

Occurrance

This alga is found to be restricted particularly in warm seas with a few exceptional species which grow in cold waters. *Bryopsis* contains mostly epilithic but sometimes free-floating algae that occupy a range of habitats including seaweed beds, shallow fringing reefs and both sheltered and well-exposed subtidal areas (Giovagnetti et al., 2018; Song et al., 2019). *Bryopsis* is widely distributed in tropical and subtropical areas and dominant in eutrophic coastal regions (Hollants, Leliaert, Verbruggen, Willems, et al., 2013). *B. pennata*, *B. hyponoides* and *B. plumosa* are among the most common cosmopolitan species within this genus (Krellwitz et al., 2001). In addition to having a wide distribution, this genus can have a high species density. For example, in Singapore it represents one of the four functional-form groups that makes up approximately 20-40% of benthic cover across coral reefs (Fong et al., 2019) and along the east coast of the United States and the Caribbean, where clusters of species from this genus dominate intertidal regions (Krellwitz et al., 2001).

Thallus Structure

The coenocytic plant body exhibits a very remarkably conspicuous development being differentiated into a comparatively inconspicuous prostrate rhizome-like portion creeping along the substrtaum (Fig. 83A) having anchored to it by the rhizoids and elaborately developed aerial erect axis with a dense and regular pinnate branching in acropetal succession.

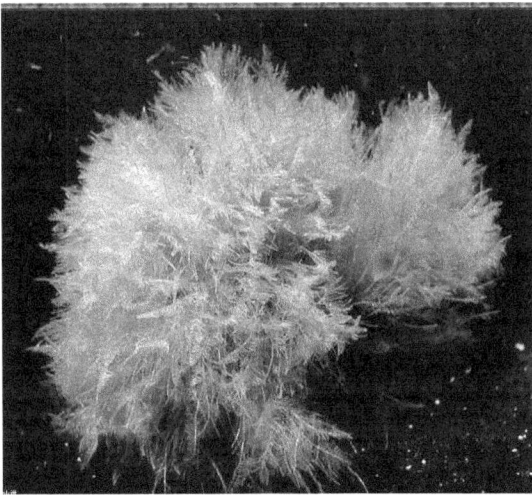

Fig. 1. *Bryopsis sp.: Thallus Structure*

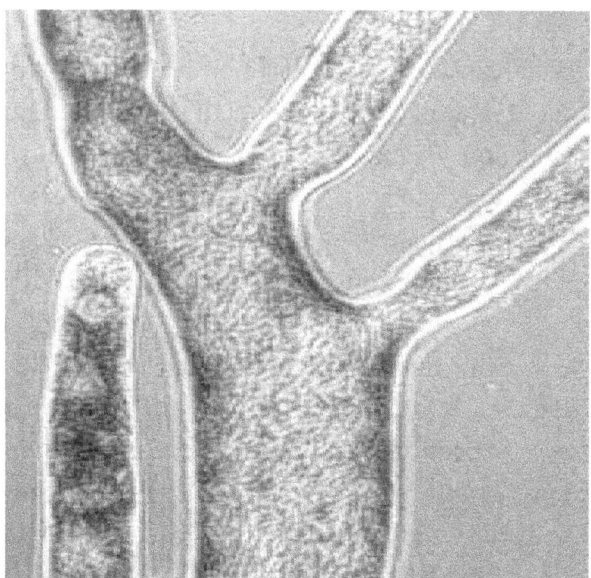

Fig.2. *Bryopsis hypnoides*: siphonal (ramified, multinucleate) thallus (Heligoland, March 1989)

The erect branches are feather-like with the axis bearing numerous branchlets. The lower region of the axis is naked and the upper part bears pinnae of variable length. Each pinna is constricted at the point of origin. The rhizome also functions as an organ of perennation. The wall of the vegetative body is composed of cellulose and pectic substances. Internal to the wall is the cytoplasmic layer lining a continuous vacuole. The cytoplasm contains numerous nuclei and small discoid to spindle-shaped chloroplasts having one pyrenoid each.

Reproduction

Vegetative reproduction is by the growth and decay of new and old parts.

Sexual reproduction is anisogamous. The plants are dioecious. Markedly differentiated pear-shaped biflagellate anisogametes are formed in gametangia in two different plants. During reproduction the pinnae develop into gametangia being separated from the parent plant by means of septa.

Macro- and micro-gametes are formed in different gametangia from the protoplasmic contents of the pinnae which behave as gametangia.

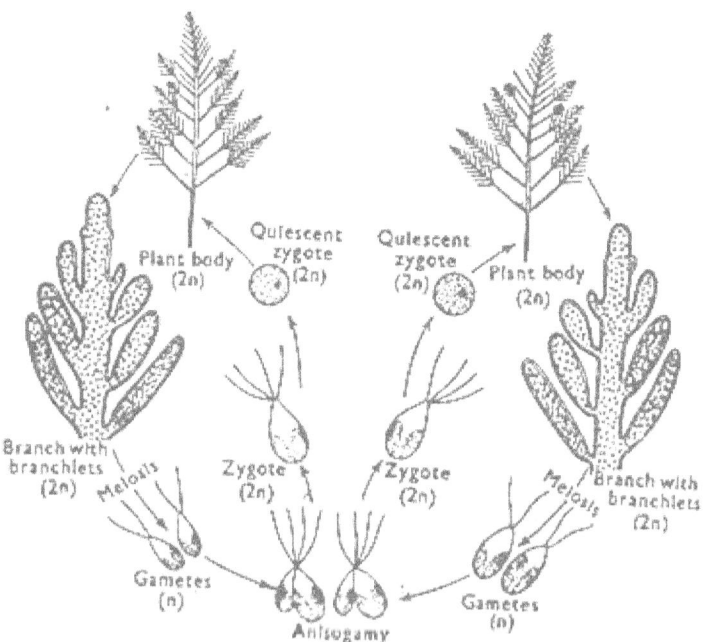

Fig.3. Life cycle of *Bryopsis sp.*

The gametangia are distinguishable at maturity from the colour. The ones producing microgametes are yellow coloured, whereas, the others that produce macrogaraetes being green. The nuclei of the gametangia divide meiotically before taking part in the formation of gametes.

The macrogametes are about three times the microgametes and have one to two chloroplasts, a pyrenoid and an eye-spot. Whereas, the microgametes have a small yellow plastid without pyrenoid.

The gametes are liberated by the gelatinization of the wall of the gametangia at the apex. They swim around for some time and fuse together resulting in the formation of a zygote. The zygote remains quadriflagellate for a very short time, soon assumes round shape and germinates directly into new plant.

The alga has an elaborate diploid phase. Meiosis being gametic the haploid phase is represented by the gametes only.

Some Indian species of Genus Brtopsis:

Bryopsis corymbosa J. Ag.; B. hypuoide's Lamour.; B. pachynema Martens.; B. pennata Lamour.; B. plumosa (Huds.) Ag.; B. tenuissima Notaris.

Special features:

1. Very conspicuously developed coenocytic plant body being differentiated into a comparatively inconspicuous prostrate portion and an elaborately developed aerial erect axis with a dense and regular pinnate branching in acropetal succession.

2. Discoid to spindle-shaped chloroplast.

3. Vegetative reproduction by the growth and decay of the thallus.

4. Sexual reproduction anisogamous.

5. Elaborate diploid phase.

6. Gametic meiosis.

Caulerpa

Systematic Position

Class: Chlorophyceae
Order: Siphonales
Family: Caulerpaceae
Genus: *Caulerpa*

Occurrance

All species of Caulerpa are marine being frequent in the quiet shallow waters of the tropics although there are some Mediterranean species. They cover extensive tracts of the sea-floor and favour relatively shallow water. Several of them are often rooted in sand or mud with their prostrate rhizome bearing colourless rhizoids with which they attach themselves with the substratum.

The production of deadly poisonous substance **Caulerpicin** has been recognized in Hawaii and in the Philippines, since freshly collected plants of **Caulerpa** eaten in salads causes health hazard.

Thallus Structure

The plant body is elaborate in form being differentiated into prostrate portion represented by creeping rhizome with rhizoids (Fig. 5A) and an aerial portion performing the function of assimilatory shoot, or assimilator which may be very variable in structure being either simple flat blade-like, or lobed or a highly complicated branched structure (Fig. 5B & G).

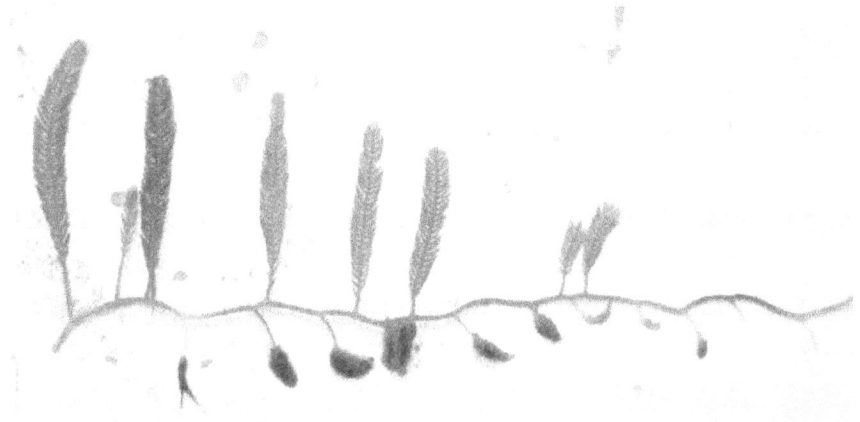

Fig.4. The plant consists of a horizontal, stem-like rhizome that produces a series of root-like rhizoids that extend downward. Branch-like structures that shoot upward compose the recognizable, feather-like fronds.

Internally, plant body is characterized by the complete absence of septation and is composed of closely apposed and intertwined coenocytic threads traversed by longitudinal and transverse skeletal strands (Fig. 5 D to F) perhaps to increase the rigidity of the plant body. These are very remarkably developed in the rhizome portion and are completely absent or poorly developed in the rhizoids.

As such the entire plant body is a single multinucleate cell with a central vacuole and lining cytoplasm with numerous discoid chloroplasts. The longitudinal walls, as well as the wall of the skeletal strands are composed of callose, pectin, and similar other materials. But cellulose is completely absent.

Reproduction

Vegetative Reproduction. Vegetatively *Caulerpa* reproduces by the method of **fragmentation.** There is sporadic separation of proliferous shoots as well. The progressive death and decay of the olderparts of the rhizome sets free the branches. Each branch or fragment becomes an independent plant. This leads to rapid, local propagation. Sometimes dispersal of detached fragments may take place. Reaching a suitable place each grows into a new plant.

Sexual Reproduction. Majority of the species are dioecious. **Iyengar (1940)** reported one monoecious species from India. Sexual reproduction is generally anisogamous. It consists in the fusion of gametes that are not alike. In some species they are distinguishable into micro and macrogametes. The macrogamete is slighty longer but decidedly broader than the microgamete. The narrower gamete is more active in its movements. Both are uninucleate, pyriform and biflagellate. The flagella are equal, whiplash and apical. Each gamete possesses a single curved chloroplast without a pyrenoid and a prominent elongate eye-spot. 'Phallus being diploid meiosis occurs at the time of gamete formation. The gametophytio structures in the life cycle. They are developed on the assimilators or rarely on the rhizome (*C. prolifera*) in special gametangial areas separated from the rest of the protoplast by a membrane. The gametes are liberated shortly after day break in a mass of mucilage through the apices of elongated papillae, the **extrusion papillae**. The papillae are developed on the assimilators simultaneously with the formation of gametes. Soon after the liberation of gametes, the fertile parts of the thallus die and decay. The gametes fuse in pairs in the open sea to produce a zygote which retains all the four flagella for a time. Soon it retracts its flagella, rounds off and secretes a wall around it. According to Miyake and Kunieda (1937) the germinating zygote increases in size. The number of chloroplasts goes up to 30 or more.

Methods of Control:

Many methods to control this plant have been tested throughout the Mediterranean. Some have tried to tear up the patches of algae but one torn leaf that gets away can generate a whole new outbreak. Divers have used pumps to pull out the plant but it seems to regenerate in the same place at a rate quicker than its original growth rate. Other eradication methods include poison, smothering the algae with a cover that lets in no light, and using underwater welding devices to boil the plant. Two species of marine snail, *Aplysia depilans* and *Elysia subornata* that attack the algae have been found. The *Aplysia depilans* snail contains a toxin itself and is avoided by other marine life. Neither has been tested in open water since its overall effect on the ecosystem has not been determined.

Some Indian species of Genus Caulerpa:

Caulerpa crassifolia (Ag.) J. Ag.; C .fastigiata Mont.; C. peltata Lamour.; C. racenosa (Fursk.) Web. van. Bosse; C. sertulariodes Gmel.; C. taxifolia (Vahl.) Ag.

Special Features of Genus Caulerpa:

1. Coenocytic plant body with remarkable differentiation of aerial and prostrate portions.

2. Development of internal skeletal strands for mechanical support of different parts of the plant body.

3. Sexual reproduction both iso- and anisogamous.

4. Fusing gametes are pyriform bearing two flagella.

5. Gametes escape from the plant body through the ends of extrusion papillae in a mass of mucilage.

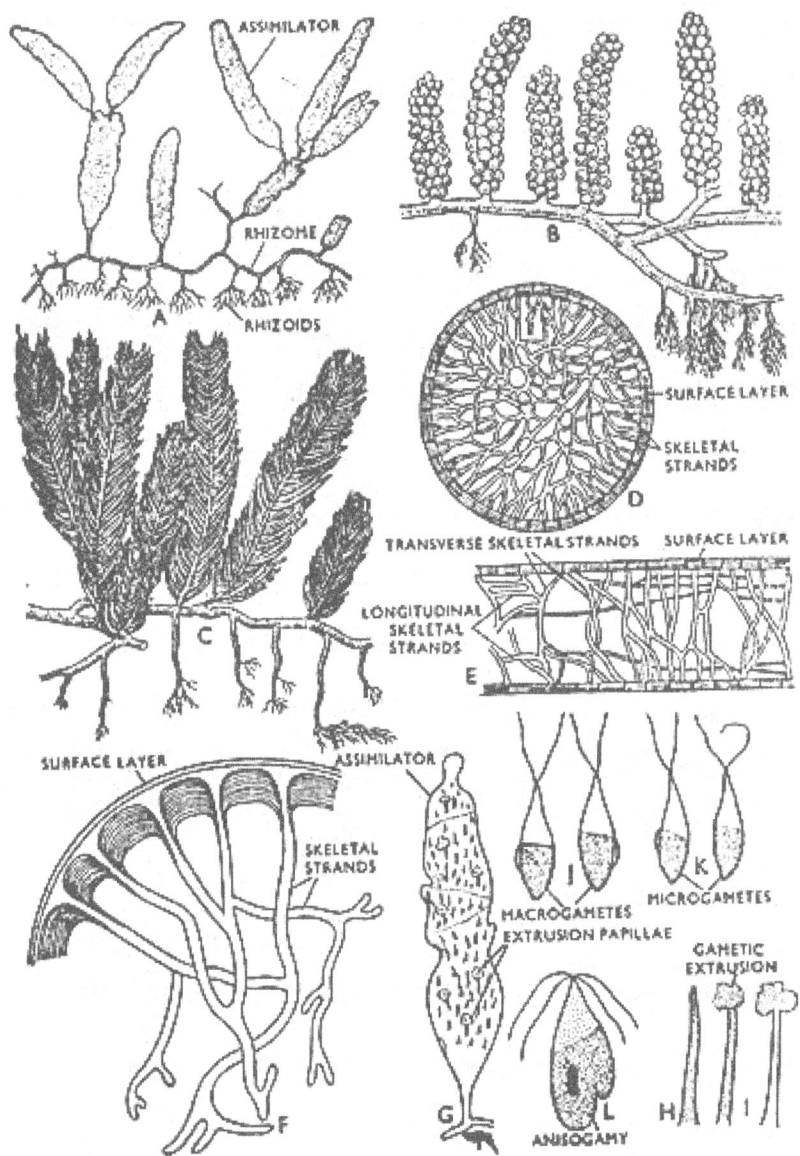

Fig. 5 . *Caulerpa* sp. A–C. Habit and variations in the aerial portion. D. Section of rhizome showing skeletal strands. E. Longitudinal section of assimilator showing structural details. F. Internal structure of rhizome magnified. G. A fertile assimilator with extrusion papillae through which swarmers are liberated. H. Unopened extrusion paiplla. I. Extrusion papillae showing gametic extrusion." J. Macrogametes. K. Microgametes. L. Anisogamy.

Family Codiaceae

The members of Codiaceae are freely branched Flamentous algae, abundantly growing in tropical waters. The thallus consists of tubular branches of spongy nature, made up of densely interwoven filaments. The filaments are characterised by the presence of constrictions with stratified annular thickenings, especially near the points of branching. They show typical siphonaceous coenocytic structure. The plastids lie at the peripheral position, and in most of the genera they lack pyrenoids. Vegetative reproduction occurs by means of detachable propagules. Sexual reproduction is anisogamous; the gametes are biflagellate and are produced in ovoid cup-shaped gametangia. The family includes more than 120 species representing some 16 genera. Rhipidodesmis, *Avraimvillea*, *Rhipilea*, *Codium* and *Udotea* are some common genera of the family. Here the life-history of *Codium* is described as the representative of the family.

Codium

Systematic Position

Class: Chlorophyceae

Order: Siphonales

Family: Codiaceae

Genus: *Codium*

Occurrence

Codium is a widely distributed non-calcareous genus with more than 45 marine species. The species of *Codium* occur from low-tide mark up to the depth of 70 meters in tropical as well as temperate waters. *Codium tomentosum*, *C. bursa* and *C. fragile* are some common tropical species of the genus.

Thallus Structure

The word *Codium* is Greek and means "skin of the animal" referring to the soft, spongy texture. Plants are green in colour and 14-18 cm tall. The surface of the thalli are finely dotted. Thallus is attached with a disc-like holdfast from which many proliferations arise with main thallus. Thallus is dichotomously branched which are 2-3 cm thick. There is a 2-4 cm gap between two divisions. Vesicles are of two types, slender and broader. Broader vesicle side walls are parallel.

Thallus Structure

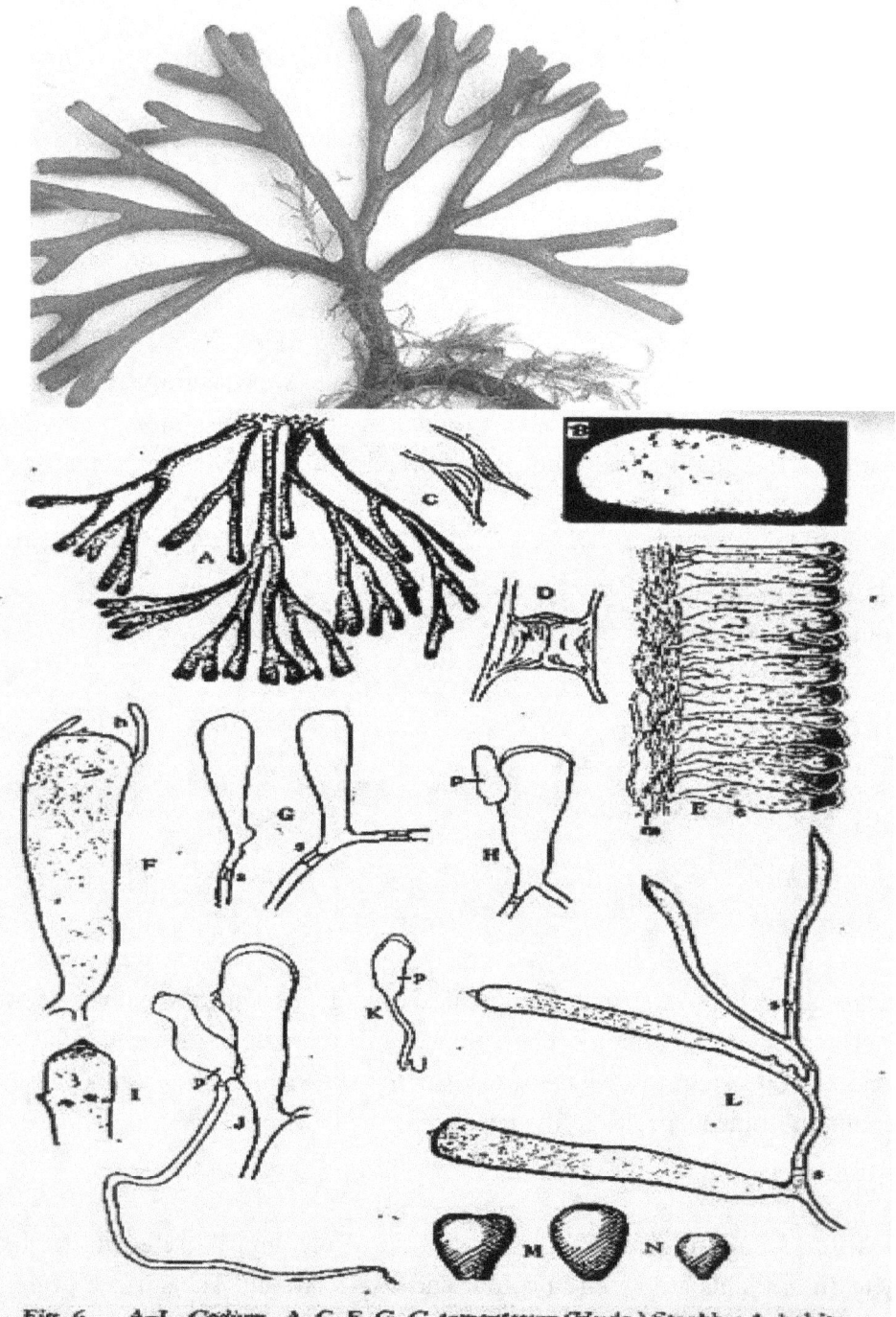

Fig. 6 . A–L, *Codium*. A, C, E, G, *C. tomentosum* (Huds.) Stackh.; A, habit; C, septum-formation; E, part of a longitudinal section; G, septum-formation

In *C. tomentosum* (Fig. 6 A) the thallus, firmly anchored by a basal disc, consists of cylindrical hairy segments which are about the thickness of an ordinary lead-pencil and repeatedly, forked; they float out into the water or hang down from the rocks at low tide. In *C. adhaerens* (Cabr.) Ag. the thallus is a flat cushion or crust attached by numerous rhizoids along the underside, while in *C. Bursa* it appears as a large rounded structure (10-20 cyn. in diameter), slightly flattened at the point of attachment to the substratum (Fig. 6 B) and becoming hollow at maturity; within the hollow may be found diverse small animals and red-coloured Myxophyceae (69,208). In *C. tomentosum* the thallus in section (Fig. 6 E) is seen to be composed of a central medulla {m) of longitudinal, rather narrow, interwoven, forked threads and a peripheral cortex (c) of large clubshaped vesicles densely grouped at the same level to form a palisade like layer. Chloroplasts are aggregated in large numbers beneath the outer surface of each cortical vesicle (Fig. 6E) and are also found in the hairs which are produced periodically on these structures (Fig. 6 F), and give the plant its tomentose character (cf. (184)); when they drop off they leave characteristic scars (Fig. 6 I). In longitudinal sections it is usually not difficult to recognise that the vesicles represent the dilated terminations of one of the forks of a longitudinal medullary thread (Fig. 6 G, L). From the base of the vesicle one or several branches arise, some of-which grow more or less transversely and help to bind the medullary ^threads together, whilst others grow lengthwise between cortex and medulla and sooner or later curve out afresh and become dilated to form another cortical vesicle (Fig. 6 L). At the base of each of the latter, as well as at the points where these lateral branches are given off, one finds the customary annular thickenings (Fig. 6 G, L, s); these are so strongly developed that they seem to meet (Fig, 6 D), although according to, Mirande (124) a delicate plasma strand traverses them through which transport of material can take place(117a); in *C. Bursa* the pores are stated to become widened in spring, but to become narrowed again in autumn.

Reproduction

Codium reproduces by vegetative and sexual methods.

Vegetative reproduction

It takes place by means of small club-shaped propagules which develop on utricles. When detached from the parent plant, the propagule grows inte a new thallus.

Sexual reproduction

Sexual reproduction is anisogamous and takes place with the help of distinct male and female gametes. Some species are strictly dioecious (e.g.. C. fragile), branch. while others are monoecious. Gametes are produced in ovoid male and female gametangia which develop laterally on utricles. A gametangium initiates as a tubular outgrowth on the lateral walls of the utricle. The outgrowth has compact cytoplasm with evenly distributed diploid nuclei and the plastids are crowded at the distal end (Fig. 7A). The young gametangium elongates and is cut off from the utricle by a strong annular thickening that develops at its base (Fig. 7 B). Some nuclei of the multinucleate gametangium degenerate and the remaining functional nuclei enlarge and divide meiotically to form haploid nuclei. Now the protoplast of the gametangium divides into many uninucleate segments. Each uninucleate segment develops into a gamete. The mature male gametangium is golden yellow and contains a few thousand small Pyriform male gametes (microgametes; Fig. 7 G), whereas the female gametangium is dark green and contains only a few hundred relatively large female gametes (macrogametes; Fig. 7 C, D, H). The male gamete contains one or two chloroplasts, whereas the female gamete has many chloroplasts. The discharge of gametes has been studied in detail in C. fragile. The internal layers of the gametangium wall are gelatinized. The apical portion of the gametangium ruptures and a large mass of mucilage is extruded through it. The mucilage mass is almost of the same shape and size as the gametangium, and it has a central canal through which gametes move out (Fig. 7 D-G). Initially gametes lack flagella and are carried out passively due to hydrostatic pressuree within the gametangium. Soon after the discharge, each gamete develops two flagella on its anterior side and swims freely with their help. Ejection of gametes occurs very rapidly and all gametes are released within a minute or so.

Fertilization takes place when motile male and female gametes come close to each other. The male gamete gets attached to the side of the female gamete and loses its flagella. The gametic union results in the formation of a diploid zygote, which propels for sometime with the help of the flagella of the female gamete. As the zygote settles on some substratum, it withdraws its flagella, assumes a spherical shape and secretes a wall (Fig. 7 A). The Zygote germinates immediately and forms a lobed germling (Fig. 7 B). The germling soon gets differentiated into an erect aerial and a horizontal disc like rhizoidal system (Fig. 7 C). The entire aerial system (medulla and cortex) is formed by the ramification of the single initial filament.

Alternation of Generation

Codium shows diplontic life-cycle with no alternation of generation. The thallus is diploid which bears micro- and macrogametangia. The gametangial nuclei divide meiotically to form haploid gametes which represent the haploid phase in the life-cycle. The micro-and macrogametes fuse to form diploid zygote and thus the diploid phase is reestablished. There is no reduction division at the time of the germination of zygote.

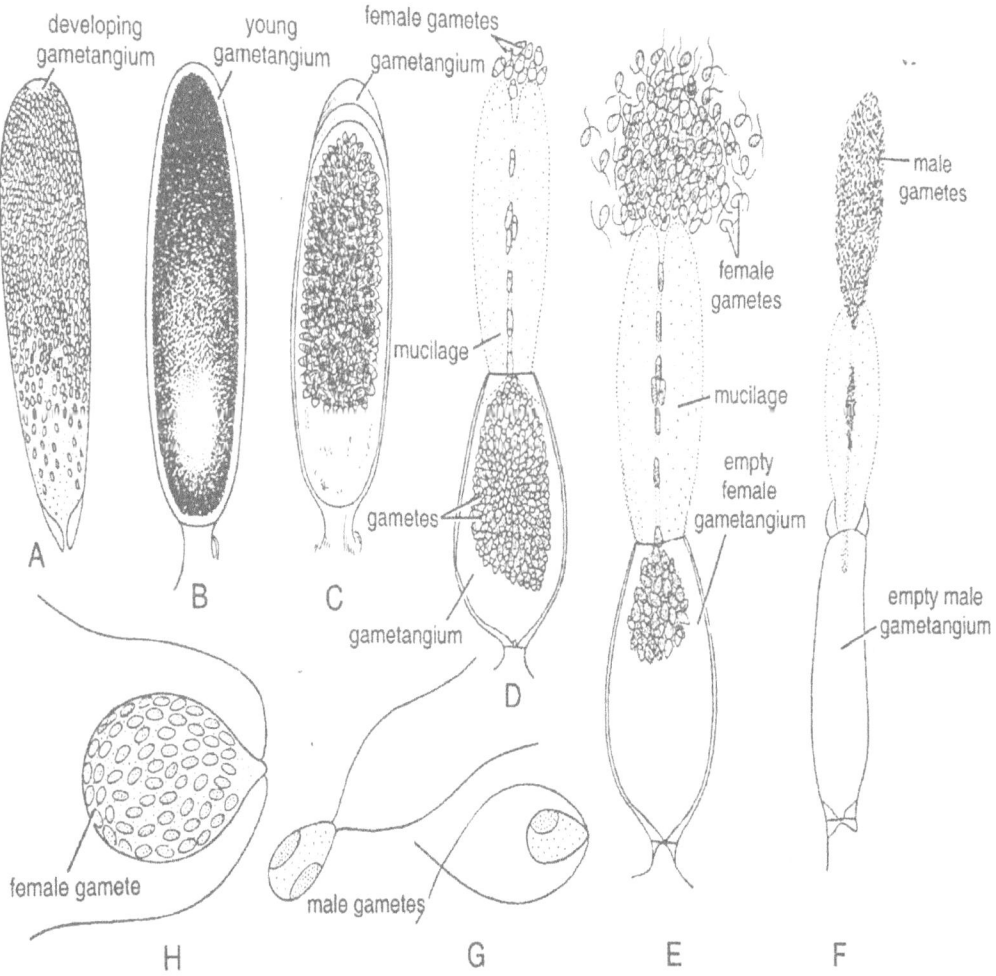

Fig. 7 A-H. *Codium*: Sexual reproduction, A-B. Young gametangia, C. Mature gametangium, D-E. Liberation of female gametes, F. Liberation of male gametes, G. Male gametes, H. Female gamete.

Family. Vaucheriaceae

Distinguishing Features

(1) Thallus is simple, branched, unseptate, coenocytic with apical growth.

(2) The product of assimilation is oil or fat in (*Vaucheria*) but starch in Dichotomosiphon.

(3) The rhloroplasts are numerous, oval or elliptical usually without pyrenoids

(4) Asexual reproduction takes place by akinetes aplanospores or multiflagellate zoospores.

(5) Sexual reproduction is highly oogamous.

The family includes 4 genera which may occur on wet soil, in fresh or brackish waters and some are exclusively marine. Out of 4 genera (*Vaucheria, Vaucheriopsis, Dichotomosiphon* and *Pseudodichotomosiphon*). *Vaucheria* has the largest numberofspecies and has been studied in great detail. The members of this family are found in fresh water habitats in temperate regions. The aseptate, siphonaceous and coenocytic thalli are found attached to the substratum by a branched hapteron. They reproduce asexually by means of zoospores or aplanospores. Sexual reproduction is of oogamous type.

Vaucheria

Systematic Position

Class: Chlorophyceae

Order: Siphonales

Family: Vaucheriaceae

Genus: *Vaucheria*

Occurrence

The genus *Vaucheria* comprises about 54 species, out of which only 9 species are Indian. Most species are terrestrial and aquatic. A few species are marine (V. piloboloides). Terrestrial species grow on moist soil and in ploughed fields where they form green velvety mats. Aquatic species occur widely in stagnant brackish water and some are seen in shallow fresh waters of ponds and ditches or near the banks of slow flowing streams. The common Indian species of *Vaucheria* are – *V. sessilis, V. geminata, V. amphibian*, etc.

Thallus Structure

The plant body i.e., thallus is filamentous. The filament is variously branched, cylindrical and tubular in structure, long and yellowish green in colour. The thallus is multi-nucleate and without any transverse wall or septation. Septa formation takes place during the formation of reproductive structures. Most terrestrial species are attached to the substratum by means of hapteron like colourless branched outgrowth called rhizoids. Rhizoids are absent or ill developed in floating species. Cell wall of the filament is thin and weak. It is composed of an outer layer of pectic substance and inner layer of cellulose. A thick layer of cytoplasm is present in the periphery just beneath the cell wall. Numerous discoid chloroplast and nuclei remain embedded in the cytoplasm. Next to the cytoplasm is the central vacuole which is filled up with cell sap. The pigments present in *Vaucheria* are – Chlorophyll-e, Chlorophyll-a, β-carotene, siphonein, siphonoxanthin, etc. The cytoplasm also contains reserve food in the form of colourless oil droplets.

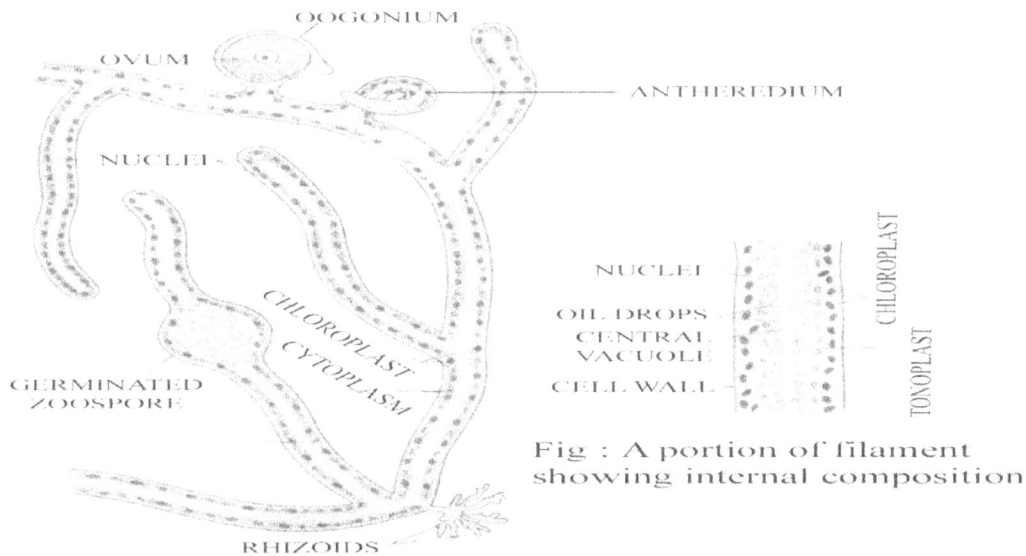

Fig.8.

Reproduction:

Vaucheria reproduces both by vegetative, asexual and sexual methods of reproduction.

1. Vegetative Reproduction: - Vegetative reproduction takes place by fragmentation. In this, the thallus accidentally breaks into short segments, each of which becomes thick walled and finally develops into new plant.

2. Asexual Reproduction: - It takes place by the production of different types of spores.

(a) Zoospores – It is the most common method of reproduction of aquatic species. The zoospores of *Vaucheria* are large, multinucleate and multi-flagellate motile spores formed singly within the elongated club shaped zoosporangia developed at the tip of the side branch. The mature zoospore escapes through a narrow aperture which is formed by the gelatinization of the wall at the distal end of the zoosporangium. The zoospore takes a short period of 5-15 minutes of rest after liberation and starts germinating.

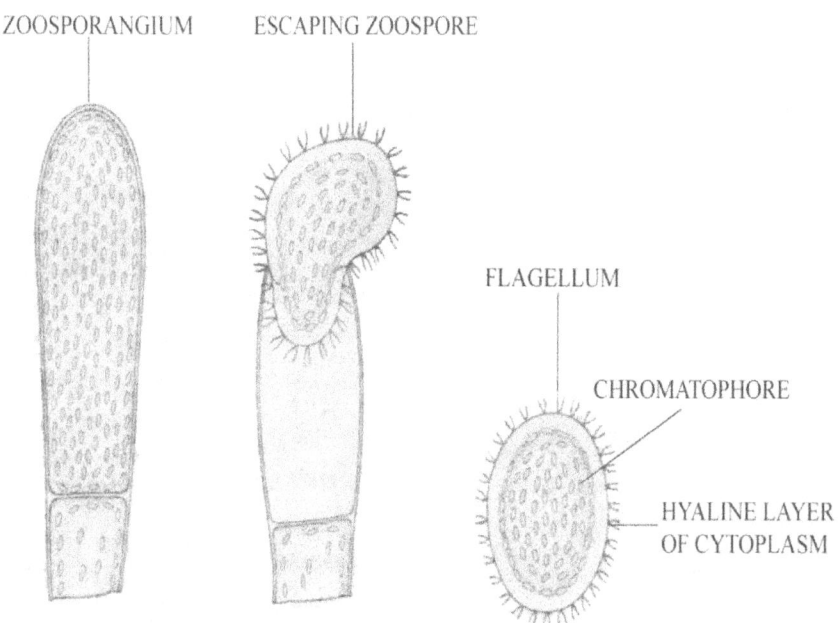

Fig. 9. *Vaucheria*: Development of Zoospore

(b) Aplanospores – These are non-motile asexual spores produced normally by the terrestrial species. These spores are developed when the terrestrial species are exposed to draught or when the species are transferred from light to darkness or from running to still water. The aplanospores are more or less rounded or elongated structures developed at the ends of short lateral or terminal branches known as aplanosporangia. Very rarely the aplanospores grows into a new plant.

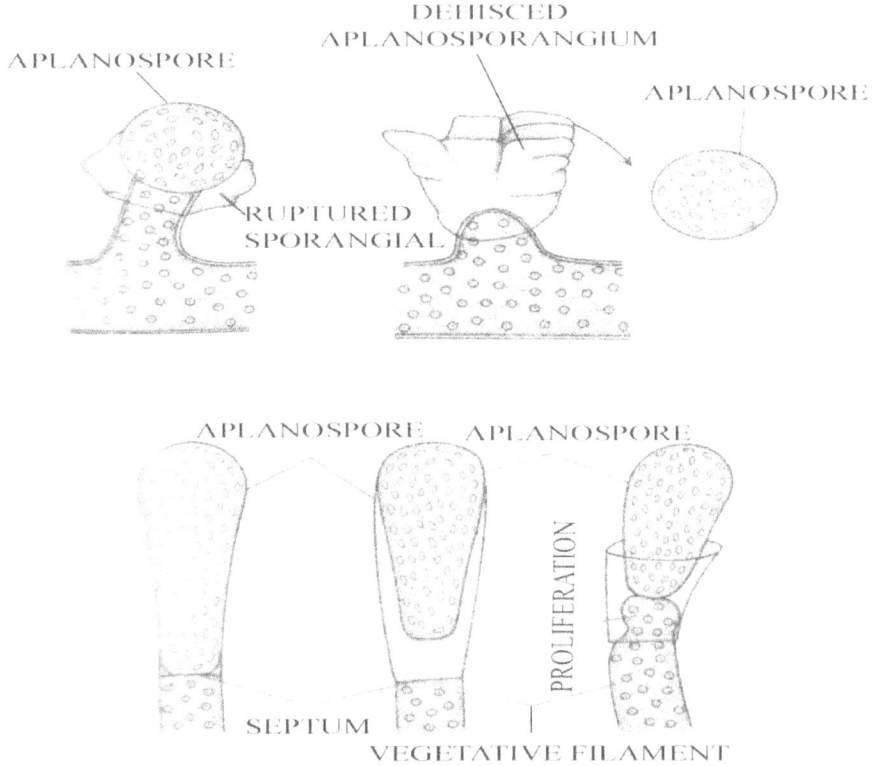

Fig. 10. *Vaucheria*: Development of aplanospore

(c) Akinetes – In some aquatic and terrestrial species when exposed to greater dessication or low temperature, the branched filaments divides into rows of short segments by thick, gelatinous cross walls. These are resting spores and are known as cysts or hypnospores or akinetes. For a time the cysts may remain connected by a parent membrane of the filament, giving it the appearance of another alga Gongrosira. This stage of Vaucheria is thus known as Gongrosira stage. With the return of conditions favourable for growth, each cysts may germinate into new plant.

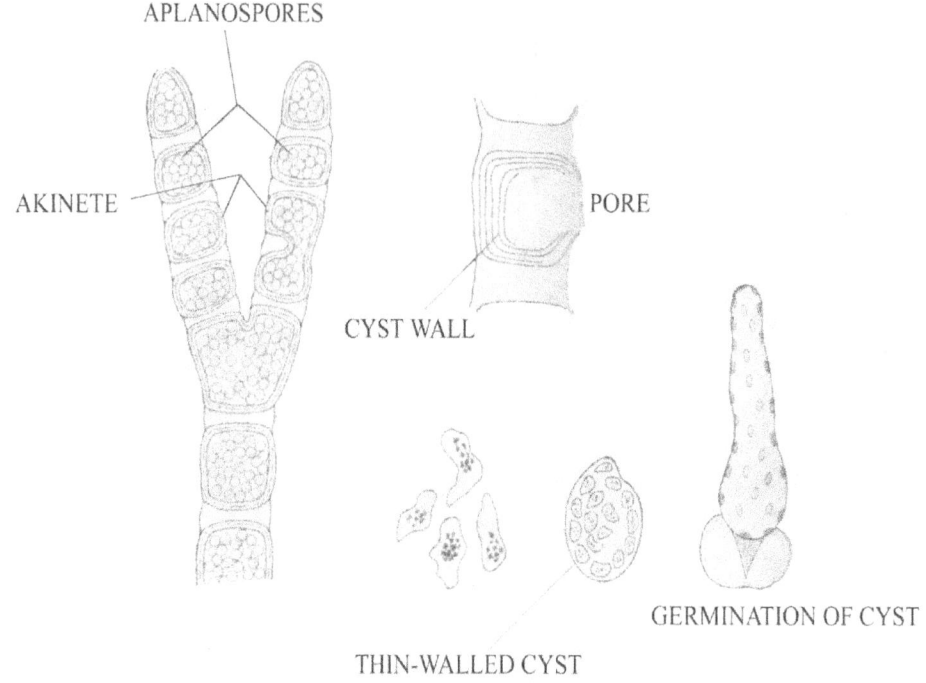

Fig.11. *Vaucheria*: Development and germination

Sexual Reproduction

The sexual reproduction of Vaucheria is of oogamous type. Most of the species of fresh water and terrestrial habitats are homothallic, while a few marine species are heterothallic. In homothallic species, both antheridia and oogonia are produced adjacent to one another, either on a common lateral branch or on adjoining branches. **(a) Antheridia**

The antheridium is slender, curved, hooked-like tubular structure opening by a terminal pore. It is formed at the end of short lateral branch. The tip of the branch producing an antheridium becomes densely filled with cytoplasm containing many nuclei and few chloroplasts. It is then cut off from the other parts of the vegetative body by a transverse wall. The protoplast of an antheridium becomes divided into number of uninucleate fragments, each of which is metamorphosed into a biflagellate sperm or antherozoid. Each antherozoid is a colourless, pear shaped structure provided with two laterally flagella of equal length. When the antheridium is fully developed, the antherozoids are liberated into the surrounding water through a terminal pore by the gelatinization of the antheridial wall at the apex.

(b) Oogonia

The oogonium is a tumid, round or oval, sessile or short stalked body and has a wall with rounded beak, which opens at maturity to receive the sperm. It is formed adjacent to the antheridium and begins to develop simultaneously with the accumulation of the colourless, multinucleate mass of cytoplasm, called waterplasm. The protoplasm i.e., the contents of the oogonium round off to form a single uninucleate egg or ovum.

(c) Fertilization

During fertilization both antheridia and oogonia open, usually simultaneously or with the difference of time ranging from a few minutes to one or two hours. By the opening and rupturing of oogonial tip a small drop of cytoplasm oozes out in the form of a twisted mass. Many sperms escape from the terminal opening of the antheridium. A few of them enter into the oogonium through the apical pore, but only one of them fuses with the egg resulting in the formation of diploid zygote (2n). After fertilization the zygote develops a thick wall (usually 3-7 layers) and undergoes a resting period of a few months. When resting period is over, it undergoes meiotic division. As a result haploid nuclei are formed and from each haploid nucleus a haploid coenocytic filament is produced.

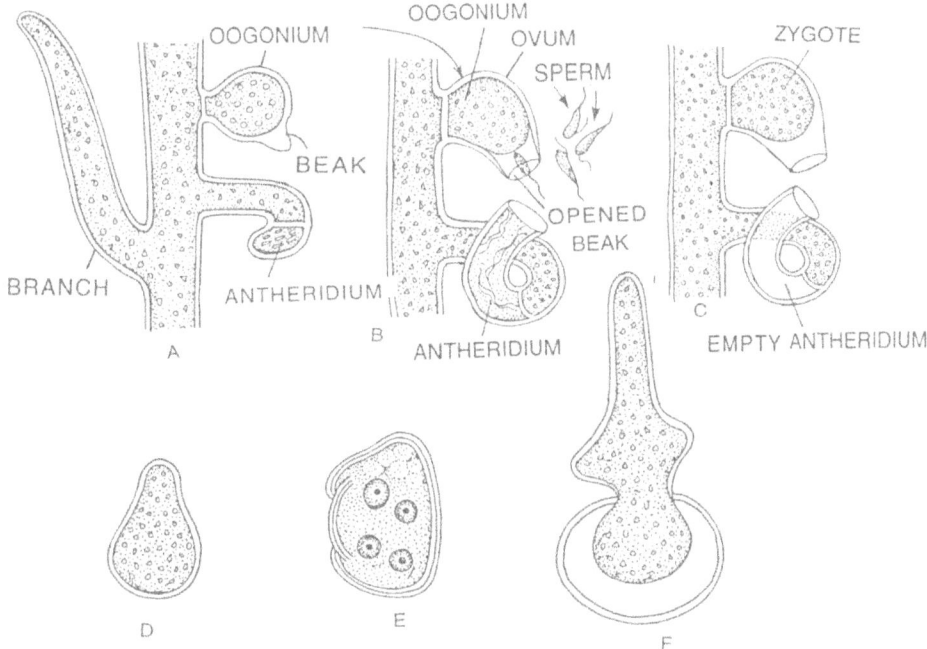

Fig (A-C) : Filament bearing Antheridium and Oogonium;
(D-F) : Different stages of germination of Zygote.

Fig.12.

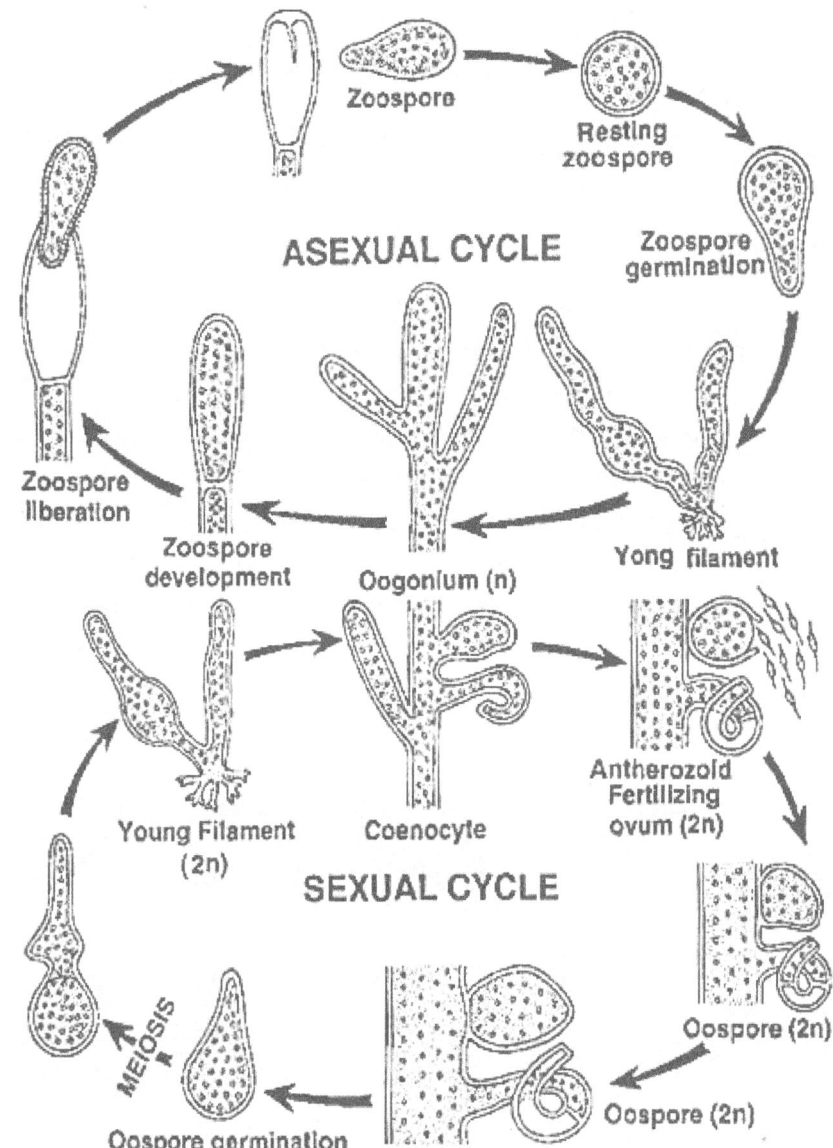

Fig.13.Vaucheria: Life cycle

16.

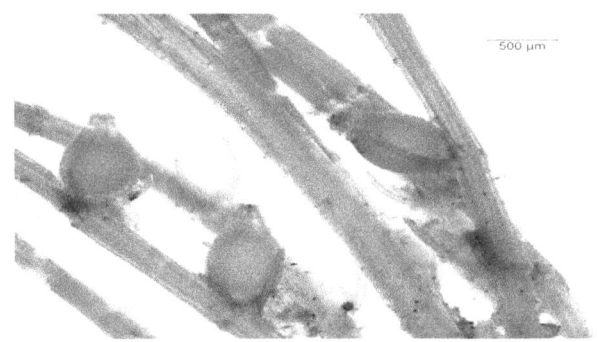

Charales: *Chara*

The systematic position of the order Charales is debated and it has been treated differently by various workers. Fritsch (1935) included it in the class Chlorophyceae Smith (1938) gave it the rank of a class, Charophyceae, and Bold (1956) and McLean and Cook (1956) included it in a separate phylum Charophyta, intermediate between algae and bryophyta. The treatment of Fritsch is followed here.

The order includes 6 genera and 300 species, mostly found in shallow freshwater lakes with sandy and muddy bottoms. Some species also grow in brackish water. The thallus is well organized and is differentiated into rhizoids and an erect branched axis. The axis is divided into nodes and internodes. At the nodes of the main axis whorls of branches of unlimited growth are present, whereas the nodes of lateral branches bear branches of limited growth. Such branched axis gives it the appearance of an angiospermic plant. Sexual reproduction is advanced oogamous. Both male and female sex organs, known as globules and nucules respectively, are macroscopic. These structures are contained within a sterile envelope, Similar to that of bryophytes. Asexual spores are, however, absent.

The order includes only one family, Characeae, represented in Inuia by five genera (*Chara, Nitella, Lychnothamnnus, Nitellopsis* and *Tolypella*) and 65 species. The life-history of *Chara* is described here as the representative form.

Chara

Systematic Position

Class: Chlorophyceae
Order: Charales
Family: Characeae
Genus: *Chara*

Occurrence:

Chara is a fresh water, green alga found submerged in shallow water ponds, tanks, lakes and slow running water. *C. baltica* is found growing is brackish water and *C. fragilis* is found in hot springs. *Chara* is found mostly in hard fresh water, rich in organic matter, calcium and deficient in oxygen. Chara plants are often encrusted with calcium carbonate and hence are commonly called stone wort. *Chara* often emits disagreeable onion like odour due to presence of sulphur compounds. *C.hatei* grows trailing on the soil *C. nuda* and *C. grovesii* are found on mountains, *C. wallichii* and *C. liydropitys* are found in plains.nThe fossils of alga have been discovered from Paleozoic era.

Thallus Structure

The thallus of Chara is branched, multicellular and macroscopic. The thallus is normally 20-30 cm. in height but often may be up to 90 cm to 1 m. Some species like C. hatei are small and may be 2-3 cm. long. The plants in appearance resemble Equisetum hence Chara is commonly called as aquatic horsetail. The thallus is mainly differentiated into rhizoids and main axis (Fig. 1).

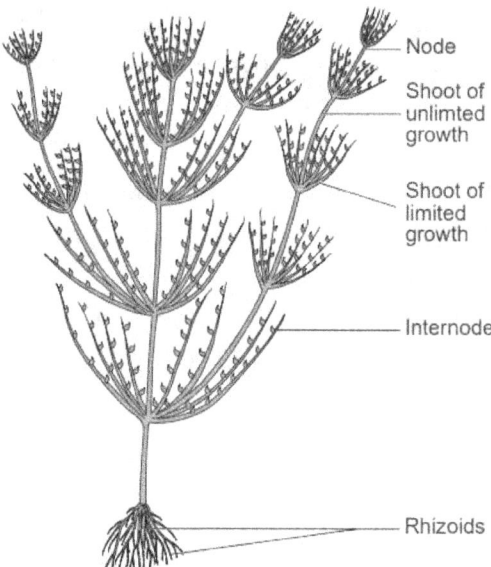

Fig. 1. *Chara*: **External features**

Rhizoids:

The rhizoids are white, thread like, multicellular, uniseriate and branched structures. The rhizoids arise from rhizoidal plates which are formed at the base of main axis or from peripheral cells of lower nodes. The rhizoids are characterized by presence of oblique septa (Fig. 2).

The tips of rhizoids possess minute solid particles which function as statoliths. The rhizoids show apical growth. Rhizoids help in attachment of plant to substratum i.e., mud or sand, in absorption of minerals and in vegetative multiplication of plants by forming bulbils and secondary protonema.

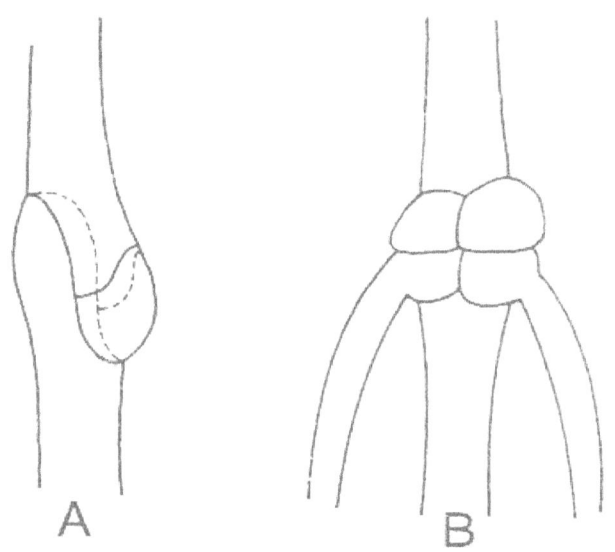

Fig. 2 A-B. *Chara*: Rhizoids; A. Showing oblique septum, B. Rhizoidal plate.

Main Axis:

The main axis is erect, long, branched and differentiated into nodes and internodes. The internode consists of single, much elongated or oblong cell. The inter-nodal cells in some species may be surrounded by one celled thick layer called cortex and such species are called as corticate species. The species in which cortical layer is absent are called ecorticate species. The node consists of a pair of central small cells surrounded by 6-20 peripheral cells. The central cells and peripheral cells arise from a single nodal initial cell.

On nodes develop these following four types of appendages:

(i) Branches of limited growth.

The branches of limited growth arise in whorls of 6-20 from peripheral cells of the nodes of main axis or on branches of unlimited growth. These are also called branchlets, branches of first order, primary laterals or leaves. These branches stop to grow after forming 5-15 nodes and hence are called branches of limited growth. The stipulodes and reproductive structures are formed on the node of these branches.

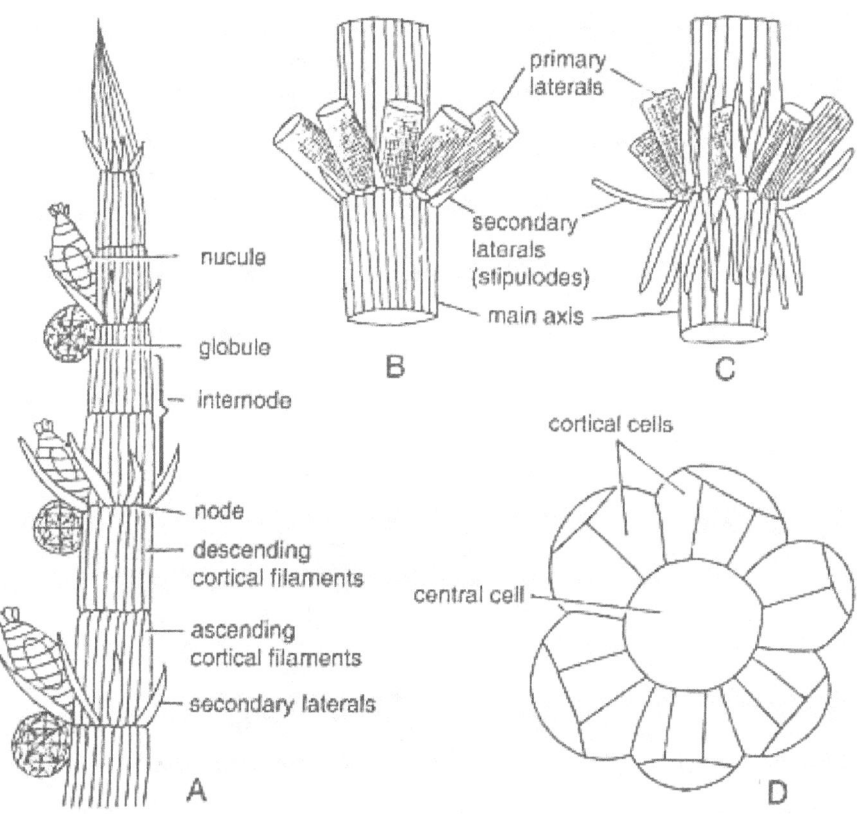

Fig. 3 A-D. *Chara*: Aerial axis; A. Branch of limited growth, B-C. Appendages at nodes, D. Transection of internode.

(ii) Branches of unlimited growth:

The branches – of unlimited growth arise from the axils of the branches of limited growth hence these are also called auxiliary branches or long laterals. These are differentiated into nodes and internodes. At nodes they bear primary laterals and these branches look like the main axis. Their growth is also unlimited like main axis.

(iii) Stipulodes:

The basal node of the branches of limited growth develops short, oval, pointed single cell outgrowths called stipulodes. In most of the species of *Chara* e.g., *C. burmanica*, the number of stipulodes at each node is twice the number of primary laterals, such species are called as bi-stipulate.

In some species of *Chara* e.g., *C. nuda* and *C. braunii*, the number of stipulodes at each node, is equal to number of primary laterals at that node, such species are called unitipulate. When stipulodes are present in one whorl at each node the species are called as haplostephanous and with two whorls on each node are called diplostephanous.

(iv) Cortex:

Many species of *Chara* e.g., *C. aspera, C. inferma* have inter-nodal cells of main axis en-sheathed by cortex cells. Such species are called corticated species. The cortex consists of vertically elongated narrow cells. The internode up to half of its length by corticating filaments developed from upper node called descending the lower half of internode is covered by filaments developed from lower node called filaments. The ascending and descending filaments meet at the middle of internode. T species without cortex e.g., *C. corallina* are called ecorticated species.

Cell Structure of Chara:

The main axis of *Chara* consists of mainly two types of cells:

(i) Nodal cells

ii) Inter-nodal cells.

The nodal cells are smaller in size and isodiametric. The cells are dense cytoplasmic, uninucleate with few small ellipsoidal chloroplasts. The central vacuole is not developed instead many small vacuoles may be present. The cytoplasm can be differentiated in outer exoplasm and inner endoplasm. The inter-nodal cells are much elongated. The cytoplasm is present around a large central vacuole. The cells are multinucleate and contain many discoid chloroplasts. The cytoplasm is also differentiated into outer exoplasm and inner endoplasm. The endoplasm shows streaming movements. The cell walls between the nodal cell and inter-nodal cells are porous to help in cytoplasmic continuity between cells.

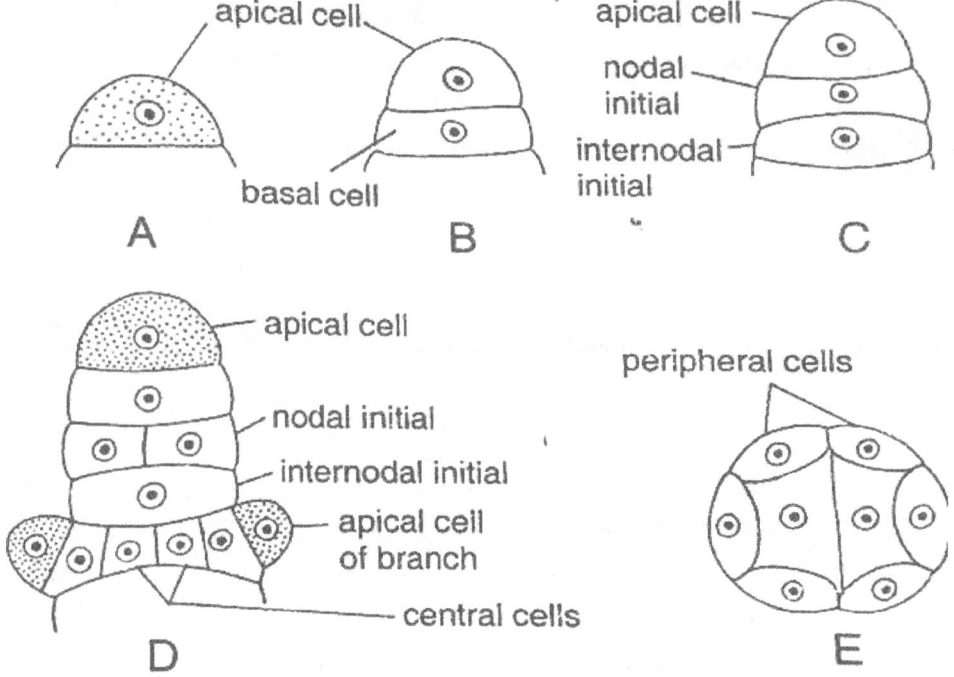

Fig. 4 A-E. *Chara*: **A-D.** Stages in apical growth, **E.** Transection of node.

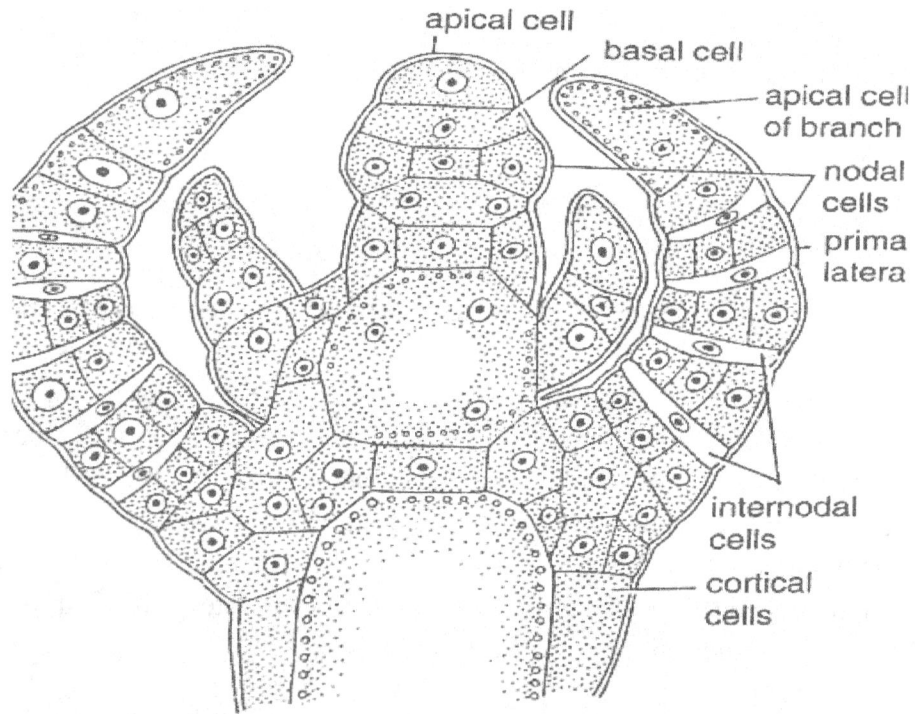

Fig. 5. *Chara*: Vertical section of the apical part.

Growth:

Growth of *Chara* takes place by a dome- shaped apical cell. The cell undergoes repeated transverse divisions and form a row of three cells. The upper one remains as apical cell, middle biconcave one forms the nodal initial and the lower one forms the internodal initial.

The nodal cell undergoes repeated vertical divisions and ultimately forms two central cells surrounded by 6-20 peripheral cells. Branches of limited growth are developed from the peripheral cells arranged in single row. The internodal initial does not divide further and elongates much more to form long internode.

Reproduction in Chara:

Chara reproduces by both vegetative and sexual means. Asexual reproduction is absent.

Vegetative Reproduction:

The vegetative reproduction takes place by the formation of following structures:

1. Bulbils:

These are small oval or spherical bodies developed on stem or root nodes. Bulbils are formed on root of *C. aspera* and stem of *C. baltica*. After detachment, they germinate and develop new plants (Fig. 6 A, B).

2. Amorphous Bulbils:

These are small cells developed and aggregated at the node, called amorphous bulbils. They are found in *C. fragilis, C. baltica* etc. On being detached from the mother plant, they germinate and develop into new plants (Fig.6 C).

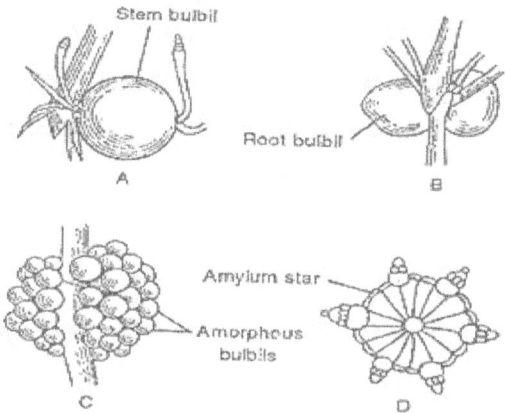

Fig. 6. Chara: A. Stem Bulbil, B. Root Bulbil, C. Amorphous Bulbil, D. Amylum stars

3. Amylum Stars:

These are multicellular aggregations of cells, looking like stars and the cells are densely filled with amylum starch; thus they are called amylum stars. The amylum stars are developed at the nodal cells of the basal region e.g., C. stelligera (Fig.6 D).

4. Secondary Protonema:

These are thread like structures developed from primary protonema or from the basal cell of the rhizoid. New plants are also developed from the secondary protonema.

Sexual Reproduction in Chara:

The sexual reproduction in *Chara* is of highly advanced oogamous type. The sex organs are macroscopic and complex in organization.

The male sex organs are called antheridium or globule and the female oogonium or nucule. Most of the Chara species are homothallic i.e., the male and male sex organs are borne on the same nodes, (Fig. 7) e.g., *C. zeylanica*. Some species e.g., *C. wallichii* are heterothallic i.e., male and female sex organs are borne on different plants.

The sex organs arise on the branches of limited growth or primary laterals, the nucule above the globule. The development of globule and nucule takes place simultaneously but species globule matures before nucule (Fig. 7).

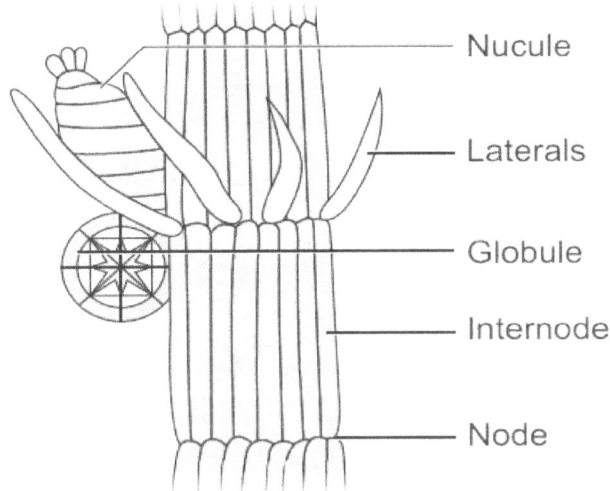

Fig.7. *Chara*: Sex Organs

Globule:

The globule is large, spherical, red or yellow structure.

Development and Structure:

The early development of globule and nucule is similar. The pripheral cell of the lower node of the primary lateral divides periclinally to form and an inner cell. The outer cell functions as antheridial initial and the lower cell again divides a periclinal division. Out of these three cells formed, the lowermost functions as inter-nodal cell the middle forms basal node, the uppermost functions as the antheridial initial (Fig. 8 A, B).

The middle basal node cell divides to make 5 peripheral cells. Out of these five peripheral cells, the upper one develops into oogonium, two lateral ones form unicellular bracteoles and two lower ones, one on either side of oogonium forms cortex or remains non-functional (Fig. 8 C, D).

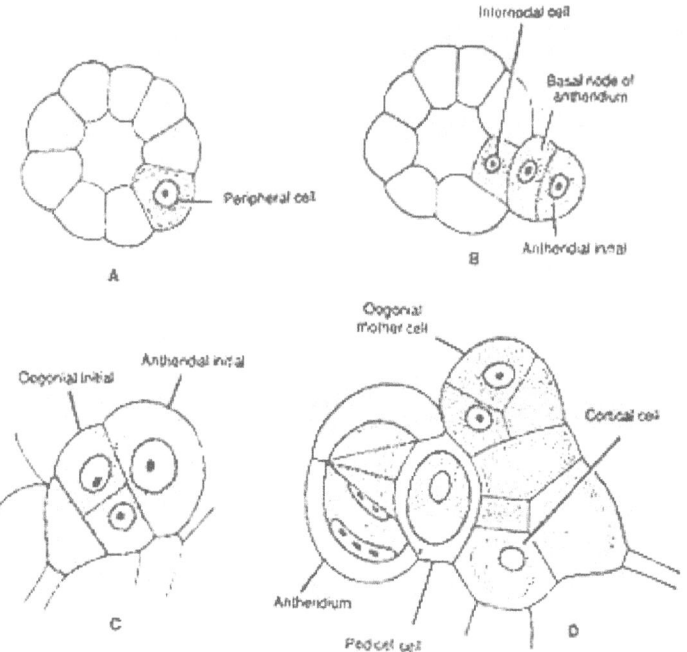

Fig. 8. (A-D). *Chara*. Early development of sex organs.

The antheridial initial divides by transverse division to make basal pedicel cell and a terminal antheridial mother cell (Fig. 9 A, B). The pedicel cell does not divide further and forms pedicel or stalk of mature antheridium. The antheridial mother cell enlarges and divides by two successive vertical division at right angle to each other to make a quadrant (Fig. 9 C-D).

All these four cells divide by a transverse wall to make eight cells or octant stage. Each eel of the octant divides periclinally and forms two layers of eight cells each.

The cells of inner or outer layer divide periclinally to make three radial layers of eight cells each. The outermost eight cells enlarge laterally to form a curved plate of eight shield cells (Fig. 9 E, F). The cells of the middle layer elongate towards centre to make eight rod-shaped manubrial cells (Fig. 9 F).

The cells of the inner layer function as eight primary capitulum cells. Each primary capitulum cell divides to form six secondary capitulum cells. Sometimes the secondary capitulum cells divide to make tertiary capitulum cells.

Each capitulum cell divides repeatedly to form 2-4 long, multicellular, branched or un-branched antheridial Filaments or sperinatogenous filaments (Fig. 9 G-H). The antheridial filament has up to 250 uninucleate cells. These cells function as sperm mother cell and each cell gives rise to a single spirally coiled, uninucleate, bi-flagellated antherozoid (Fig. 9 J-L).

The mature globule thus is made up of 8 curved shield cells, 8 elongated manubrial cells, 8 centrally located primary capitulum cells and 48 secondary capitulum cells. The secondary capitulum cells give rise to many antheridial filaments. Each sperm mother cell forms a single bi-flagellated antherozoid (Fig. 9 K).

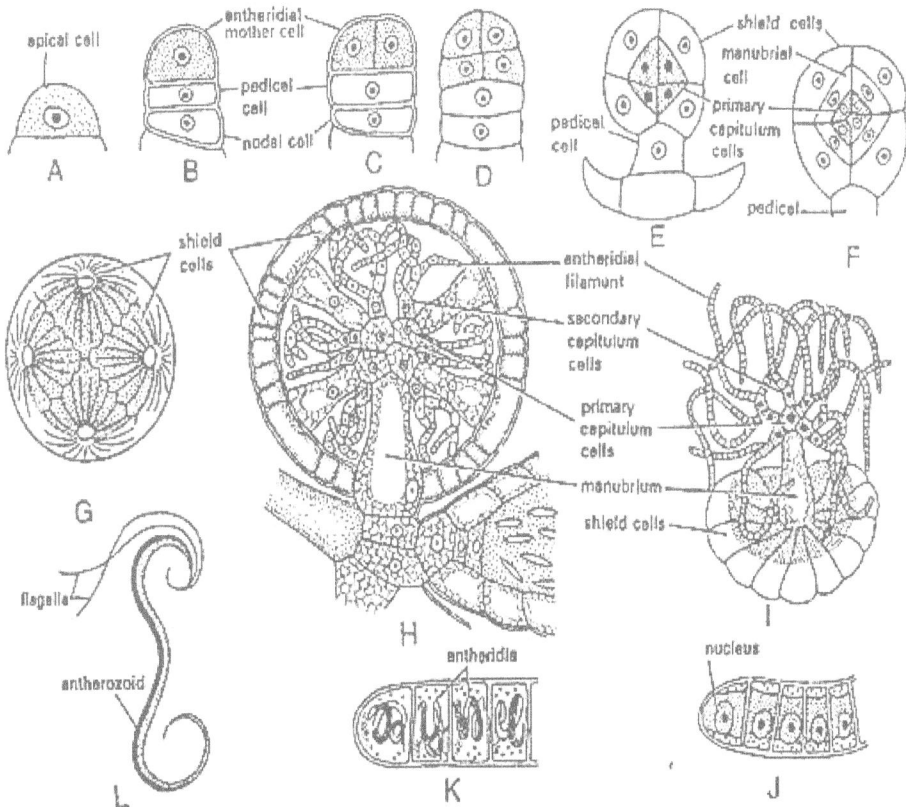

Fig. 9. (A–K). *Chara*. Development of globule and antherozoids.

Liberation of Antherozoids:

At maturity the shield cells of antheridium separate from each other exposing antheridial filaments in water. The sperm mother cell gelatinizes to liberate the antherozoids.

Nucule:

The nucule of Chara is large, green, oval structure with short stalk. It is borne at the node of the primary lateral. It lies just above the globule in homothallic species.

Development and Structure:

The upper peripheral cell of the basal node of the antheridium functions as the oogonial initial. The oogonial initial divides by two transverse divisions to make three celled filament. It has lower pedical cell, the middle nodal cell and the upper oogonial mother cell (Fig. 10 A-C). The pedicell does not divide further and makes pedicel of the

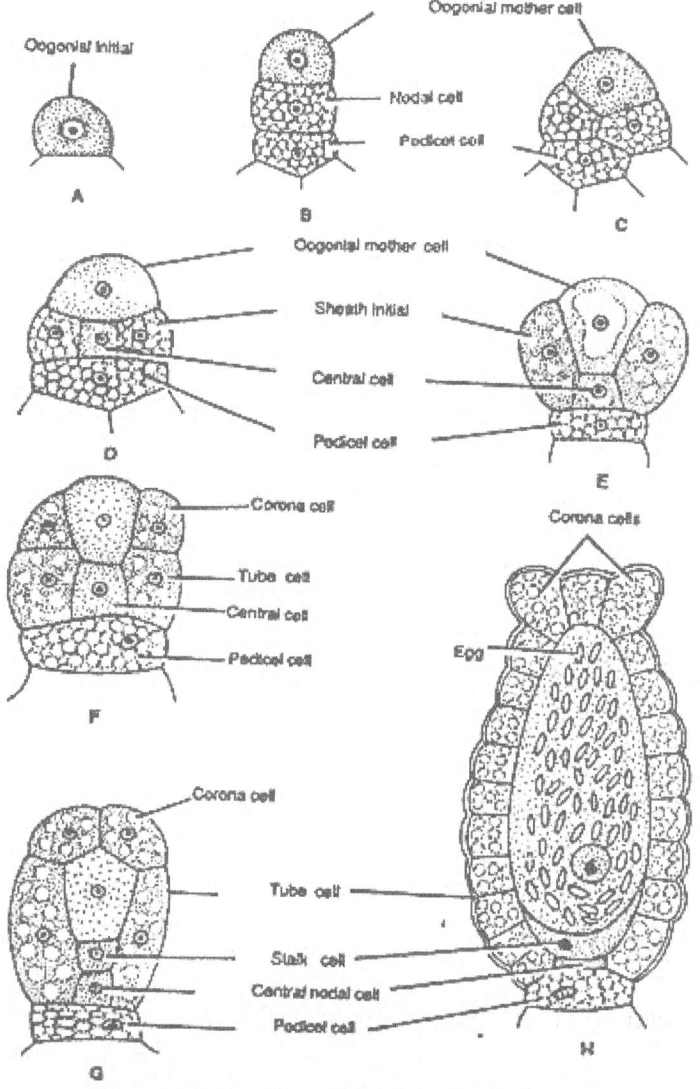

Fig. 10. (A–H). *Chara*. Development of oogonium

oogonium. The middle nodal cell under many vertical divisions to make five sheath cells or peripheral cell which surrounds the central cell (Fig. 10 D-F). The central cell does not divide and functions as the node of the oogonium.

The oogonial mother cell elongates vertically and divides by transverse division to make lower small stalk cell and an upper large oogonium. The oogonium contains uninucleate ovum or egg (Fig. 10 G, H). The peripheral cells or five sheath initials elongate and divide by transverse division to make two tiers of five cells each.

The five upper tier cells from coronary cells which form the corona of nucule. The five lower tier functions as tube cells, the tube cells elongate and get spirally twisted in clockwise directions on oogonium. The nucleus migrates on lower side and receptive spot develops at the tip of oogonium. Large amount of starch and oil get deposited in oogonium.

The mature nucule is attached to the node by the pedicel cell. The nucule is surrounded by five tube cells. The tips of tube cells from corona at the top of nucule. The oogonial cell possesses a single large egg or ovum. The nucule contains large amount of starch and oil. The receptive spot is present at the upper part of nucule.

Fertilization:

When the oogonium is mature, the five tube cells get separated from each other forming narrow slits between them. Antherozoids are chemotactically attracted towards ovum. The antherozoids enter through these slits and penetrate gelatinized wall of the oogonium. Many antherozoids enter oogonium but one of those fertilizes the egg to make a diploid zygote. The zygote secretes a thick wall around itself to make oospore.

Oospore:

The mature oospore is hard, oval, ellipsoid structure which may be brown e.g., C. inferma, black e.g., C. corallina or golden brown e.g., C. flauda. The oospore inside contains a diploid nucleus and many oil globules in cytoplasm. On maturity of oospore the inner walls of tube cells get thickened, suberised and silicified. The oogonial as well as oospore walls become thick. The oospore nucleus moves towards the apical region. In advanced stage the outer walls of the envelope or sheath cells fall off and the inner parts remain attached to mature oospore in form of ridges.

Germination of Oospore:

The oospore germinates when favourable conditions appear. The diploid nucleus present in apical colourless region divides by meiosis forming four haploid daughter nuclei (Fig. II A-B). At this stage a septum divides oospore into two unequal cells. The upper smaller apical cell contains a single nucleus and the large basal cell contains three nuclei. The three nuclei of basal cell degenerate gradually. The oospore apical cell divides by longitudinal division to make a rhizoidal initial and protonemal initial (Fig.

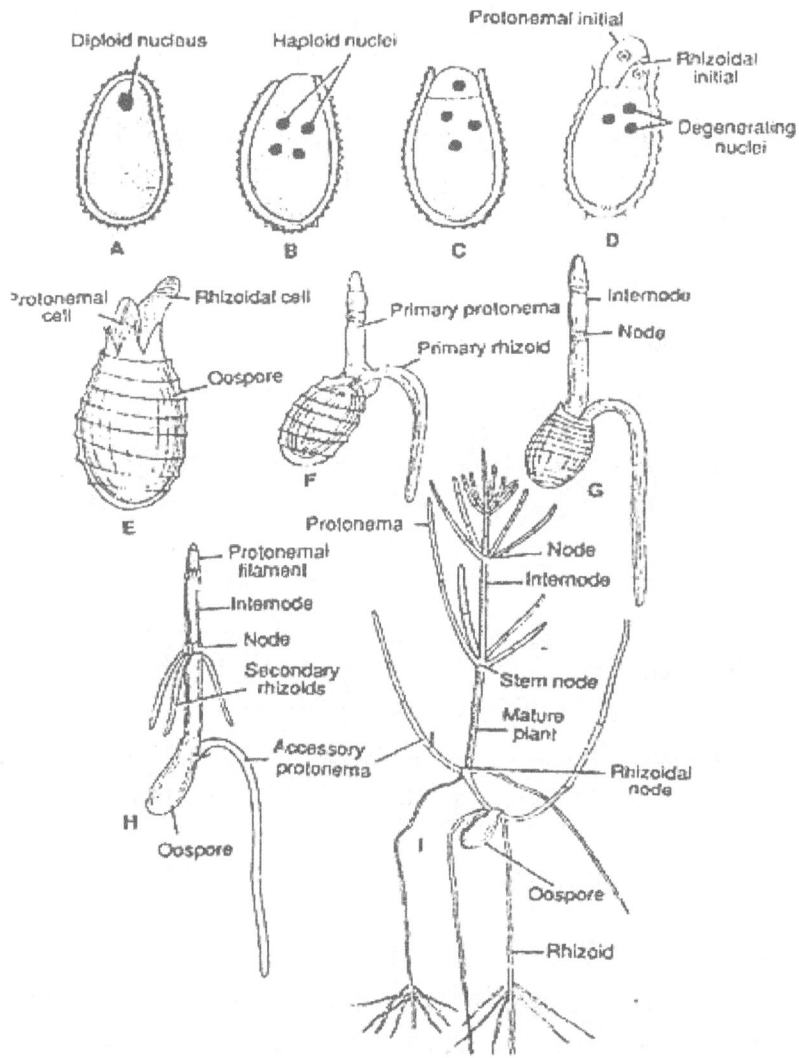

Fig. 11. (A–I). *Chara.* Germination of oospore and development of new plant

11 D).

The rhizoidal initial shows positive geotropism and forms primary rhizoid, the protonemal initial shows negative geotropism and forms primary protonema. The primary protonema differentiates into nodes and internodes. The peripheral cells of the basal node give rise to rhizoids and secondary protonema. The peripheral cells of the upper nodes give rise to lateral branches (Fig. 11 E-I).

Life Cycle of Chara:

The plant body of Chara is haploid. The vegetative reproduction takes place by the formation of amylum stars, bulbils and secondary protonema. Asexual reproduction is absent. The sexual reproduction is advanced oogamous type.

The male and female sex organs are globule and nucule respectively. After fertilization a diploid spore is formed. At the time of germination diploid oospore nucleus divides to make hapoid nuclei and haploid Chara plant. Thus the life cycle of Chara a predominantly haploid type (Fig. 12, 13).

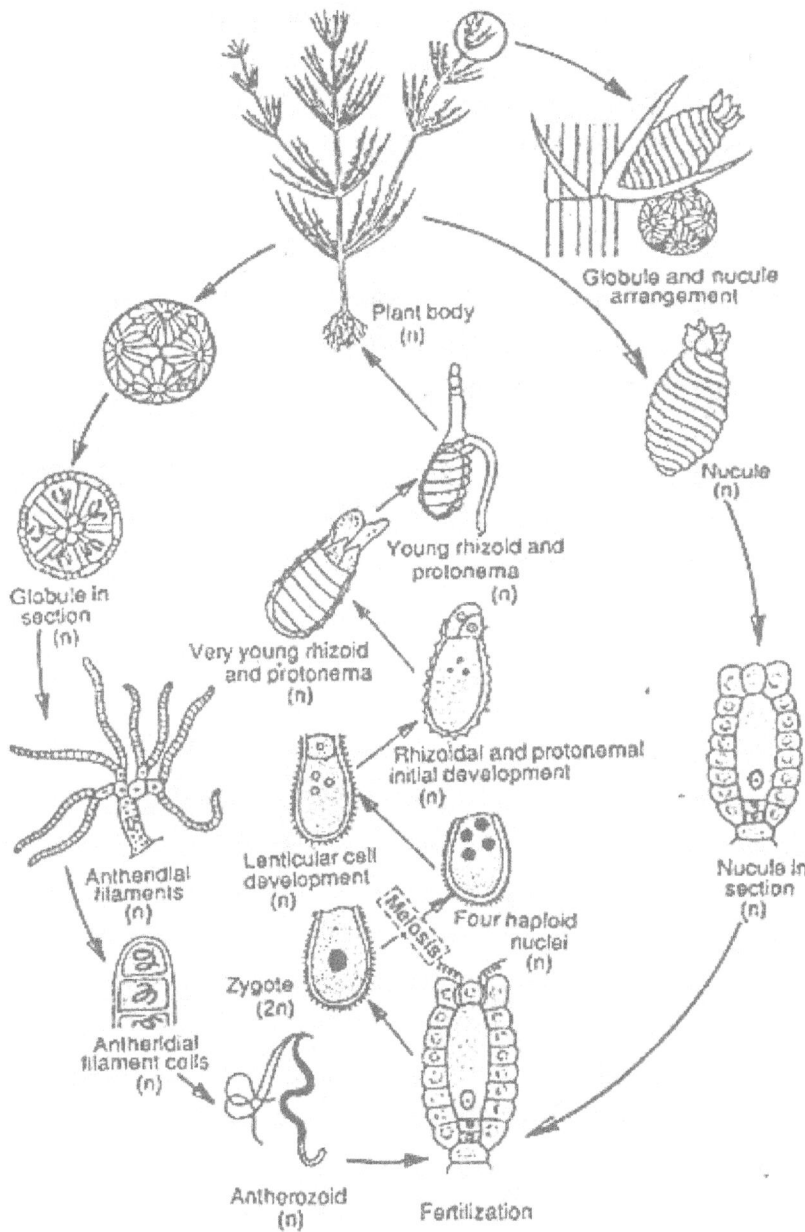

Fig. 12. *Chara.* Diagrammatic life Cycle

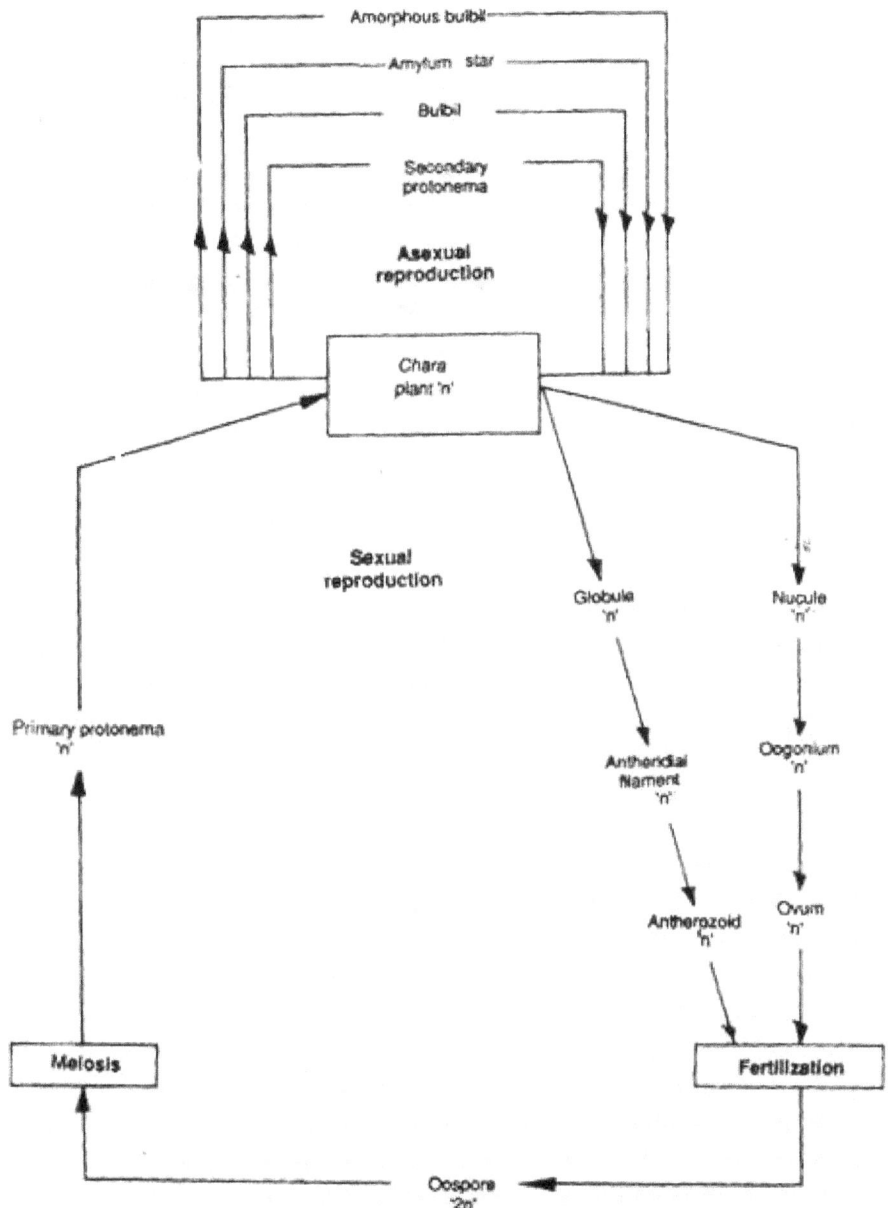

Fig. 13. *Chara*. Graphic life Cycle

17.

Bacillariophyceae: General Characters and Classification

Diatoms (Class:Bacillariophyceae) are unique microscopic algae containing silica and having distinct geometrical shapes. They are unicellular, eukaryotic and photosynthetic organisms.Their cell size ranges between 5 µm–0.5 mm. They occur in water moist places where photosynthesis is possible. Diatoms are either planktonic (free-floating) or benthic (attached to a substratum) in Nature (Figure1). The individuals are solitary or some-times form colonies. Diatoms are mostly non-motile; however,some benthic diatoms have a specialized raphe system1thatsecretes mucilage to attach or glide along a surface. They are also known to form biofilms, i. e., layers of tightly attached cells of microorganisms. Biofilms are formed on a solid surface and are often surrounded by extra-cellular fluids.

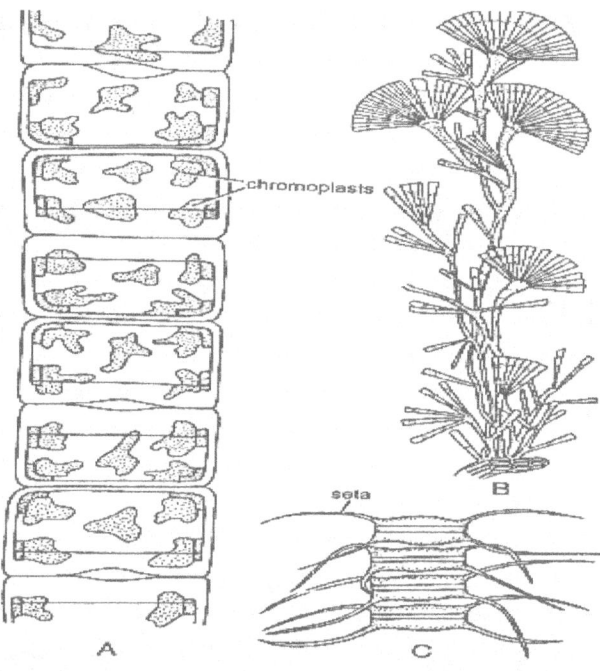

Fig.1: A-C. Some diatoms: A. *Melosira*, **B.** *Licmophora flabellate*, **C.** *Chaetoceras diodema*

Diatoms are a major group of organisms in terms of diversity, abundance and productivity of marine and freshwater ecosystems. Diatoms are solely responsible for about 20–25% of global oxygen production, i.e., approximately every fourth breath of oxygen we inhale. Diatoms alone account for around 40% of the phytoplankton on Earth which implies 20–25% global net primary production. This is more than the primary production of all tropical rainforests combined! They also play a significant role in silica cycle and also contribute to other biogeochemical cycles, especially in carbon fixation in the carbon cycle by converting CO_2 into biomass in a large proportion.

General Characteristics of Bacillariophyceae

They are commonly unicellular and free- living but some members form colonies of various shapes like filaments, mucilaginous colonies etc.

2. Microscopic cells are of different shapes. They may be oval, spherical, triangular, boat- shaped etc.

3. Plant bodies are either bilateral or radial in symmetry.

4. The cells are surrounded by a rigid cell wall, called frustule, consisting of upper epitheca and lower hypotheca; arranged in the form of a box with its lid.

5. The cell wall is composed of pectic substances impregnated with high amount of siliceous substance.

6. The wall may have secondary structures like spines, bristles etc.

7. Vegetative cells are diploid (2n).

8. The cells generally have many discoid or two large plate-like chromatophores. Some cells possess stellate chromatophore.

9. The photosynthetic pigments are chlorophyll a, chlorophyll c along with xanthophylls like fucoxanthin, diatoxanthin and diadinoxanthin.

10. Reserve food is oil, volutin and crysolaminarin.

11. Some vegetative cells show gliding movement.

12. Motile structure (antherozoid) has single pantonematic flagellum.

13. Vegetative multiplication takes place by cell division, which is very common. Some of the cells become very much reduced in size.

14. They produce characteristic spore, the auxospore which develops to regain the normal size.

15. Sexual reproduction takes place by isogamy and oogamy.

Distribution / habitat

This Bacillariophyceae specification consists of widespread diatoms in freshwater and seawater, as well as in moist soils. The number of diatoms is very large, estimated at 16,000 species. Due to the large number, diatom phytoplankton become an important producer component in marine waters. Diatoms are living alone and some form colonies. Some live freely on the surface of the water, some other species live on the substrate.

Cell Structure

Cell Structure of Diatoms:

The cell consists of cell wall and protoplast. The cells are covered by a siliceous wall, the frustule. It consists of two overlapping halves, the theca. The upper one is epitheca and lower one is hypotheca. The common region of the connecting bands, where both the theca remain fitted together, is the girdle. [When the diatoms are observed from the valve side i.e., valve side is uppermost, called the valve view, but when viewed from the connecting band, it is the girdle view]. Depending on symmetry, the cells are divided into two orders: Pennales (bilaterally symmetry) and Centrales (radially symmetry).

In some pinnate diatoms (Cybella cistula, Pinnularia viridis etc.) an elongated slit is present on their valves, called raphe. The raphe is interrupted at its midpoint by thickening of the wall called central nodule. Similar thickening is also present at the ends called polar nodules. Some members like Tabellaria fenestrate etc. of the order Pennales, do not have raphe, called pseudoraphe.

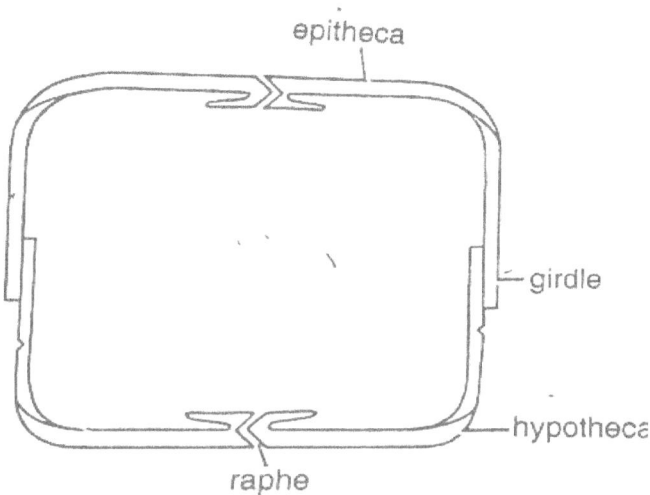

Fig. 2. *Pinnularia viridis*: Transverse section.

Besides raphe or pseudoraphe, the cell walls have other types of openings, called pores and locules. Based on electron microscopic studies, Hendey (1971) observed four basic types of secondary structures. These are: Punctae (small perforations on valve surface), Canaliculi (tubelike narrow channels which run through the valve surface), Areolae (large boxlike depressions) and Costae (riblike structures on the valve surface). The cell wall is mainly made up of pectic substances, impregnated with silica. The content of silica varies from 1% (*Phaeodactylum tricornutum*) to about 50% on the basis of dry weight of the cell.

Both the theca consist of two portions:

(a) Valve — the upper flattened top and

(b) Connecting band or cingulum (pl. cingula) — the incurved region.

Each cell contains one nucleus with one or more ribbon-shaped plastids or discs, containing abundant golden brown pigments. Food substances in the form of droplets of oil. The shape of the diatom cells extends, coated by a cell wall (shell) consisting of two interlocking hemispheres. This cell wall is made of pectin and silica layers. When the diatoms die, the remaining silica shells are translucent. The shell on a diatom is equipped with a small hole that allows the cell to connect to the water environment.

Protoplast:

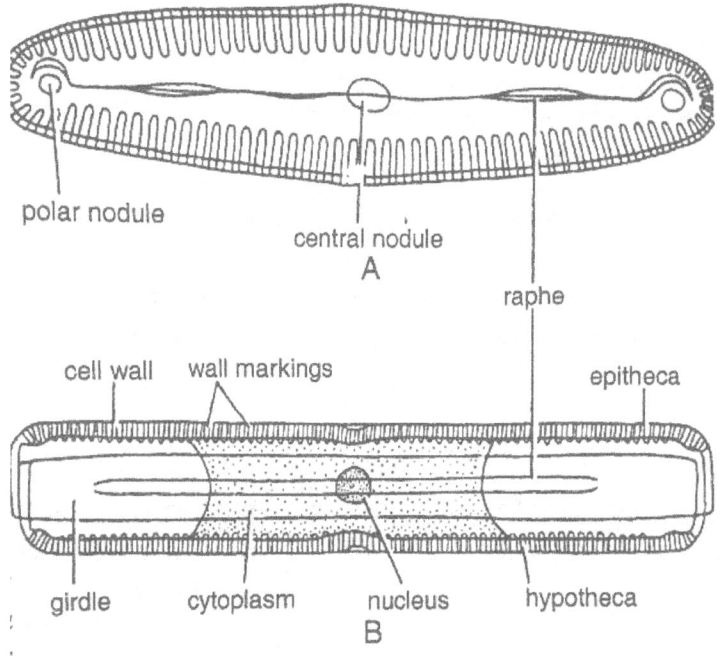

Fig. 3 A-B. *Pinnularia*: A. Valve view, B. Girdle view.

The entire content present inside the cell wall is the protoplast. The cell membrane encloses a large central vacuole surrounded by cytoplasm. The cytoplasm contains single nucleus, mitochondria, golgibodies and chloroplasts. The chloroplasts may be of different shapes like stellate, H-shaped, discoid etc. In some species the chloroplasts contain pyrenoids. The photosynthetic pigments are chlorophyll a, c_1 and c_2, β-carotene, fucoxanthin, diatoxanthin and diadinoxanthin. The latter two are present in small quantity. (The golden-brown colour of diatom cells is due to the presence of xanthophylls like fucoxanthin, diatoxanthin and diadinoxanthin. The term diatomin is used for the mixture of chlorophyll and carotenoids, particularly carotene and several brown xanthophylls pigments.) The reserve food of diatoms is chrysolami- narin and oil droplets (they do not store in the form of starch).

Reproduction of Diatoms:

Diatom reproduces by vegetative and sexual means.

1. Vegetative Reproduction:

Vegetative reproduction performs with the help of cell division (Fig. 4.). It takes place usually at midnight or in the early morning.

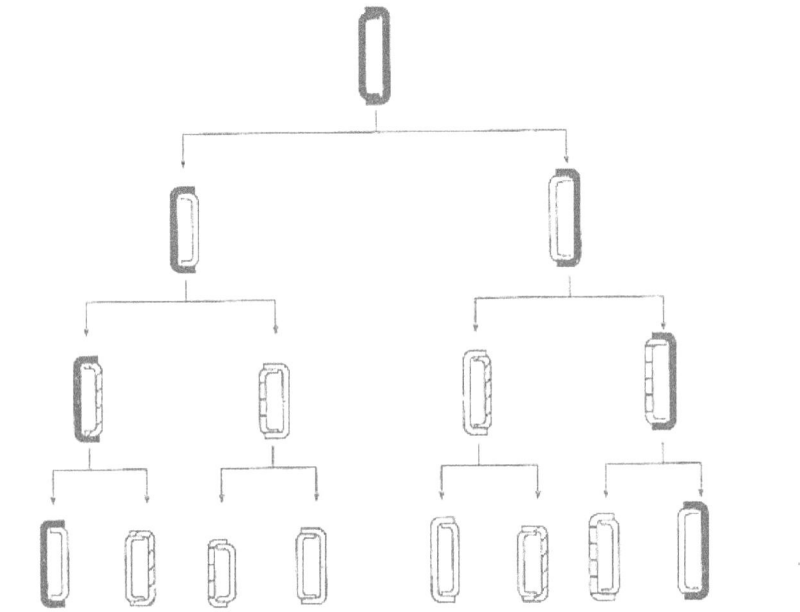

Fig.4. Cell division in diatoms (diagrammatic) showing reduction in cell size in successive generations except one (extreme right)

During cell division the protoplast of the cell enlarges slightly, thus the cell increases in volume and slightly separates both the theca (epitheca and hypotheca). Then the protoplast undergoes mitotic division and gets separated along the longitudinal axis through the median line.

Thus one half of protoplast remains in epitheca and the other one in hypotheca. One side of the protoplast thus remains naked. Now both the theca i.e., epitheca and hypotheca of mother cell behave as epitheca of the daughter cells.

Thus new silicious valves are deposited towards the naked sides of the protoplast and always behave as hypotheca of the daughter cells. Connecting bands are developed between the theca. Later on, the daughter cells get separated.

During cell division, both the theca i.e., epitheca and hypotheca of the mother cell behave as epitheca of the daughter cells. So at the side where the hypotheca behaves as epitheca, the cell becomes reduced in size. Thus with continuous cell division some cells gradually become reduced in size.

2. Sexual Reproduction:

The pattern of sexual reproduction differs in both orders — Pennales and Centrales. During this process, auxospore is formed in both the groups. During cell division, those cells become reduced in size, are able to regain their normal size through the formation of auxospore, so it is a "restorative process" rather than multiplication.

Auxospore Formation in Pennales:

It takes place through gametic union, autogamy and parthenogenesis.

These are of the following types:

1. Production of one auxospores by two conjugating cells. In this process two uniting cells come very close to each other (Fig. 5.) and become covered by a mucilaginous sheath. The diploid nucleus of each cell undergoes meiosis.

Out of four nuclei, three degenerate and only one survives. The surviving nucleus behaves as gamete (n). The gametes come out from the parent frustules and unite together, to form a zygote (2n).

After a short period of rest the zygote elongates considerably and functions as an auxospore. The auxospore projects out from the parent frustules along with mucilage and elongates in a plane parallel to the long axis of the parent diatom.

The auxospore is enclosed in a pectic membrane, the perizonium. The auxospore then develops new frustule inside the perizonium. Thus new diatom cell is formed which regains the normal size. It is found in Cocconis placentula, Surirella saxonica etc.

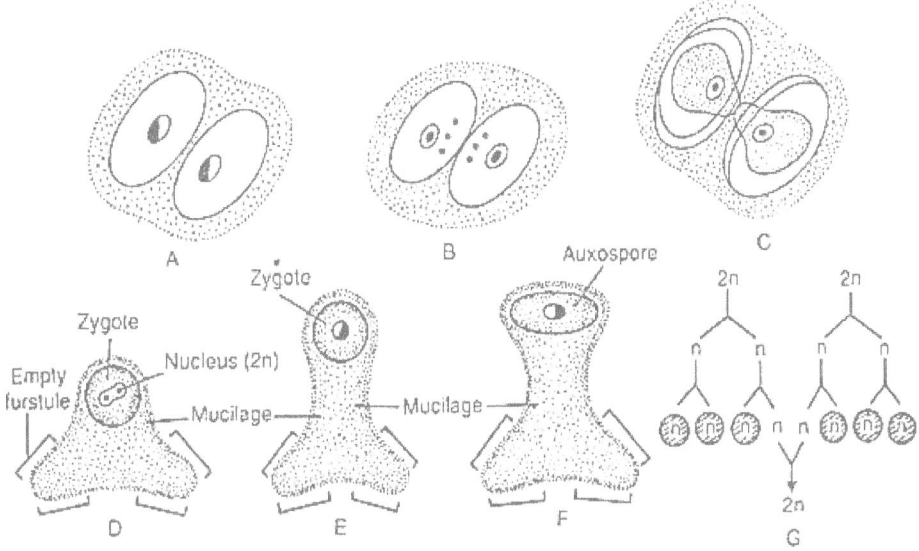

Fig.5.: A-F. Production of one auxospore by two conjugating cells of *Cocconeis placentula*, and G. Nuclear behaviour during reproduction

2. Production of Two Auxospores by Two Conjugating Cells:

This is a very common process of auxospore formation. In this process the conjugating cells come very close to each other and get enclosed by mucilage (Fig. 3.104). The nucleus (2n) of each cell undergoes meiotic division and forms four nuclei.

Out of four nuclei, two degenerate, the rest two survive. The cytoplasm then divides either equally or unequally and along with one nucleus they behave as gametes. Thus two gametes are formed in each cell.

The pattern of union between the gametes varies from species to species. Both the gametes of a cell may be active and fuse with the gametes of other cell, thus two zygotes are produced in a single cell or out of two, one becomes active and fertilises with the opposite one and thus one zygote is produced in each cell.

The zygotes elongate and function as auxo- spores. The auxospores develop the perizonium around themselves and both of them develop new frustules on their outer sides i.e., inside the perizonium. Thus two diatom cells of normal size are formed. It is found in Cymbella lanceolata, Gomphomema parvulum etc.

3. Production of One Auxospore by One Cell:

This process of auxospore formation is called Paedogamy (Pedogamy). In this process the diploid nuclei of a vegetative cell undergo meiosis and form four haploid nuclei. Out of the four nuclei two partially degenerate. Each of the rest two along with the cytoplasm and one partially degenerated nucleus, behaves as gamete. Later on, the union between the two sister gametes takes place and forms the zygote.

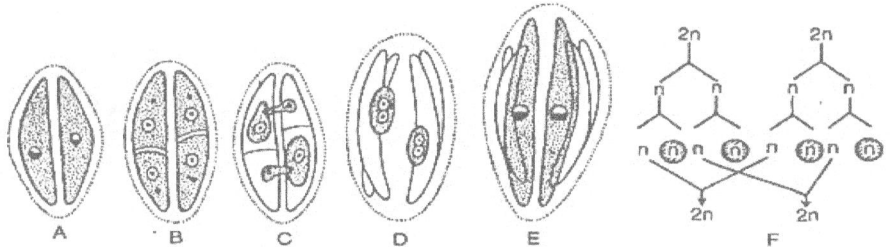

Fig.6.: *Cymbella lanceolata* : A-E. Production of two auxospore by two conjugating individuals, and F. Nuclear behaviour during reproduction

The zygote comes out from the parent frustule and behaves as an auxospore. The auxospore then gets covered by perizonium and develops wall inside the perizonium. Thus one diatom cell of normal size is formed.

4. Production of One Auxospore by Autogamy:

In this process the diploid nucleus undergoes first meiotic division. Thus two haploid nuclei are formed. The two nyclei in the protoplast come side by side, fuse together and form diploid (2n) nucleus. This is called autogamous pairing.

The protoplast along with diploid (2n) nucleus comes out from the parent frustule and behaves as an auxospore. The auxospores are then covered by perizonium. New wall develops on the auxospore inner to the perizonium. Thus a new individual of normal size is developed. This is found in Amphora normani.

5. Production of Auxospore by Parthenogenesis:

The diatom cells come together and are covered by a common mucilage envelop (Fig.7.). The diploid nucleus undergoes two sequential mitotic divisions. Meiotic division does not take place here. One nucleus in each mitotic division degenerates.

Thus only one diploid (2n) nucleus along with protoplast remains, and comes out from the mother cell and behaves as an auxospore.

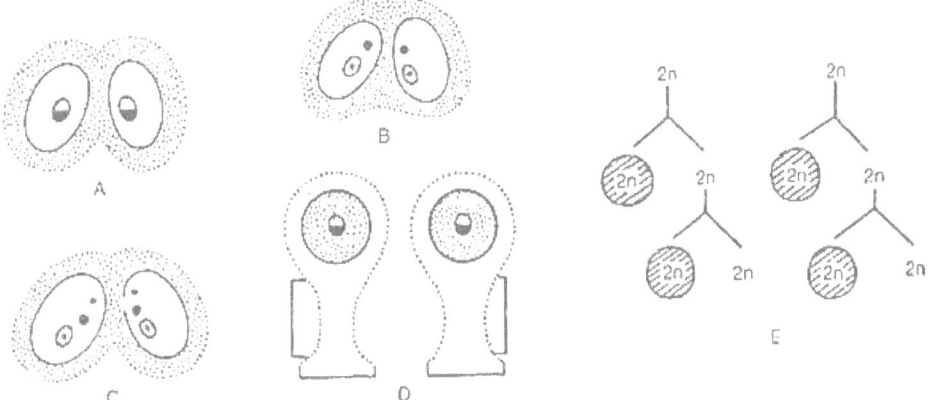

Fig.7.: A-D. Production of auxospore by parthenogenesis in Pennales, and E. Nuclear behaviour during reproduction

The auxospore is then covered by perizonium and secretes new wall around itself. Thus normal size cell is formed.

6. Production of Auxospore by Oogamy:

In this process (Fig. 8) the nucleus (2n) of female cell which behaves as oogonium, undergoes meiosis and forms four nuclei. The protoplast is also divided into two unequal parts, each containing two nuclei.

The lower half is larger and behaves as functional ovum and the upper smaller one as non-functional ovum. The functional ovum contains one functional nucleus and one non-functional nucleus, which gradually degenerates at maturity.

The male cell (2n) behaves as antheridium, also undergoes meiosis and forms four nuclei. The protoplast also divides into two parts. Thus two microgametes are formed. Each of which contains two nuclei, of which one is functional and other is non-functional. The microgametes are naked, globular and non-flagellate. After coming out, the male gamete fertilizes the egg and forms the zygote (2n). Later it functions asan auxospore and forms new individual of normal size. It is found in *Rhabdonema adriaticum*.

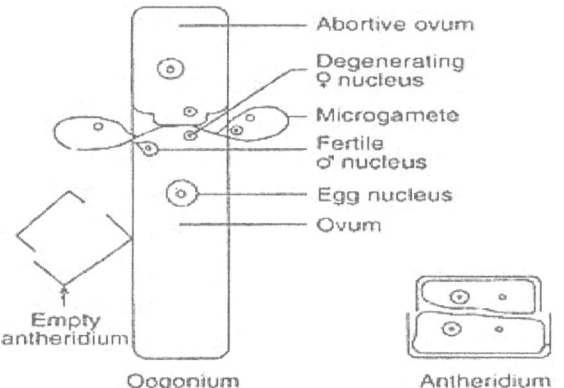

Fig. 8.: Production of auxospore by oogamy in Pennales

Auxospore Formation in Centrales

It takes place by autogamy and oogamy:

1. Auxospore Formation by Autogamy:

The protoplast of the vegetative cell (Fig.9) secretes mucilage which separates both the theca. The nucleus (2n) then undergoes meiosis and forms four nuclei. Of the four nuclei two degenerate and the other two undergo fusion to form diploid (2n) nucleus again.

This is called autogamy. The protoplast with 2n nucleus functions as an auxospore. The auxospore forms fresh frustule inside the perizonium covering and forms cell of normal size. It is found in Melosira nummuloides.

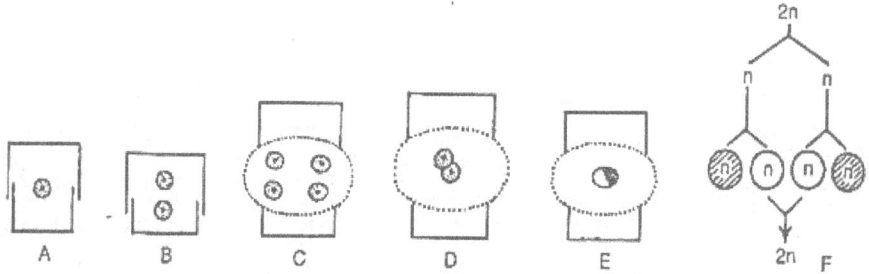

Fig.9 : A-E. Auxospore formation by autogamy in Contrales, and F. Nuclear behaviour during reproduction

2. Auxospore Formation by Oogamy:

Oogamy takes place by the fusion of egg and sperm developed inside the oogonium and antheridium respectively (Fig. 10).

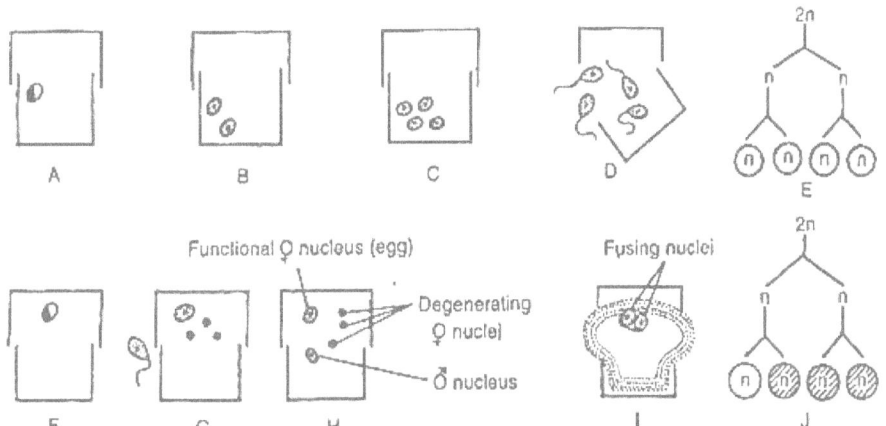

Fig.10 : Auxospore formation by Oogamy in Centrales : A-D. Formation of sperms, E. Nuclear behaviour during sperm formation, F-G. Formation of egg, (F. Single oogonium, G. Sperm approaching oogonium), H. Male nucleus entered inside the oogonium, I. Fusing male and female nuclei and J. Nuclear behaviour during Egg-formation

Oogonium:

Single vegetative cell behaves as an oogonium. The protoplast of oogonium undergoes meiotic division and forms four nuclei. Of the four nuclei three degenerate and the remaining one functions as an egg.

Antheridium:

The pattern of development of sperms varies in different species. In species like Melosira varians the protoplast undergoes meiotic division and forms four haploid nuclei. Each haploid nucleus with some protoplast metamorphoses into an uniflagellate (tinsel type) sperm. In others the number of sperms may go up to 8 or even 128.

Fertilisation:

After coming out of the antheridium only one sperm enters inside the oogonium and fertilises the egg. The resultant zygote undergoes mitotic division but one nucleus degenerates in each division. The remaining nucleus with its protoplast behaves as an auxospore. The auxospore then develops new wall inside the perizonium covering and forms new cell of normal size like the mother. It is also called firstling cell.

From the above processes of sexual reproduction in both pennales and centrales, it becomes clear that the sexual process in diatom does not lead to multiplication but is to regain the normal size.

3. Resting Spores:

These spores are formed during unfavourable conditions. Some members reproduce by the formation of thick-walled resting spores, the cysts or statospores. They are formed in *Melosira*.

****

18.

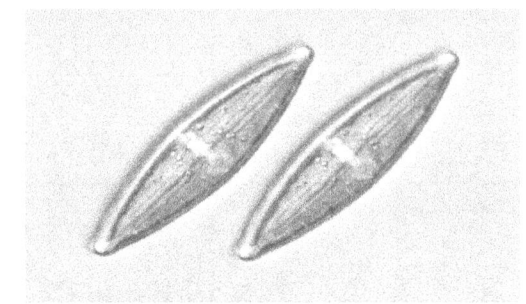

Pennales: *Navicula*

The members of the order Pennales are characterised by the presence of a bilateral symmetry. A pseudoraphe or a raphe is usually present in the median line of the valve. They show gliding movement. Sexual reproduction, if present, is of amoeboid isogamous type. AUxospores are normally formed by conjugation. They occur mostly in fresh water habitats. The order is classified according to the presence or absence and the number and morphology of the raphe on the valves.

Family. Naviculoideae

In the members of this family the raphe is present on both the valves. It occupies the apical axis and is usually without a keel.

Navicula

Systematic Position:

Class: Bacillariophyceae
Order: Pennales
Family: Naviculoideae
Genus: *Navicula*

Occurrence

Navicula, a fresh water diatom, is cosmopolitan and ubiquitous in distribution, found in ponds, lakes, etc. However, a few species occur in terrestrial habitats. The species of Navicula are very important as primary producers in the food web of aquatic ecosystems. The most common species of the genus occurring in India is *N. halophile.*

Thallus Structure:

Thallus is represented by an isobilaterally symmetrical diploid unicells.

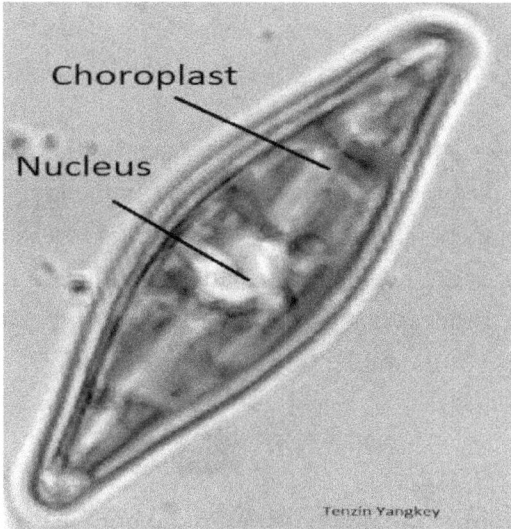

Fig.1: *Navicula*

Structurally, it can be differentiated into two parts:

A siliceous cell wall (called frustule) and the protoplast. The cell wall is made of pectic substances which are impregnated with silica (SiO_2). Cell wall consists of two overlapping halves called epitheca and hypotheca. Epitheca remains fitted over the hypotheca as a lid over the box.

Each theca is further divided into two parts:

The main surface called valve and the incurved margin known as connecting band or cingulum. The two connecting bands of the two thecas are fitted together. The connecting band of the epitheca overlaps that of the hypotheca and the two bands remain united in the overlapping region (called girdle) by a connecting cement present between the (Fig. 1).

A frustule can be seen in two views:

Valve view and girdle view. In top view or valve view it appears as boat shaped and in girdle view or side view more or less rectangular. The valve view shows marking or striations which spread out laterally in two parallel series, one on either side of the axial strip. The axial strip pears a longitudinal cleft known as raphe.

The raphe extends from one end of the valve to the other. It also bears three enlargements or rounded nodules, one central nodule and two polar nodules. Due to presence of raphe Navicula shows gliding movement. This movement is caused **"by streaming cytoplasm by circulation within the raphe, and by the extrusion of the mucilage."**

Just inner to the cell wall is present a plasma membrane which encloses the cell protoplast It is differentiated into a single nucleus and cytoplasm. The cytoplasm forms a thick layer just below the cell wall and encloses a large central vacuole. The cytoplasm includes mitochondria, Golgi bodies, and two large parietal brownish yellow chromatophores. Pyrenoids are absent. The photosynthetic pigments are chlorophyll-a, Chlorophyll-c, p-carotene, fucoxanths, diatoxanthin and diadinoxanthin. Reserve food material is in the form of chrysolaminarin and oil droplets.

Reproduction

Vegetative Reproduction:

It takes place by the mitotic cell division or fission. Successive cell division takes place very rapidly at night. Presence of aluminium-silicate in water is essential for cell division to occur. As the cell division starts, the cell protoplast increases in diameter. The cell also increases in size. The diploid nucleus divides mitotically and produces two daughter nuclei. Two chromatophores divide. The single chromatophore spits longitudinary in such a manner that one chromatophore comes to lie in each half. Now the protoplasm cleaves into two uninucleate portions by division in longitudinal plane parallel to valve surface. One daughter protoplast now lies in epitheca and the other in hypotheca.

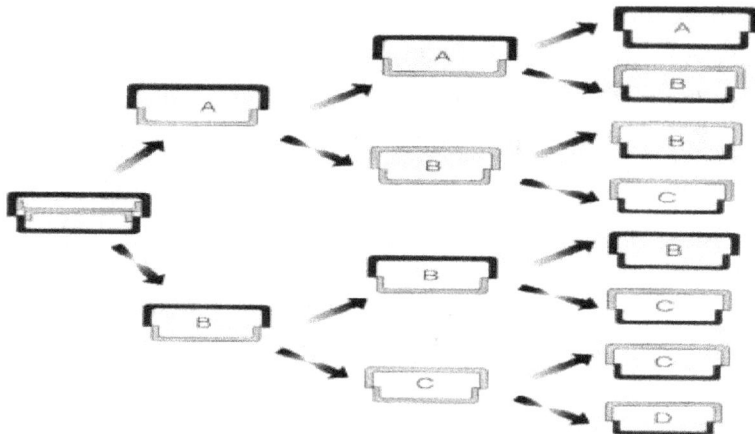

Fig.2. *Navicula*: **Diagrammatic representation of the vegetative cell**

Now both the daughter protoplasts with one daughter nucleus secrete the new siliceous wall on the two fresh protoplasmic surfaces exposed along the cleavage plane. The new valves developed always become hypotheca while the older theca (which may be epitheca or hypotheca of the parent cell) becomes the epitheca of the new or daughter cell.

When this cell again divides, it produces a daughter cell which is again smaller than the present parent. Thus, in a population of diatom cells during successive divisions there is normally a progressive decrease in the average cell size (Fig. 2). It is called Macdonald-Pfitzer law. The smaller cells of later series of division loose their vitality and capacity of division.

Sexual Reproduction: (fig.3.)

It takes place by the formation of auxospores. The successive decrease of cell size in vegetative reproduction is prevented by the auxospore formation. The auxospore formation is actually a **'restorative process'** because the reduction in the original size of the cells, during the cell division is restored. During the process only those cells which have diminished sufficiently in size can act as **'sex cells'** or conjugating cells.

Those cells which do not decrease in size by cell division apparently do not show sexual reproduction. Majority of the species of *Navicula* are monoecious but, *N. haplophila* is dioecious. Two sex cells come together, pair up longitudinally (called gamontogamy) and secrete a common mucilaginous envelope. The diploid nucleous of each cell undergoes meiosis to form 4 haploid nuclei. Out of these two nuclei degenerate and only two remain functional. The protoplasm of each cell now cleaves into two portions each obtaining one haploid nucleus. The functional nuclei ultimately metamorphose into gametes.

The parent cell fuses (cytogamy) and the fusion of gametes occurs in a copulatory jelly. In *N. haplophila* the two gametes formed in one cell (conjugant) are amoeboid and the two gametes formed in the other are passive or immobile. The amoeboid gametes emerge through the open valves of the parent frustule and dip into open shell of the other to fuse with opposite gametes to form two zygotes in one shell. The other is empty. Thus, *N. haplopila* shows physiological anisogamy. Two diploid fusion cells or zygotes escape from the enclosing pustules. They remain doranant for some time. Later the zygote elongates (more in the longitudinal plane) and functions as auxospore which develops a silicified membrane called perizonium around its protoplasm. It may be secreted by the auxospore or by the remains of the zygotic membrane. The auxospore secretes new pustules around itself around the perizonium. The reconstituted new cell is of normal size and after sometime begins to divide vegetatively to form new generations. The valves of the old pustules are often seen attached to the newly formed pustules.

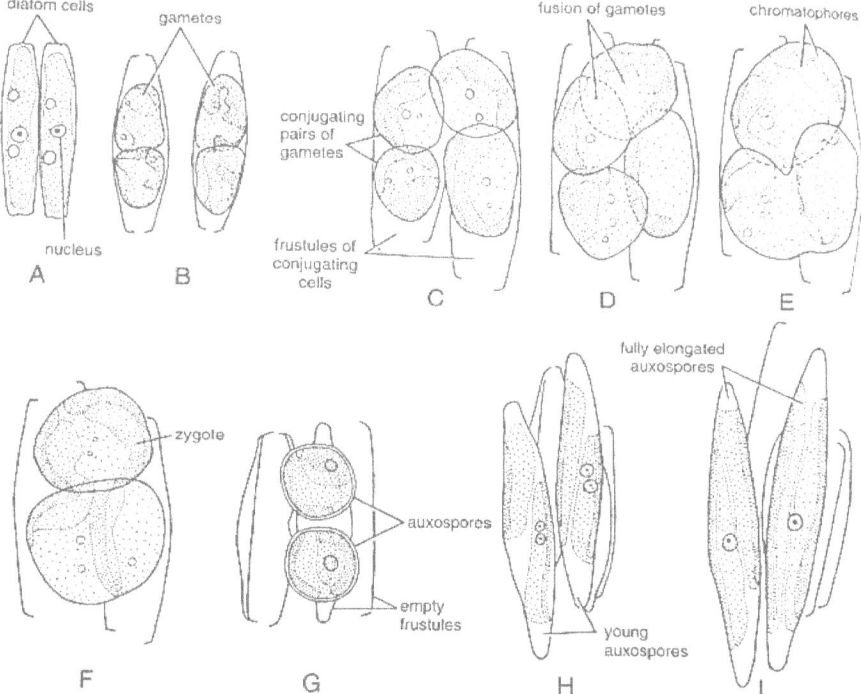

Fig. 3 A-L. *Navicula*: Auxospores; A-G. Conjugation of gametes and formation of the auxospore, H. A stage in the enlargement of auxospore,. Fully elongated auxospore (After Subrahmanyam, 1946).

19.

Phaeophyceae (Brown Algae): Description and Classification

It is a large group of algae consisting of 240 genera and over 1,500 species out of which 32 genera and 93 species are reported from India. They are commonly known as brown algae, due to the presence of a golden brown xanthophyll pigment, fucoxanthin ($C_{40}H_{54}O_6$) in the chromatophores.

About 99.7% members are marine and a few grow in fresh water. The fresh water members are *Pleurocladia, Heribaudiella, Pseudobodanella, Lithoderma* and *Sphacelaria*. Members like *Pleurocladia lacustris* grow both in fresh water and marine habitats.

The members of Phaeophyceae are popularly called brown algae. The brown colour of this group of algae is due to the possession of a pigment called fucoxanthin.

The main characteristics of phaeophyceae are:

1. The algae of this family are commonly known as **brown algae.**
2. The members of phaeophyceae are mostly **marine.**
3. Most of them are **large sized** and **multicellular;** simple forms are absent.
4. In addition to the golden brown **carotene pigment** it also possesses **chlorophyll a, chlorophyll c.**
5. The reserve food material is present as *Laminarin* and Mannitol.
6. It possesses double layered **cell wall**; the inner layer of **cellulose** and outer layer of **phycocolloids and fucoxanthin.**
7. Many of the cells possess a characteristic **fucosan vesicle.**
8. The plant body is attached to the substratum by a **hold fast, has a stalk, a stipe and leaf like photosynthetic part.**
9. **Reproduction** occurs both by **asexual** and **sexual methods**.
10. **Asexual reproduction** occurs by **fragmentation, zoospores** and **aplanospores.**

11. Sexual reproduction takes place by **isogamy, anisogamy or oogamy**.
12. The large brown algae are called **trees of seas or Kelps**.

Classification of Phaeophyceae (Brown Algae):

Fritsch (1935, 45) classified the Class. Phaeophyceae into nine orders. This was also followed by Mishra (1966).

Thallus Structures

Whatever their form, the body of all brown algae is termed a *thallus*, indicating that it lacks the complex xylem and phloem of vascular plants. This does not mean that brown algae completely lack specialized structures. But, because some botanists define "true" stems, leaves, and roots by the presence of these tissues, their absence in the brown algae means that the stem-like and leaf-like structures found in some groups of brown algae must be described using different terminology. Although not all brown algae are structurally complex, those that are typically possess one or more characteristic parts.

Fig.1.Two specimens of *Laminaria hyperborea*, each showing the rootlike holdfast at lower left, a divided blade at upper right, and a stemlike stipe connecting the blade to the holdfast.

A *holdfast* is a rootlike structure present at the base of the alga. Like a root system in plants, a holdfast serves to anchor the alga in place on the *substrate* where it grows, and thus prevents the alga from being carried away by the current. Unlike a root system, the holdfast generally does not serve as the primary organ for water uptake, nor does it take in nutrients from the substrate. The overall physical appearance of the holdfast differs among various brown algae and among various substrates. It may be heavily branched, or it may be cup-like in appearance. A single alga typically has just one holdfast, although some species have more than one stipe growing from their holdfast.

A *stipe* is a stalk or stemlike structure present in an alga. It may grow as a short structure near the base of the alga (as in *Laminaria*), or it may develop into a large, complex structure running throughout the algal body (as in *Sargassum* or *Macrocystis*). In the most structurally differentiated brown algae (such as *Fucus*), the tissues within the stipe are divided into three distinct layers or regions. These regions include a central pith, a surrounding cortex, and an outer epidermis, each of which has an analog in the stem of a vascular plant. In some brown algae, the pith region includes a core of elongated cells that resemble the phloem of vascular plants both in structure and function. In others (such as *Nereocystis*), the center of the stipe is hollow and filled with gas that serves to keep that part of the alga buoyant. The stipe may be relatively flexible and elastic in species like *Macrocystis pyrifera* that grow in strong currents, or may be more rigid in species like *Postelsia palmaeformis* that are exposed to the atmosphere at low tide.

Many algae have a flattened portion that may resemble a leaf, and this is termed a *blade*, *lamina*, or *frond*. The name *blade* is most often applied to a single undivided structure, while *frond* may be applied to all or most of an algal body that is flattened, but this distinction is not universally applied. The name *lamina* refers to that portion of a structurally differentiated alga that is flattened. It may be a single or a divided structure, and may be spread over a substantial portion of the alga. In rockweeds, for example, the lamina is a broad wing of tissue that runs continuously along both sides of a branched *midrib*. The midrib and lamina together constitute almost all of a rockweed, so that the lamina is spread throughout the alga rather than existing as a localized portion of it.

Fig.2. Species like *Fucus vesiculosus* produce numerous gas-filled pneumatocysts (air bladders) to increase buoyancy.

In some brown algae, there is a single lamina or blade, while in others there may be many separate blades. Even in those species that initially produce a single blade, the structure may tear with rough currents or as part of maturation to form additional blades. These blades may be attached directly to the stipe, to a holdfast with no stipe present, or there may be an air bladder between the stipe and blade. The surface of the lamina or blade may be smooth or wrinkled; its tissues may be thin and flexible or thick and leathery. In species like *Egregia menziesii*, this characteristic may change depending upon the turbulence of the waters in which it grows. In other species, the surface of the blade is coated with slime to discourage the attachment of epiphytes or to deter herbivores. Blades are also often the parts of the alga that bear the reproductive structures.

Gas-filled floats called *pneumatocysts* provide buoyancy in many kelps and members of the Fucales. These bladder-like structures occur in or near the *lamina*, so that it is held nearer the water surface and thus receives more light for photosynthesis. Pneumatocysts are most often spherical or ellipsoidal, but can vary in shape among different species. Species such as *Nereocystis luetkeana* and *Pelagophycus porra* bear a single large pneumatocyst between the top of the stipe and the base of the blades. In contrast, the giant kelp *Macrocystis pyrifera* bears many blades along its stipe, with a pneumatocyst at the base of each blade where it attaches to the main stipe. Species of *Sargassum* also bear many blades and pneumatocysts, but both kinds of structures are attached separately to the stipe by short stalks. In species of *Fucus*, the pneumatocysts develop within the lamina itself, either as discrete spherical bladders or as elongated gas-filled regions that take the outline of the lamina in which they develop.

Life cycle

Most brown algae, with the exception of the Fucales, perform sexual reproduction through sporic meiosis. Between generations, the algae go through separate sporophyte (diploid) and gametophyte (haploid) phases. The sporophyte stage is often the more visible of the two, though some species of brown algae have similar diploid and haploid phases. Free floating forms of brown algae often do not undergo sexual reproduction until they attach themselves to substrate. The haploid generation consists of male and female gametophytes. The fertilization of egg cells varies between species of brown algae, and may be isogamous, oogamous, or anisogamous. Fertilization may take place in the water with eggs and motile sperm, or within the oogonium itself.

Certain species of brown algae can also perform asexual reproduction through the production of motile diploid zoospores. These zoospores form in plurilocular sporangium, and can mature into the sporophyte phase immediately.

In a representative species *Laminaria*, there is a conspicuous diploid generation and smaller haploid generations. Meiosis takes place within several unilocular sporangium along the algae's blade, each one forming either haploid male or female zoospores. The spores are then released from the sporangia and grow to form male and female gametophytes. The female gametophyte produces an egg in the oogonium, and the male gametophyte releases motile sperm that fertilize the egg. The fertilized zygote then grows into the mature diploid sporophyte.

In the order Fucales, sexual reproduction is oogamous, and the mature diploid is the only form for each generation. Gametes are formed in specialized conceptacles that occur scattered on both surfaces of the receptacle, the outer portion of the blades of the parent plant. Egg cells and motile sperm are released from separate sacs within the conceptacles of the parent algae, combining in the water to complete fertilization. The fertilized zygote settles onto a surface and then differentiates into a leafy thallus and a finger-like holdfast. Light regulates differentiation of the zygote into blade and holdfast.

The outline of the classification is:

Class. Phaeophyceae

Order.

1. **Ectocarpales** Examples: *Ectocarpus, Punctaria, Iyengaria, Holothrix*
2. **Tilopteridales** Example: *Tilopteris*
3. **Cutleriales** Example: *Cutleria*
4. **Sporochnales** Example: *Sporochnus*
5. **Desmarestiales** Example: *Desmarestia*
6. **Laminariales** Examples: *Alaria, Chorda, Macrocystis, Laminaria*
7. **Sphacelariales** Examples: *Sphacelaria, Halopteris*
8. **Dicotyotales** Examples: *Dicotyota, Hormosira, Ascoseria, Cystophora*
9. **Fucales** Examples: *Sargassum, Fucus*

****

20.

Ectocarpales: *Ectocarpus*

The order Ectocarpales includes 60 genera of brown algae which are found in the cold sea water of the temperate and polar regions. Most of the members of this order are heterotrichous with a prostrate creeping disc, which functions as holdfast and an erect monosiphonous filamentous foliose system. These algae show trichothallic growth with intercalary divisions confined to certain areas in the filament. Asexual reproduction is by zoospores which are produced in uni- or plurilocular sporangia. Sexual reproduction is isogamous but in some species it is oogamous. Reproductive structures are borne laterally or in uniseriate rows.

Fritsch (1935) divided the order into 12 families- Ectocarpaceae, Myrionemataceae, Elachistaceae, Leathesiaceae, Megbloeaceae, Aenotrichaceae, Spermatochnaceae, Splachnidiaceae, Punctariaceae, Asperococcaceae, Encoeliaceae and Dictyosiphonaceae. But most algologists restrict Ectocarpales to a single family Ectocarpaceae characterized by simple, heterotrichous members exhibiting isogamous sexual reproduction and 1Somorphic alternation of generations.

Family. Ectocarpaceae

The members of the family are characterised by simple heterotrichous thallus organisation, isogamous type of sexual reproduction with isomorphic alternation of generation. The life-history of Ectocarpus is described here as a representative of the family.

Ectocarpus

Systematic Position:
Class: Phaeophyceae
Order: Ectocarpales
Family: Ectocarpaceae
Genus: *Ectocarpus*

Occurrence

Ectocarpus is word-wide in distribution particulary in colder seas and Polar Regions. *Ectocarpus* is very common on sea shore of Atlantic Ocean. *Ectocarpus* is found attached on sea rocks. Some species of Ectocarpus are epiphytic e.g., **E. coniferus**, and **E. breviarticulatus** grow on larger algae like *Fucus* and *Laminaria*. *E. dermonematus* is endophytic species. *E. fasciculatus* is epizoic species growing on fins of faster. In India *Ectocaupus* is represented by about 100 species.

Structure of thallus

Genetically the thalli may be haploid or diploid. But both the types are morphologically alike. The thallus consists of profusely branched uniseriate filaments. It shows heterotrichous habit. There are two systems of filaments. These are prostrate and projecting system. The filaments of the projecting system arise from the filaments of prostrate system

a) Prostate system:

The prostrate system consists of creeping, leptate, irregularly branched filaments. These filaments are attached to the substratum with the help of rhizoids. This system penetrates the host tissues in epiphytic conditions. Prostrate system is poorly developed in free floating species.

Projecting system: The projecting system arises from the prostrate system. It consists of well branched filaments. Each °ranch arises beneath the septa. The main axis and the branches of the projecting system are **uniseriate.** In this case, rens are joined end to end in a single series. The branches terminate into an acute point to form a **hair.** In some species the older portions of main axis are **ensheathed** (corticated). This sheath is formed of a layer of descending rhizoidal branches.

Cell Structure

The cells are small. They are cylindrical or rectangular and uninucleate.

1. The cell wall is thick It is composed of three layers composed of pecticcellulose. **Algin and fucoidan** are also present in the cell wall. These are characteristic gelatinous substances of tne walls of brown algae.

2. The chromatophores may ribbon-like with irregular outline or disc shaped. The dominant of *Ectocarpus* is **fncoxanthin**. It gives this algae golden brown colour. The other photosynthetic iigments are chlorophyll-a, -c, beta.carotene and other xanthophylls.

3. "Pyrenoid-like bodies-are associated with the chromatophores.

4. **All** other eukaryotic organelle are present.**Growth**The growth of filaments in projecting system may be **intercalary** or

a) **Intercalary:** In some species, an intercalary meristem ir present it the base of the hair. It is called **trichothallic meristem.** It increases the length of the terminal hair and vegetative cell of the branch. This growth is called trichothallic growth.

b) The growth in the prostrate system is **apical**

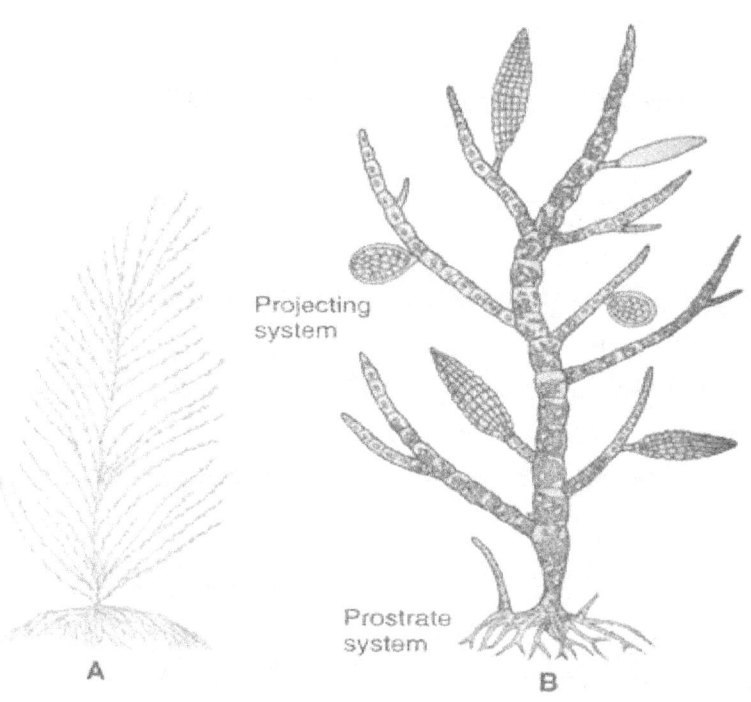

Fig.1. Thallus structure

Reproduction

Ectocarpus reproduces by both asexual and sexual methods.

Asexual reproduction

The asexual reproduction takes place by the formation of biflagellate **zospores.** These zoospores may be haploid produced in one-celled **unilocular sporangia.** Or they may be diploid formed in many celled **plurilocular sporangia.** Both kinds of sporangia are present on the same diploid **sporophyte** plant. The sporangia are borne terminally and singly on lateral branches.

(a) **Unilocular Sporangia**

A unilocular sporangium develops from a terminal cell of a short lateral branch.

1. Ile **sporangial initial** enlarges in size. It becomes globose or ellipsoidal. The number of chromatophores also increases in it.

2. The nucleus of the sporangium divides meiotically to produce four haploid nuclei. These nuclei undergo repeated mitotic divisions to produce 32-64 daughter nuclei. The cytoplasm of the sporangium divides.

3. A small amount of cytoplasm surrounds a nucleus and athromatophore to produce daughter protoplasts. Each daughter protoplast metamorphoses into a **meiozoospore** (produced by meiosis). Meiozoospore is pyriforrn and biflagellate. The flagella are laterally inserted and are of unequal size. The larger one directed forward and the smaller one is directed backward.

4. An apical pore is formed in the gelatinous mass of sporangia. The meiozoospores come out of this pore.

5. These are separated from each other after few moments. They swim freely in all directions. A new sporangium may be produced within the old sporangial wall after the liberation of zoospores.

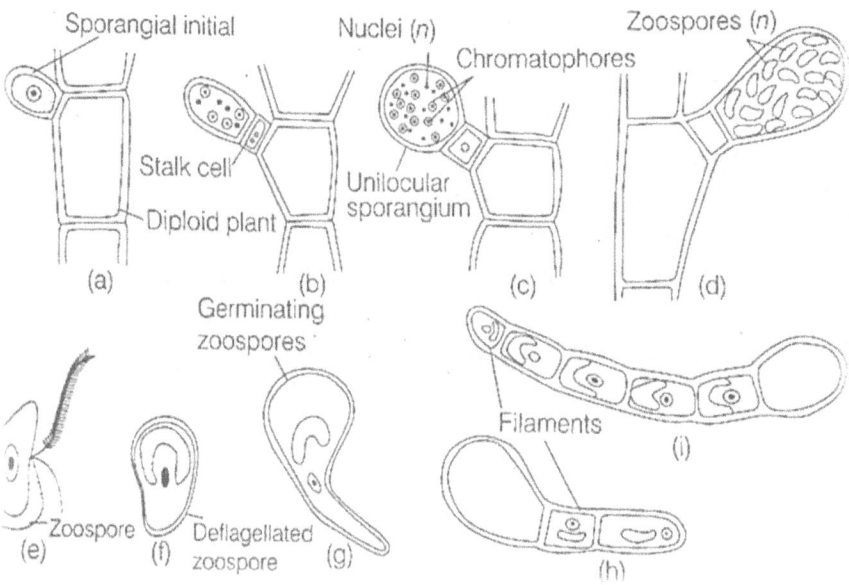

Fig. 3 A-L Ectocarpus: Asexual reproduction: A-D. Successive stages in the development of unilocular sporangium. **E.** A zoospore, **F-I.** Stages in the germination of zoospore.

(b) **Plurilocular Sporangia**

The plurilocular sporangia are stalked or sessile. These are elongated, cone-like multicellular structures. These also develop from a terminal cell of a short lateral branch.

    1. The sporangial initial enlarges in size. It undergoes repeated transverse mitotic divisions. It produces a vertical row of 6-12.
    2. These cells then divide by vertical and transverse divisions repeatedly. They form a cone-like structure. This cone consists of hundreds of small cubical cells. These cells are arranged in 20-40 transverse **tiers.** Each cell represents a sporangium.

3. Protoplast of each cell metamorphoses into a single **zoospore** (produced by mitosis). The zoospore is pear-shaped, diploid and biflagellate. The flagella are of unequal size and are laterally inserted.

4. The zoospores are liberated through a terminal or a lateral pore. This pore is formed in the wall of the sporangium.

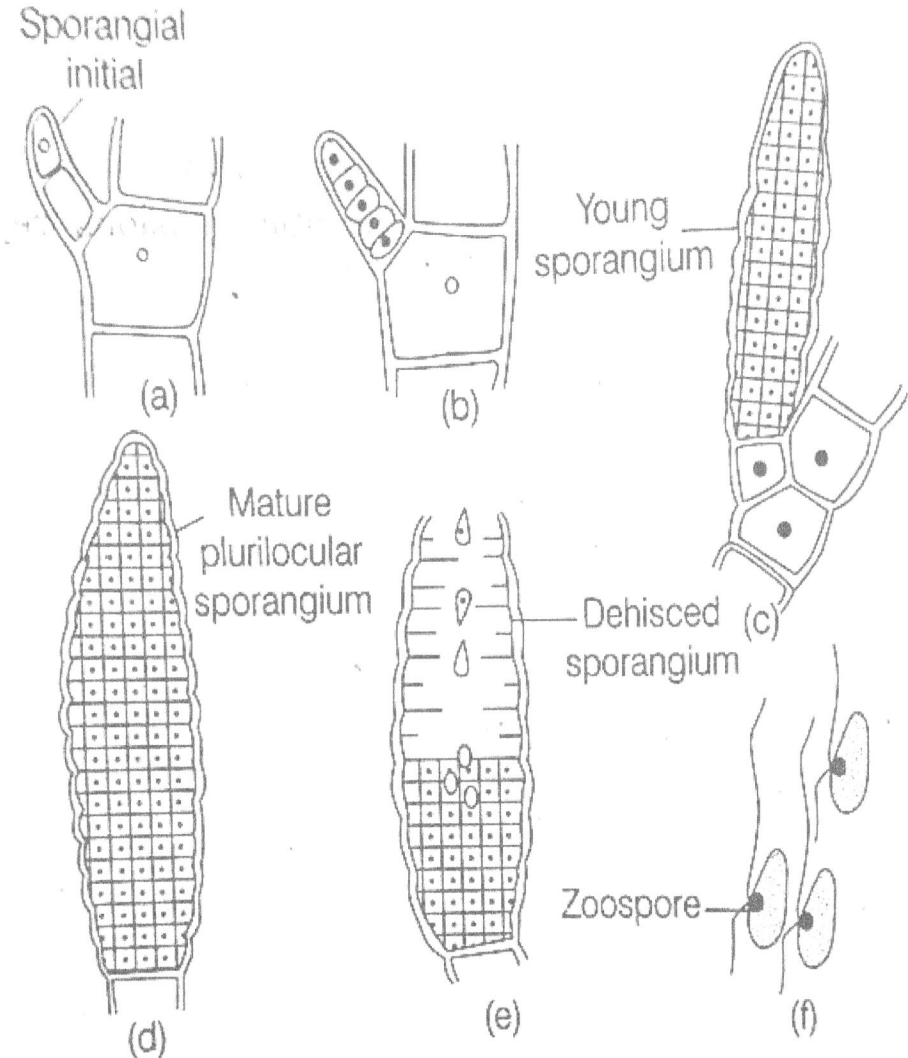

Fig. 4 A-F. Ectocarpus: Asexual reproduction; A-C. Development of plurilocular sporangium, D. Mature plurilocular sporangium, E. Liberation of zoospores from sporangium, F. Zoospores.

Germination of Zoosporesa: The zoospores formed in unilocular sporangia (zoospores) swarm for scientism. They then come to rest on some solid object. They withdraw their flagella and secrete a membrane around then. They germinate and form a small **germ tube.** This tube is separated from the zoospore cell through a septum. This germ tube divides and redivides. It forms the prostrate system of plant. The system arises from the filaments of the prostrate system. The new plant form is haploid. Therefore, it is gametophyte. The zoospores develop into a gametophytic plant. Therefore, these spores are also called as **gonozoospores**.

Germination of mitozoospores: The zoospores produced in plurilocular sporangia are zoospores. They develop in the same manner as the unilocular zoospores. But they are diploid. Therefore, they develop into a diploid **sporophyte.** Therefore, the zoospores are also called as **neutral spores.** They are reduplication of sporophyte generation.

Sexual Reproduction:

The sexual reproduction is both isogamous and anisogamous type. Oogamy is absent. Anisogamy is very common. Anisogamy may be of two types: morphological anisogamy (*E. secundus*) and physiological anisogamy (*E. siliculosus*). The gametes are produced inside the plurilocular gametangia, developed on haploid plants.

Plurilocular Gametangia:

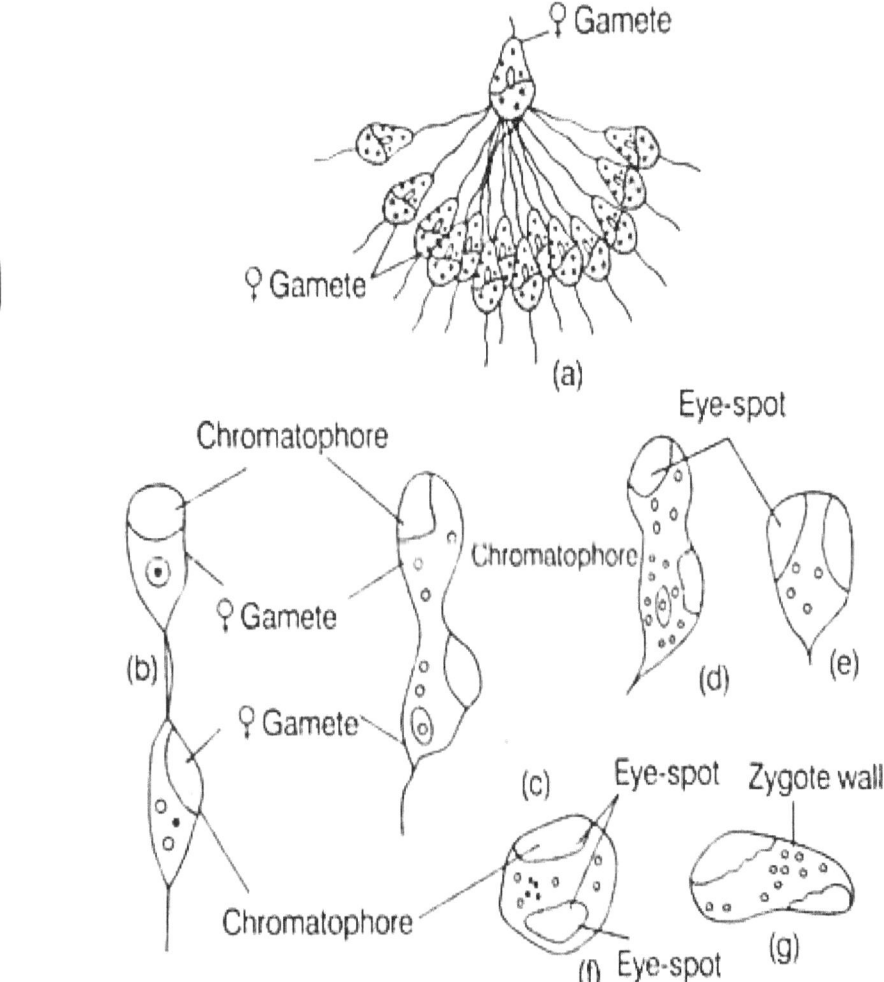

Fig. 5. Physiological anisogamy in *Ectocarpus siliculosus*. (a) Clump formation (b-e) Formation, stages in gametic union, (f). Naked zygote with 2 chloroplasts and 2 eye-spots: (g) Zygote at later stage (After Berthold)

They are large, elongated, sessile or short stalked, multicellular structures. Morphologically, both plurilocular gametangia and plurilocular sporangia are alike. The plurilocular gametangia produce haploid gametes; on the other hand plurilocular sporangia produce diploid zoospores.

Though both are morphologically more or less alike, the gametes are slightly smaller in size than the zoospores. The development of plurilocular gametangia is alike with the development of plurilocular sporangia. The gametes are liberated from the gametangia following the same procedure as that of zoospore liberation from the plurilocular sporangia.

Fertilisation:

Majority of the species show physiological anisogamy (Fritsch, 1945), but morphological anisogamy is observed in *E. secundus*. In physiological anisogamy both the uniting gametes are morphologically similar but in morphological anisogamy female gamete is larger than the male gametes.

During fertilisation, many male gametes encircle the female gamete and get entangled by the anterior large flagellum. This stage is called clump formation. Out of many, only one male gamete fuses with the female gamete and the remaining gametes go astray and gradually get destroyed.

The uniting gametes then form zygote through plasmogamy and karyogamy.

Germination of Zygote:

The zygote undergoes germination without any reduction division and rest. On germination it develops into a sporophytic (2n) plant. The sporophytic plant again develops unilocular and plurilocular sporangia.

Common Indian species:

Ectocarpus arabicus, E. filife, E. enhali, E. coniger, E. zeylanicus, E. rhodochortonoides etc.

Life Cycle of Ectocarpus:

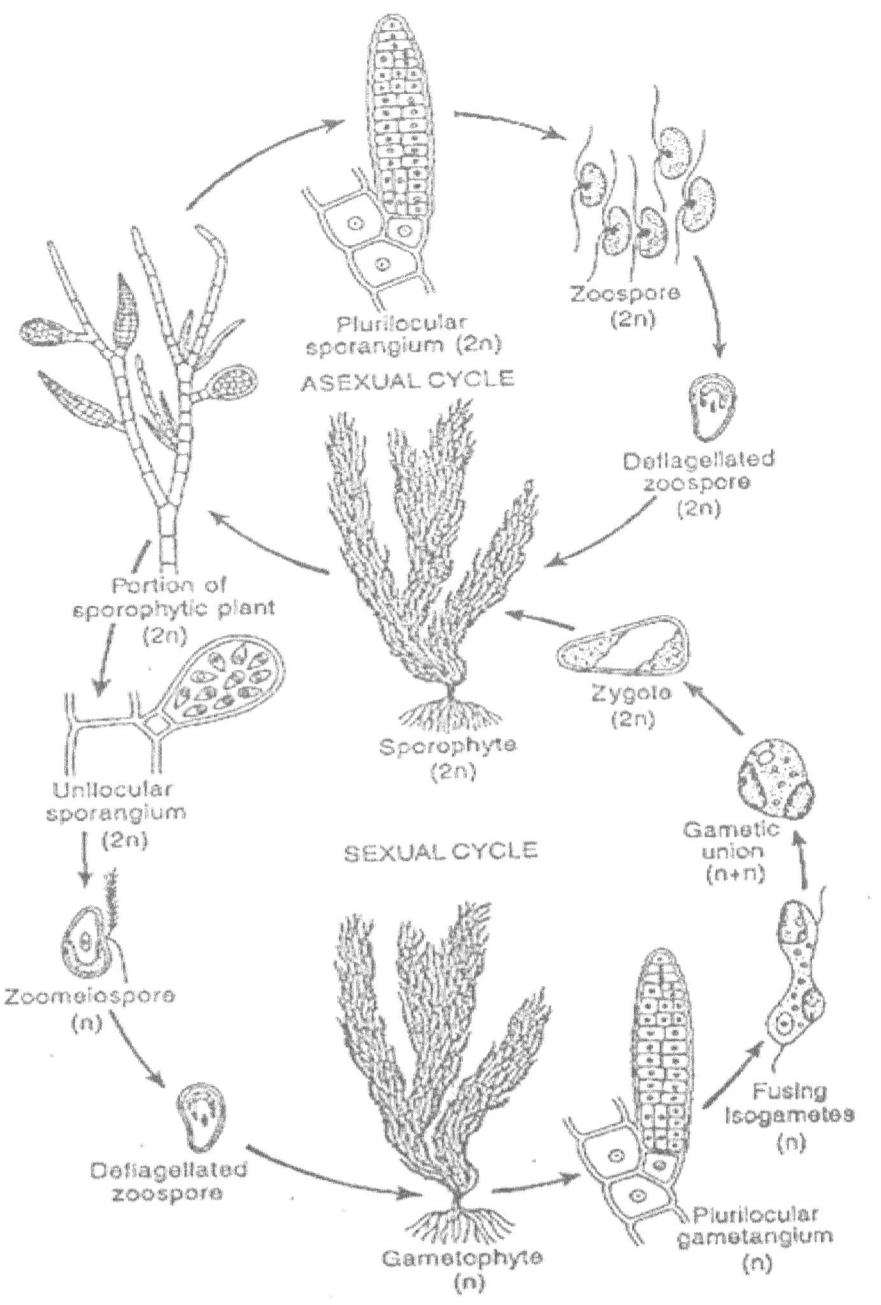

Fig.6. Life cycle of *Ectocarpus sp.*

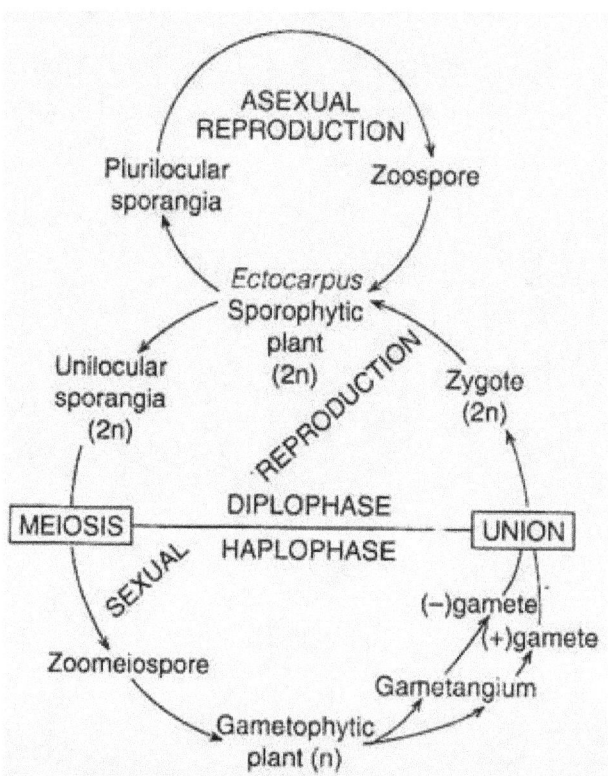

Fig.7. Graphic Life cycle of *Ectocarpus sp.*

21.

Fucales: *Fucus, Sargassum*

There are 40 genera and 3,000 species in the order Fucales. They are cosmopolitan in distribution. Species occurring in the arctic and north temperate seas differ considerably from those of the antarctic and south temperate waters. They are mostly found attached to the substratum by discoid holdfasts. The plant body is complex both morphologically and anatomically. These algae reproduce by vegetative and sexual methods; asexual spores are absent. Sexual reproduction is oogamous and the sex organs are borne in special cavities, known as conceptacles. The special branches bearing conceptacles are called receptacles. Conceptacles bear microsporangia or macrosporangia which function as antheridia and oogonia respectively. Both micro-and macrosporangia are unilocular. The members of Fucales do not show distinct alternation of generation. The sporophytic thallus is diploid and the haploid stage is confined only to gametes.

The order Fucales has been divided into seven families:

(1) Fucaceae

(2) Himanthaliaceae

(3) Cystoseiraceae

(4) Sargassaceae

(5) Hormo-siraceae

(6) Durvilleaceae

(7) Ascoseiraceae.

Family. Fucaceae

The plant body is usually flattened and dichotomously branched; axes subterate to alate with mid-rib, but not foliar.

Fucus

Systematic Position

Class – Phaeophyceae
Order – Fucales
Family – Fucaceae
Genus – *Fucus*

Occurrence

The genus *Fucus* comprises more than 100 species and is an extremely marine alga, widely distributed along the sea coasts of temperate and arctic regions. They are attached on the rocks under cold water between the high and low tide marks by their flat, discoid holdfasts. They are commonly called as "*rock weeds*".

Thallus Structure

The plant body i.e., thallus is more or less 30 cms in height. The vegetative body of the *Fucus* plant i.e., sporophyte is complex in organization. It consists of flat, dichotomously branched ribbon-like leathery and dark brown coloured thallus.

The thallus is composed of three parts –

(a) Holdfast – It is a basal flat and broad disc shaped structure which anchors the thallus to the substratum under water.

(b) Stipe – It is relatively short and stem-like portion. It is the lowermost part of the thallus.

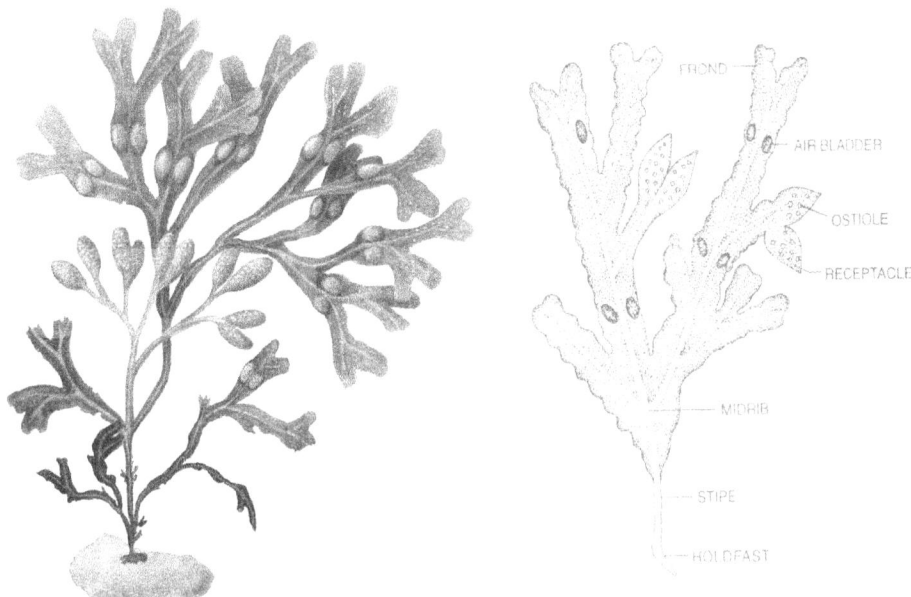

Fig. 1. *Fucus*: Thallus Structure.

(c) Frond or Blade – It is the expanded, ribbon-like dichotomously branched, leathery, parenchymatous and somewhat foliose structure. Usually, the thallus is dark brown and slimy to touch. In many species of *Fucus*, the stipe is continued to form a mid-rib in the frond, but the mid-rib never reaches the apex of the frond. The margin of the frond may be entire, serrated or smooth.

Oranization of the Frond:

The frond is a foliose-like and consists of different structures –

(a) Air Bladders – The thallus of some species is provided with numerous hollow, bladder-like, air-filled expansions in pairs here and there along the dichotomy. These are called *air bladders or air vesicles or pneumatocysts*. These air bladders give buoyancy to the submerged thallus.

(b) Receptacles – The swollen tips of the mature fertile branches of the thallus are called *receptacles*. These receptacles lack mid-rib and are covered with small scattered pimple-like outgrowths with minute openings called *ostiole*. These are flask shaped structure and are called conceptacles. Conceptacles contain *antheridia* and *oogonia*.

Internal Structure of the Frond:

The transverse section of the frond shows the following structures –

(a) Epidermis – It is the outermost single layered peripheral structure, consisting of thin walled, small, palisade-like cells containing numerous chromatophores. Its function is photosynthetic.

(b) Cortex – It is many layered and is present just beneath the epidermis. The cortex is composed of thin-walled, parenchymatous cells. Cortex is often differentiated into outer cortex and inner cortex. Its function is both photosynthetic and storage of food.

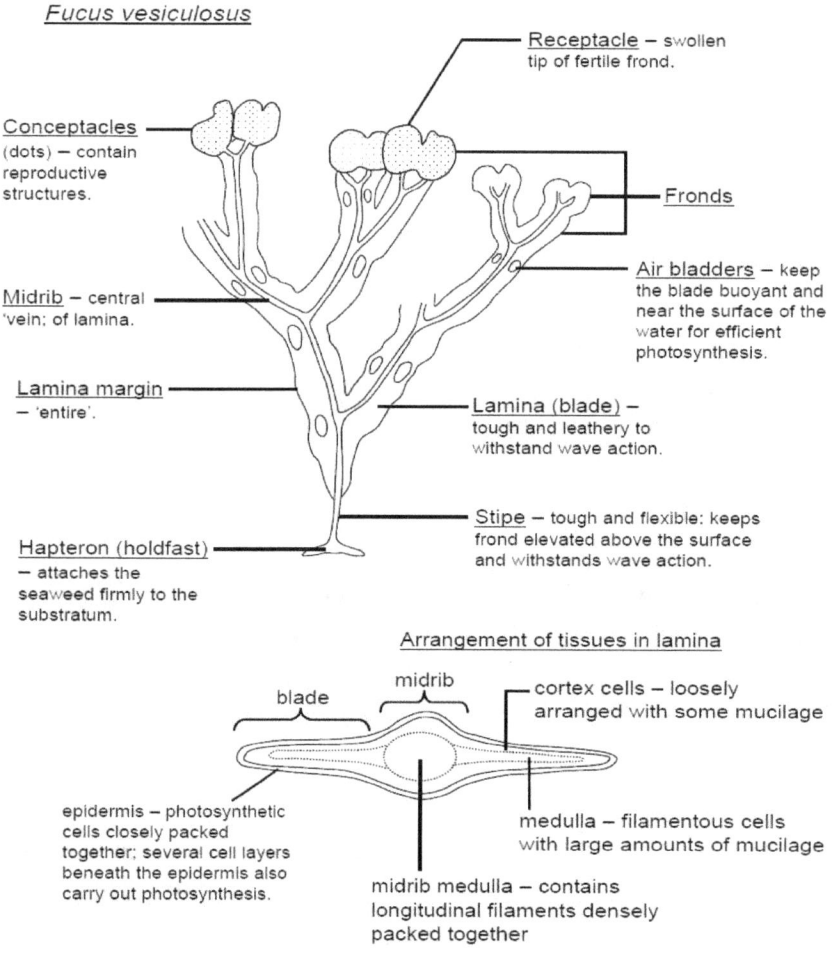

Fig.2.

(c) Pith or Medulla – It is situated in the centre and consists of several-layered hypha-like elongated, thin walled colourless, inter-wooven cells. The walls of medulla cells have sieve plates.

Reproduction:

Fucus reproduces both by vegetative and sexual methods of reproduction.

1. Vegetative Reproduction:

Vegetative reproduction takes place by fragmentation of the vegetative body into many parts, each of which grows into a new thallus. Sometimes adventitious branch develops on the stipe, which on separation forms a new plant.

2. Sexual Reproduction:

Sexual reproduction is of *oogamous* type. The species of *Fucus* are *homothallic* (*F. spiralis*) or *heterothallic* (*F. serratus*). The sex organs i.e., *antheridia* and *oogonia* in both homothallic and heterothallic species are produced in a special type of flask-shaped chamber called conceptacles.

(a) Antheridia – The male conceptacle bears numerous antheridia and paraphyses. Antheridia are small, club-shaped, unicellular and provided with a small stalk-like structure. Each antheridium has a delicate wall which is composed of 2 layers – outer firm layer called *exochite* and an inner gelatinous layer called *endochite*. It contains *a single diploid nucleus (2n)*. The diploid nucleus of each antheridium divides first by meiotic division to form *haploid nuclei (n)*, which again divides and redivides forming *64 or more haploid nuclei*. Nuclear division is followed by cleavage of the protoplasts. Each uninucleate protoplast directly metamorphoses into a single pear-shaped, biflagellate *antherozoid* or *sperm*. At maturity the tip of the antheridium breaks which allows the mass of antherozoids to escape within the conceptacle.

(b) Oogonia – The female conceptacle bears numerous oogonia and paraphyses. Oogonia are oval or somewhat ellipsoidal in shape and are provided with one-celled short stalk. It contains *a single diploid nucleus (2n)*. The oogonium divides first by meiotic division to form *four haploid nuclei (n)*, followed by mitotic division to form *8 haploid nuclei*. At this stage the cleavage of the oogonial protoplast takes place and 8-

uninucleate bits of cytoplasm are formed, which metamorphosed into *eight eggs (n)*. The eggs are surrounded by three layered oogonial wall – *exochite, mesochite* and *endochite*. Eggs come out within the conceptacle by rupturing of the exochite.

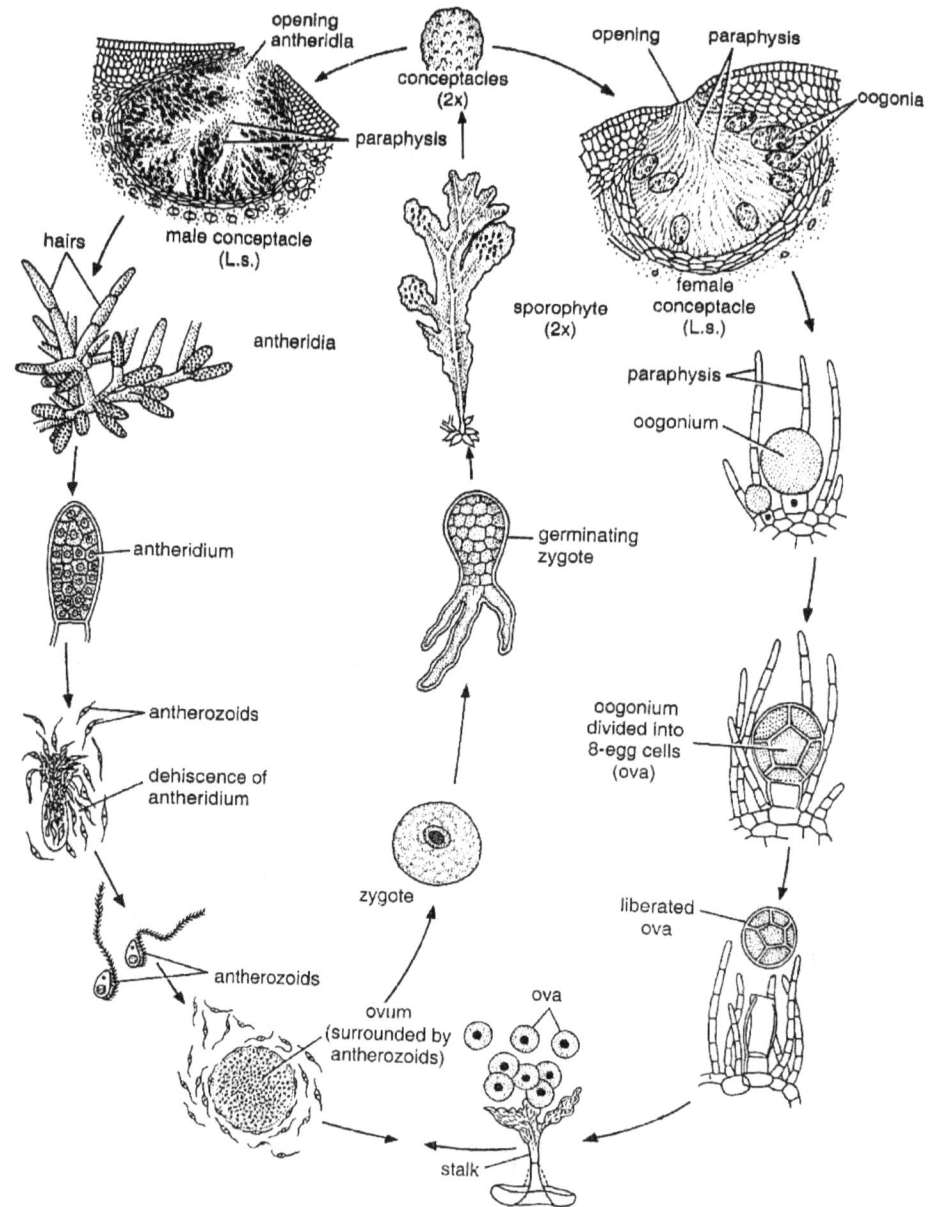

Fig.3.*Fucus*: Life Cycle

(c) Fertilization: - At the time of fertilization, both sperms and eggs come out from the conceptacle through the ostiole and remain free in the surrounding water. Numerous free swimming antherozoids gather around the egg, but only one sperm penetrates the egg and fuses with it. As a result of fusion, *a diploid zygote (2n)* is formed. After fertilization, the zygote takes a rest of few hours and starts germinating, producing diploid sporophytic *Fucus* plant.

Family. Sargassaceae

The plants possess flat costate branches or transitional stages to cylindrical branches bearing numerous macroscopic spine like projections, turbinate foliar organs, or costate leaves, usually with eryptostomata and buoyant air bladders. Conceptacles are in ordinary branches or in special receptacular branchlet systems. Sexual reproduction is oogamous. Each oogonium produces a single egg.

Sargassum

Systematic Position

Class: Phaeophyceae
Order: Fucuales
Family: Sargassaceae
Genus: *Sargassum*

Occurrence

This genus is widely distributed specially in warmer regions mainly in tropical and subtropical seas of the Southern hemisphere. The plants form huge floating masses in the Atlantic Ocean of the African continent that is known as Sargasso sea. The algae grow abundantly both in East and West coasts of India, Australia and Ceylon. This algae grows attached to the rocks in little bushes in the intertidal zone or in the shallow puddles of the zone.

Thallus Structure

The thallus of *Sargassum* is differentiated into holdfast and the main axis. This holdfast is discoid or stolon-like structure and helps in attachment of thallus to the substratum. *Sargassum* plants are highly differentiated algae in the organisation of the thallus. The main axis or stipe or 'stem is erect, elongated, cylindrical or flattened. In some species, e.g. *S. fulipendula* the thallus can be more than a metre in length. The main axis and primary laterals bear flat leaf-like branches known as secondary laterals or leaves'. The leaf-like laterals are flat and simple with blade, veins and petiole-like structure. The branching is always monopodial. The leaf is a short sterile lateral organ provided with midrib. The margins of the leaves are entire, serrate or dentate. On the surface and margins of the leaves are small pores known as ostioles. These pores are open in small flask-shaped sterile cavities called cryptostomata or sterile conceptacles. These cavities bear hairs and paraphysis inside it. The laterals specially those of the lower branch lets modify into air bladders. These are globular or spherical, air-filled structures. They help in floating of plants by increasing buoyancy. According to some algologists, the air bladders also help in respiration. In some species, the air bladders terminate into leaf-like structures. Another modification of these laterals is in the form of highly branched or swollen structures bearing reproductive bodies called receptacles. The receptacles bear reproductive structure in special flask-shaped cavities called as conceptacles.

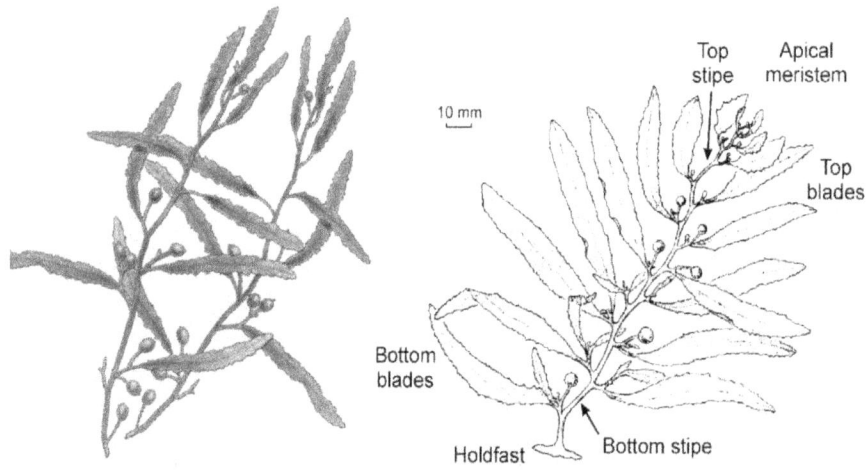

Fig.4. Sargassum: Thallus Structure

Growth

The growth in *Sargassum* is apical. Increase in diameter of the axis is initiated by the activity of a lateral meristem zone or meristoderm.

Internal Structures

Internal structure of Main axis (Fig.5.)

The main axis is circular in outline and internally it is differentiated into three regions.

1. Meristoderm is single cell thick outermost layer and is made of compactly arranged columnar cells. The meristoderm functions as protective layer and as assimilatory layer due to the presence of chromatophores in cells. The meristoderm is covered with a thin layer of mucilaginous cuticle.

2. Cortex is the largest part present between meristoderm and medulla. It is made up of loosely arranged cells elongated parenchymatous cells with intercellular spaces. The cortex cells contain reserve food material and form the storage region of the main

3. Medulla It is made up of thick walled narrow and elongated cells. Sometimes these cells have scalariform thickenings. The function of medulla is to transportwater and metabolites to different parts of the thallus.

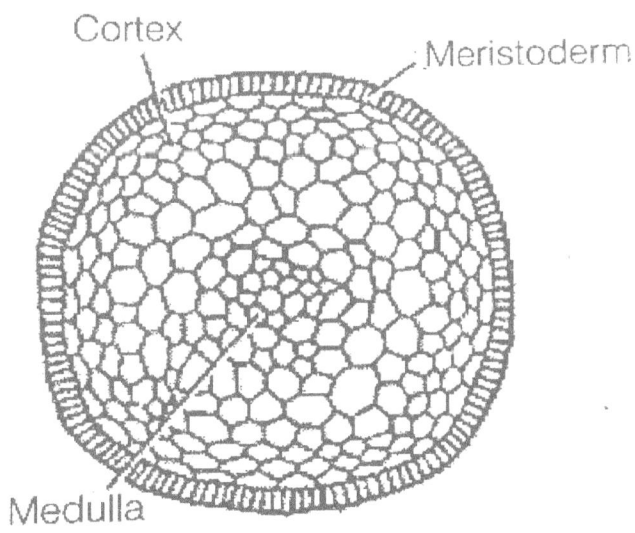

Fig.5. *Sargassum*: T.S of main axis

Internal structure of the leaf.

Just like the main axis, the leaf is also differentiated into **meristoderm, cortex** and **medulla** (Fig. 6). The meristoderm consists of small radially elongated, meristematic cells. These cells have plenty of reserve food and chromatophores. The meristoderm is followed by **cortex** which consists of thin walled cells, rich in reserve food material. The cortex is the thickest in the midrib region and gradually narrows towards the blades. The central part of the leaf is made up of thick walled cells which make up **medulla**. In structure and function it is similar to the medulla of the main axis. On both the surfaces of the leaf are present many sterile conceptacles, known as cryptoblasts or eryptostomata. The cryptoblast is a flask-shaped structure and it opens on the surface of the 'leaf by a small pore, known as ostiole. The wall of the cryptoblast is made up of sterile cells which have chromatophores. Many unbranched multicellular filaments arise from the wall of the conceptacles and protrude through the ostiole. These filaments are called paraphysis.

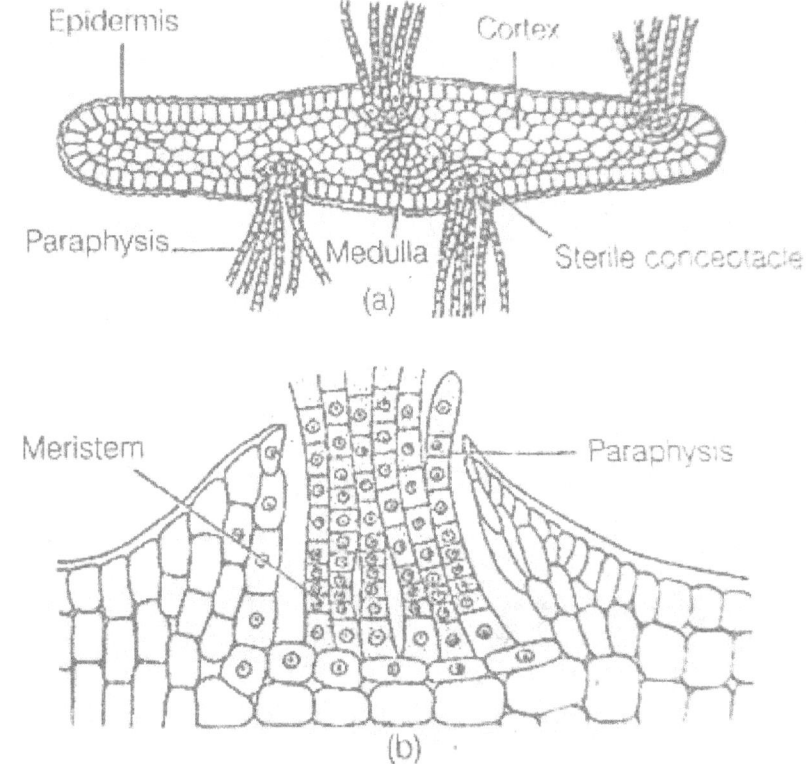

Fig.6. *Sargassum*: (a) T.S of leaf, (b) T.S through conceptacle

Internal structure of Air Bladder

The structure of air bladder is also like main axis and leaf. It is differentiated into **meristoderm** and **cortex** but medulla is absent. The meristoderm is made of radially elongated narrow cells. Inner to meristoderm is 4-8 layered parenchymatous cortex. The central part of the bladder is made of large hollow cavity. The air bladder helps in buoyancy and gaseous exchange.

Reproduction in Sargassum:

The reproduction takes place by vegetative and sexual methods. The asexual reproduction is absent.

(A) Vegetative Reproduction in Sargassum:

Sargassum multiplies profusely by vegetative fragmentation. The thallus breaks into fragments due to mechanical injury or death and decay of older parts. The species like S. hystrix and S. natans growing in Sargasso sea are completely sterile as they do not form any reproductive structures. In these species the fragmentation is the only method of multiplication.

Sexual reproduction

It is oogamous. The male sex organs are calledantheridia and the female are oogonia. The sex organs develop inspecial flask-shaped cavity called conceptacle. These conceptacles are present in specially modified laterals called receptacles. The male and female sex organs develop in separate male and female conceptacles.

In homothallic or Monoecious species, the male conceptacle and female conceptacles are produced on same receptacle, but antheridia and oogonia are not produced in same conceptacles. Sargassum species are mostly monoecious.

In dioecious plants, the male and female conceptacles are produced on separate male and female plants.

Development of antheridium

Any cell of the fertile layer of conceptacle can function as antheridial initial. This cell is dense cytoplasmic and develops a papilla-like outgrowth. It divides by transverse division to,make lower stalk cell and upper antheridial cell. The antheridial cell rounds off to make antheridium. The stalk cell elongates and pushes the antheridium to one side. The growing stalk cell divides again to make basal cell and the antheridial cell. This process is repeated many times and results in formation of many antheridia and a sterile paraphysis.

Antheridium

The antheridia are oval structures with two-layered cell walls. The outer wall is called exochite and the inner is called endochite. At young stage the antheridia are inside conceptacles and on maturity the antheridia are detached from stalk and come out of ostiole. The antheridium has one diploid nucleus which divides first by meiotic division and later by mitotic divisions. This results in formation of 32-64 haploid nuclei. The protoplast of antheridium also divides in equal number of segments. Each protoplast segment with haploid nucleus develops into an antherozoid. The antherozoid is pear-shaped structure with two lateral flagella. The flagella are heterokontic, one being acronematic and the other pantonematic. The antherozoids are liberated in water after gelatinisation of the antheridial wall.

Development of Oogonium (Fig. 8.)

Any cell of the fertile layer of the female conceptacle can function as oogonial initial. The oogonial initial divides by transverse division to make small, lower stalk cell and the large, upper oogonial cell. The stalk cell further does not divide or elongate, so the oogonial cells are almost sessile. The oogonial cell enlarges and makes spherical oogonium. The oogonia wall has three layers, the outer exochite, middle mesochite and the inner endochite. On maturity of the oogonium the exochite ruptures, the mesochite forms the gelatinous stalk and the oogonial nuclei- and protoplast remains surrounded by endochite. The diploid oogonial nucleus undergoes meiotic and mitotic divisions to form 8 nuclei. The seven of these eight nuclei degenerate and only one remains functional. This nucleus with protoplasm forms single ovum or oosphere.

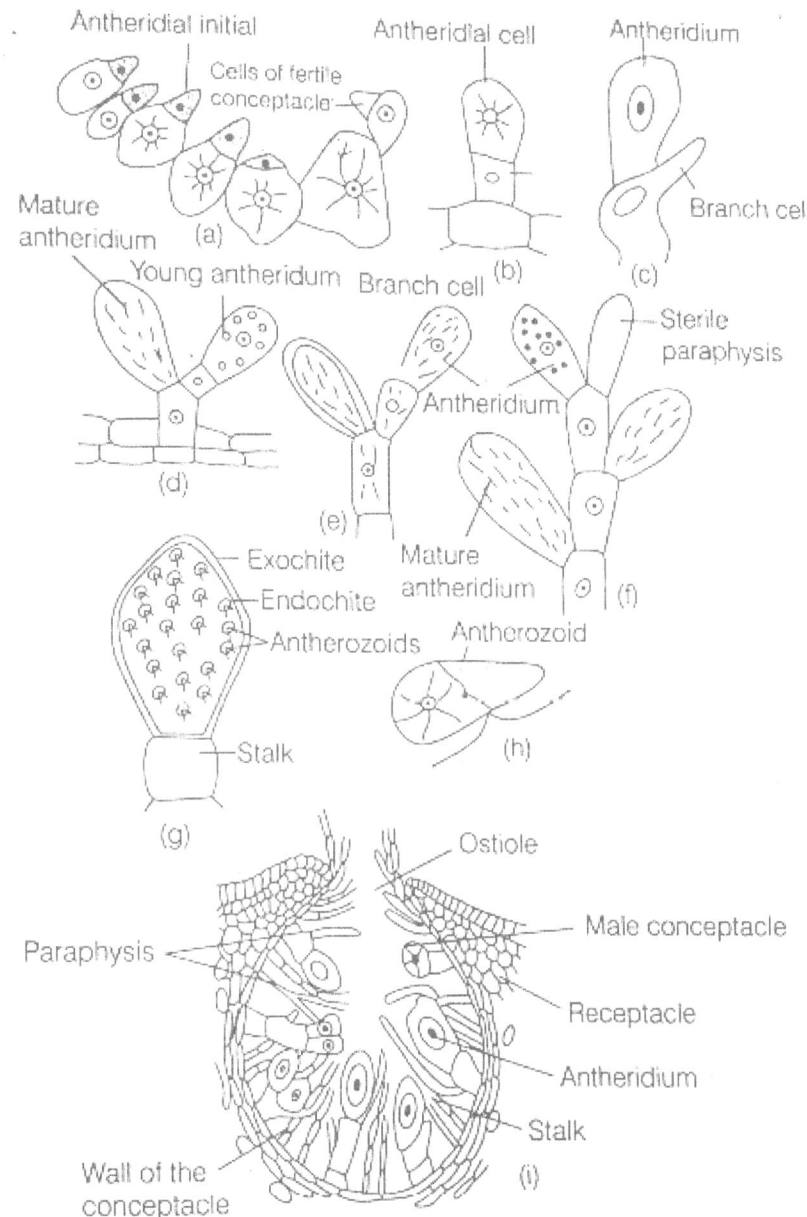

Fig. 7.Sargassum: Development of antheridium; (a-1) Stages in the development of antheridium, (g) An antheridium, (h) A male gamete, (i) VS passing through male conceptacle

The cells of female conceptacle which do not form oogonia develop into long hair like paraphyses.

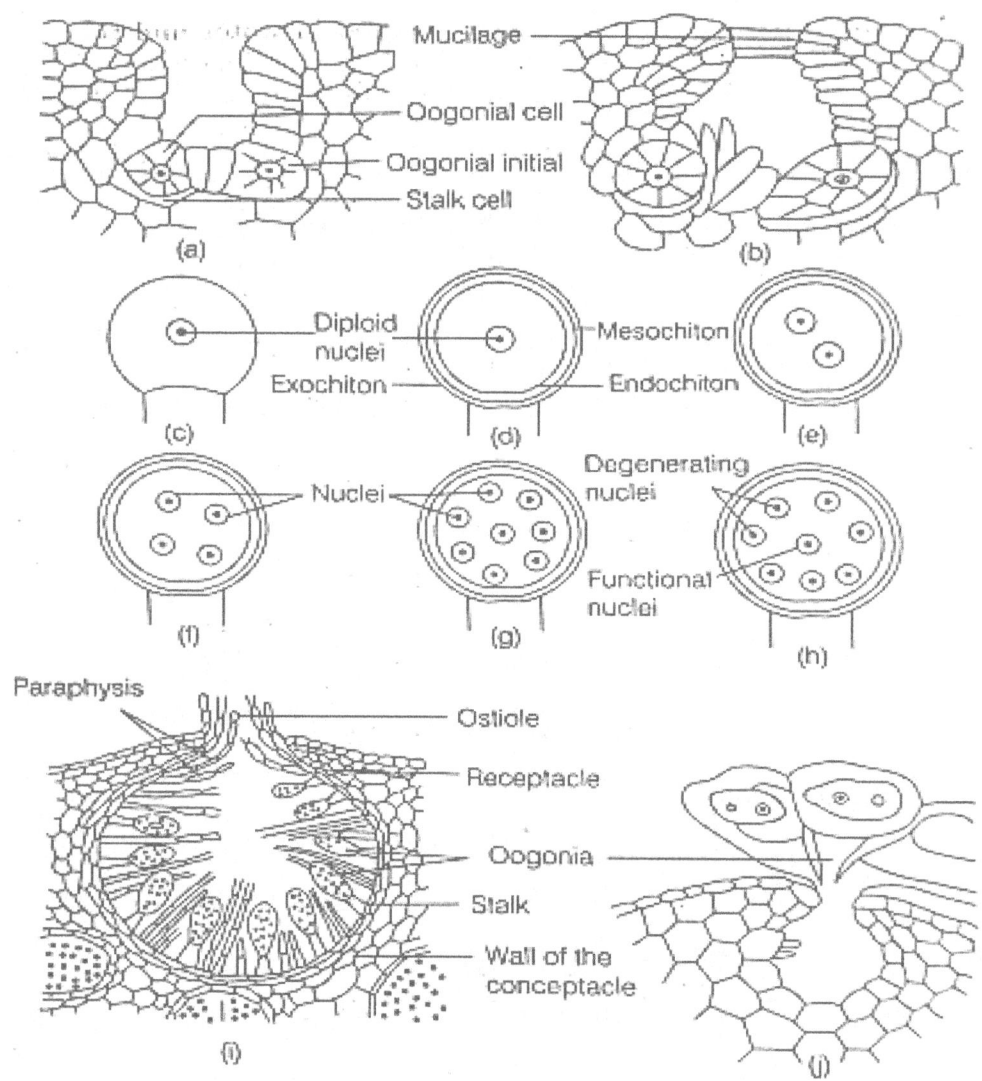

Fig. 8. *Sargassum*: Development of oogonium; (a-h) Stages in the development of oogonium, (i) VS of receptacle through female conceptacle (j) liberated oogonia.

Fertilization: (Fig. 9.)

The antherozoids are released in water and the oogonia remain attached to the conceptacle base by mucilaginous stalk. The oogonia protrude out of the ostiole. A large number of antherozoids surround the oogonium and attach to oogonial wall with the help of anterior flagellum. Only one antherozoid penetrates the oogonial wall. The male and female nuclei fuse to form a diploid zygote.

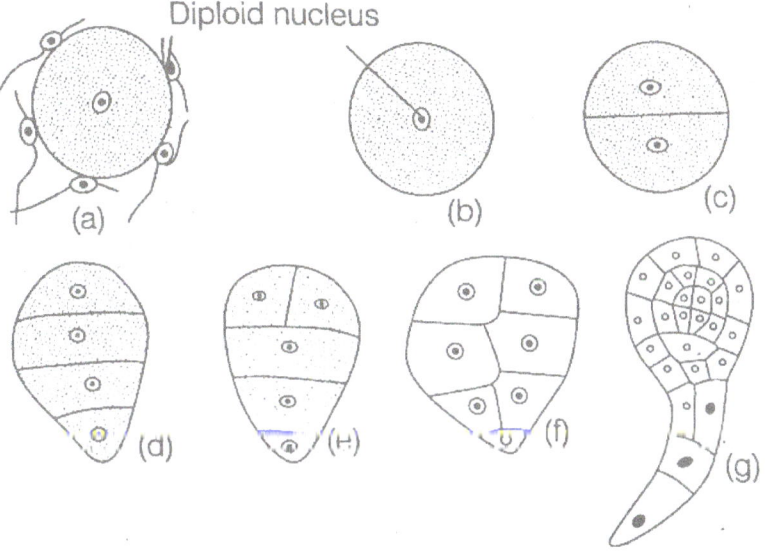

Fig. 9. *Sargassum*: **Fertilization and Germination of Zygote.**

Germination of Zygote: (Fig. 9.)

The zygote germinates immediately after fertilization when the oogonium still remains attached to the wall of conceptacle by a mucilaginous stalk. After some time the zygote is liberated by gelatinization of the oogonial wall. After liberation the zygote gets attached to any substratum in sea water. The zygote first divides by transverse division to make a lower cell and upper cell (Fig. 9 C-F).

The lower cell forms the rhizoids. The upper cell first divides by transverse division and later by anticlinal and periclinal divisions. It results in the differentiation of three layers—the meristoderm, cortex and medulla. The divisions of upper cell result in formation of a diploid, sporophytic Sargassum plant.

Life Cycle of Sargassum:

The life cycle of Sargassum is diplontic type and there is no alternation of generation. The thallus is diploid sporophytic. It forms diploid antheridia and oogonia. The reduction division in antheridia and oogonia forms haploid antherozoid and oognial nuclecus. The gametes only are haploid structure in the life cycle. After fertilization a diploid zygote is formed which divides to make a diploid sporophytic thallus (Fig. 10)

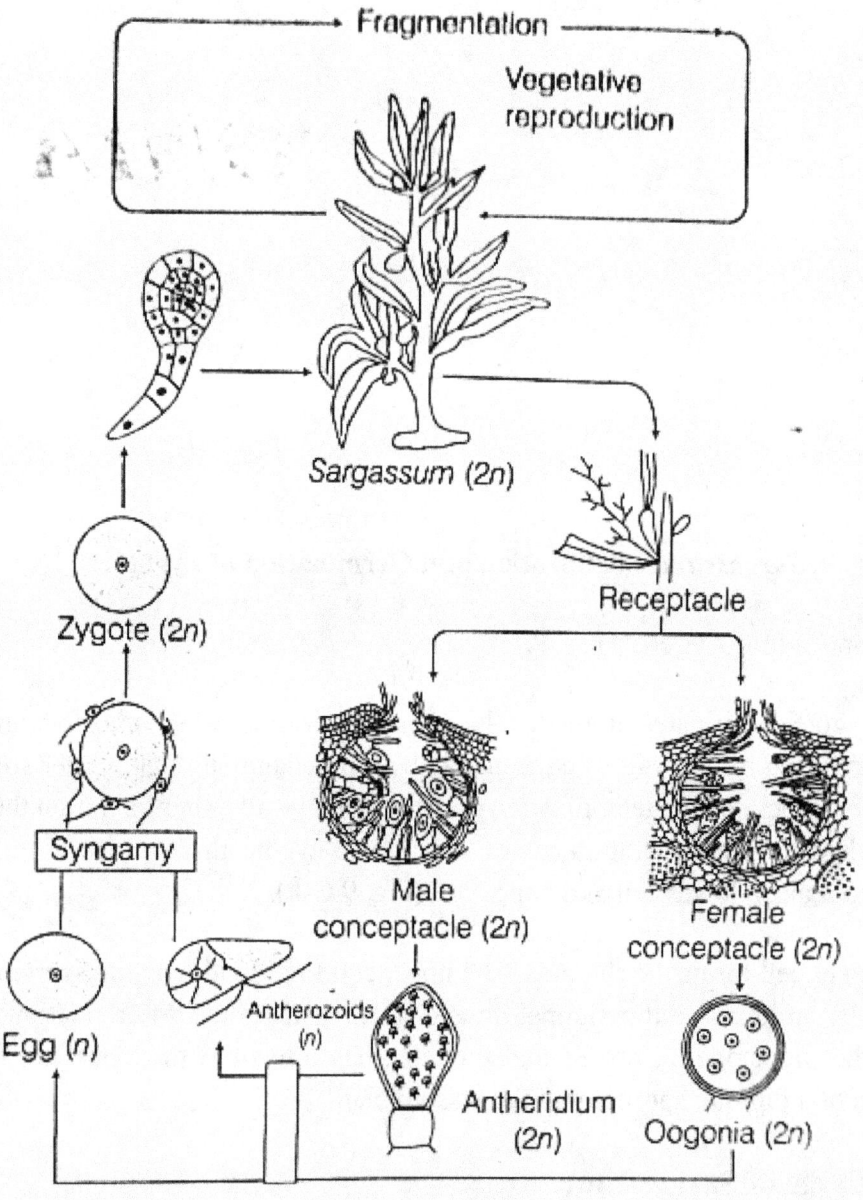

Fig. 10. *Sargassum*: Life cycle

22.

Dictyotales: *Dictyota*

There are 20 genera and 180 species in the order Dictyotales. The members occur both in temperate and tropical seas but attain their greatest development in the warmer oceans of the tropics and subtropics. Plant body is erect, flat, ribbon-like, well branched and attached to the substratum with the help of a rhizomatous disc. Growth is apical and the branching is usually dichotomous. Asexual reproduction takes place by non-motile tetraspores formed in unilocular sporangia. Sexual reproduction is oogamous type, and the plants exhibit isomorphic alternation of generations.

Family. Dictyotaceae

Plants are of moderate to large size, simple or branched, the divisions growing from apical cells or marginal rows of apical cells, forming fan-like foliaceous, strap-shaped branching thalli, usually composed of cortical and medullary layers of cells. Sexual and asexual generations are distinct, the plants of the two phases similar in form.

Dictyota

Systematic Position

Class: Phaeophyceae

Order: Dictyotales

Family: Dictyotaceae

Genus: *Dictyota*

Occurrence

The genus Dictyota is represented by about 35 species, found chiefly in the warmer Seas of the tropics. In India, the genus is represented by 12 species which occur along the coasts of Okha, Dwarka, Tuticorin, Mumbai, Porbandar and Mahabalipuram. *D.*

dichotoma is a common species found along the Indian coast. It grows attached to the rocks in the intertidal zone.

Thallus Structure

Thallus flat and leaf-like, to 300 mm long (usually 100-150 mm) and 5-30 mm broad; fronds thin and translucent, olive to yellow-brown, occasionally with a bluish iridescent, and ± regularly dichotomously forked, but lacking a midrib, and the reproductive structures are scattered over the fronds. Very variable, with narrow spirally twisted plants being found in mid-tidal pools, broader less twisted and more regularly branched plants being found in lower tidal pools, and broader very regularly branched, less twisted plants in the sub tidal (below). Often growing epiphytically. It is likely that more than one species is represented in Britain and Ireland.

Internal Structure

The thallus in section is seen to consist of three layers of cell. There is an upper and a lower assimilatory or photosynthetic layer between these two is the median layer consisting of large, colourless medullary cells which contain fucosan vesicles and reserve food in the form of globules but lack chromatophores.

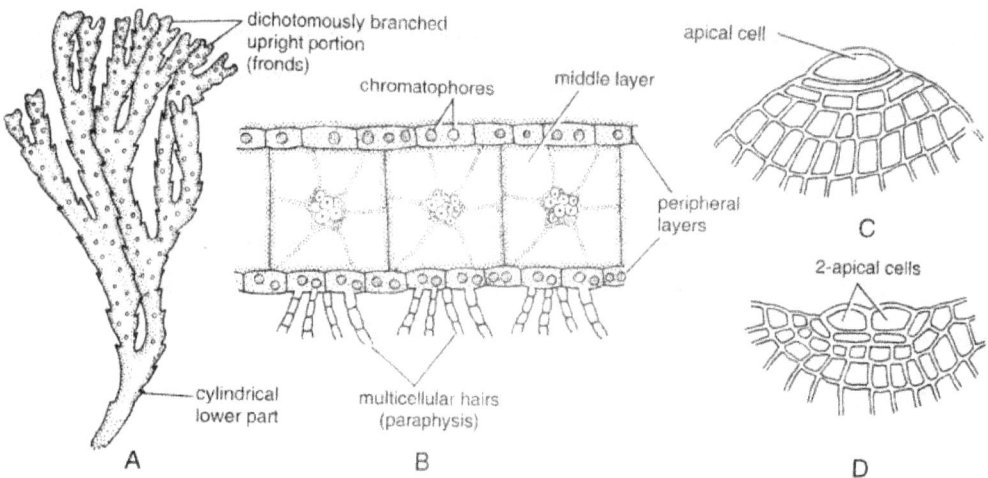

Fig. 1 A-D. *Dictyota*: A. Thallus structure; B. Internal structure. C-D. Apices of branches in surface view.

There is a single large vacuole in the centre. The septa between the medullary cells are characterised by the presence of pits. This layer serves as a food storage region. The cells of the peripheral photosynthetic or epidermal layers are small and contain abundant chromatophores. They constitute the photosynthetic region of the thallus.

Here and there groups of epidermal cells on both the surfaces of the thallus develop into colourless, multicellular mucilage hairs. The hairs are shed when the plant enters thereproductive phase.

Reproduction in Dictyota:

Dictyota reproduced by vegetative, asexual and sexual methods.

Vegetative reproduction

It take place by the decay of the older parts of the thallus, followed by subsequent fragmentation. In some cases, specially modified structures, called gemmae or brood buds, are developed on the body of the thallus for the purpose.

Asexual reproduction

It takes place with the help of large, non- motile spores, called tetraspores. The tetraspores are produced inside tetrasporangia, which are usually grouped together forming ill-defined sori on both the sides of the thallus.

The tetrasporangia are without any involucral covering and are produced on diploid asexual plants only. The solitary nucleus of a tetrasporangium undergoes a reduction division and gives rise to four, slightly elongated, naked tetraspores, each having a haploid set of chromosomes.

On maturity, the tetraspores come out through an apical opening on the tetrasporangial wall and germinate directly giving rise to male and female sexual plants in equal proportions, which are morphologically alike with the asexual plants. In some cases, a tetrasporangium may fail to produce tetraspores, and then it germinates quickly and reduplicates the same diploid generation.

Sexual reproduction

Sexual reproduction is oogamous. The gametophytes are normally heterothallic. The sex organs are differentiated in groups of sori on the male plants, each sorus being enclosed by a well-defined involucre (a sterile jacket layer of cells).

Each white and glistening antheridial sorus consists of about 100-200 antheridia, and each antheridium forms about 1500 uniflagellate sperms. The oogonial sori are deep brown in colour and are also invested by individual rudimentary involucre.

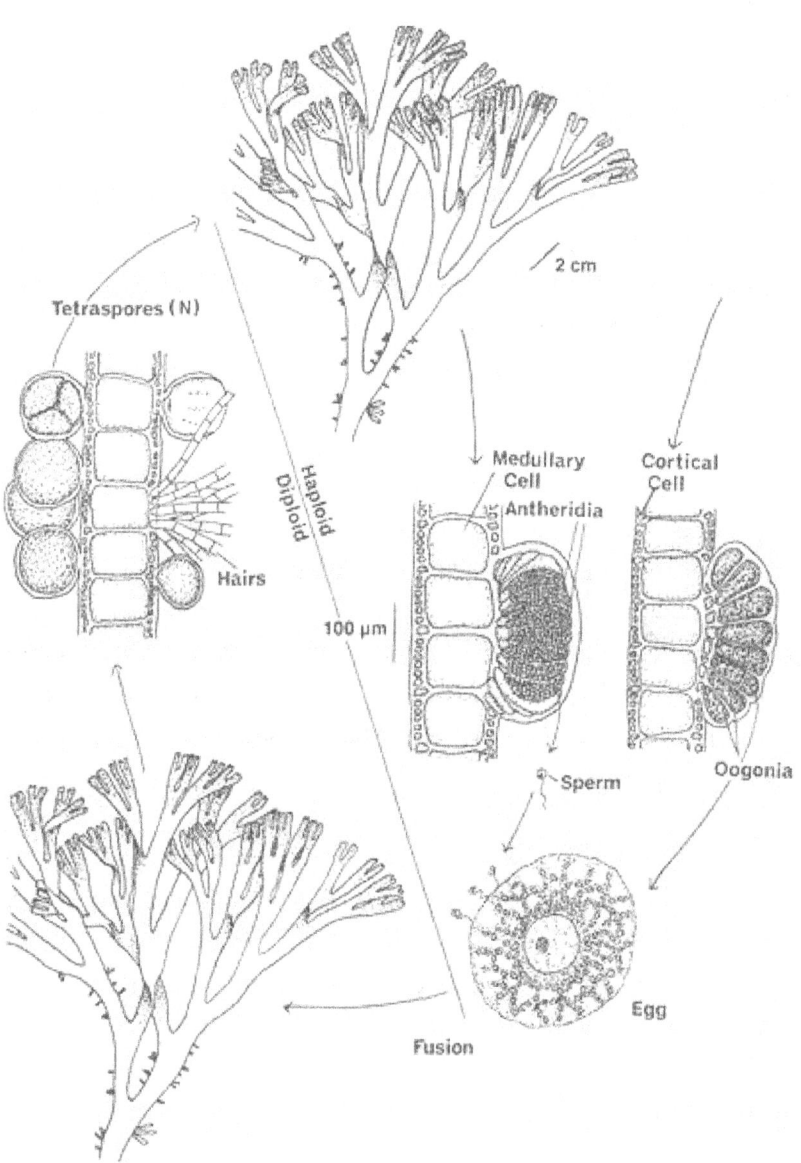

Fig.2.Dictyota: Life Cycle

Each oogonial sorus contains about 25-50 oogonia, each producing a single egg. On maturity, the apex of the oogonium becomes gelatinized and dehisces, thereby liberating the egg. There is a marked periodicity, usually correlated with the fortnightly sequence of spring and neap tides of the lunar month, in the development of sex organs and gametes.

When sperms and eggs are discharged in sea-water, the latter secrete a substance which attracts the sperms and fertilization is effected very soon (often in course of half an hour's time). The fertilized egg or oospore readily germinates into a diploid asexual plant. In rare cases, an unfertilized egg secretes a wall around it and germinates parthenogenetically.

It is to be noted that Dictyota possesses an isomorphic alternation of generations.

****

23.

Laminariales: *Laminaria*

The members of order Laminariales, commonly known as kelps, are of very large size and complex structure. Some plants reach up to 50-60 m. Most of the members show parenchymatous construction. The plant body may be differentiated into stipe and blade (lamina) and the meristematic region is located in between them. Asexual reproduction takes place generally by zoospores formed in unilocular filamentous and dioecious sporangia. The gametophytic body is microscopic. Sexual reproduction is oogamous type. Plants exhibit heteromorphic alternation of generation. There are about 30 genera and 100 species, mainly inhabitants of the colder oceans of the arctic and antarctic regions. The members are distinguished from the Fucales by their intercalary growth and heteromorphic alternation of generation. The order Laminariales is usually divided into four families.

1. Chordaceae,
2. Laminariaceae
3. Lessoniaceae
4. Alariaceae.

Family. Laminariaceae

The sporophytic plant body exhibits a high degree of morphological and anatomical differentiation well divisible into holdfast, stipe and blade. The sporangia are arranged in distinct groups (sori) on both the surfaces of blade. The gametophytes are much reduced, heterotrichous and dioecious. Members possess diplohaplontic heteromorphie alternation of generations.

Laminaria

Systematic Position

Class: Phaeophyceae

Order: Laminariales

Family: Laminariaceae

Genus: *Laminaria*

Occurrence

Laminaria (Gr. a thin plate or leaf) with some 40 species are exclusively grown on marine habitat in comparatively shallow water, mainly along the rocky shores, just below the tidal region. They are distributed widely in the Atlantic and Pacific coasts, especially where the water is cool. They are popularly called as "devil's apron". Some common members are *L. digitata, L. saccharina, L. japonica, L. flavicans* and *L. flexicaulis*.

Thallus Structure:

The plant body of *Laminaria* is diploid and attains a length of 2 to 12 metre or more in length. The sporophytic plant body is differentiated into holdfast, stipe and blade (lamina).

Holdfast:

It is an organ of attachment in variable forms. It may be like a solid disc or a cluster of profusely branched cylindrical root-like organs, also called haptera.

Stipe:

It is an unbranched and cylindrical or slightly flattened stalk-like structure, developed on the holdfast and bears terminal blade (s).

Blade (Lamina):

Blade may be simple (*L. saccharina*) or a cluster of vertically divided segments (*L. cloustoni, L. digitata*). Blades are flat, long, ribbon-shaped with wavy or smooth margin and tough and leathery in texture. Growth is intercalary and meristematic zone is situated between the stipe and blade.

Most of the members are perennial (L. ephemera etc.) and live about 5 years or more, but in annual member (*L. ephemera*), a new sporophyte develops in each spring.

Internal Structure of *Laminaria*: (Fig.1)

1. Stipe:

In transverse section, the stipe shows the following regions:

(i) Meristoderm,

(ii) Cortex, and

(iii) Medulla.

i) Meristoderm:

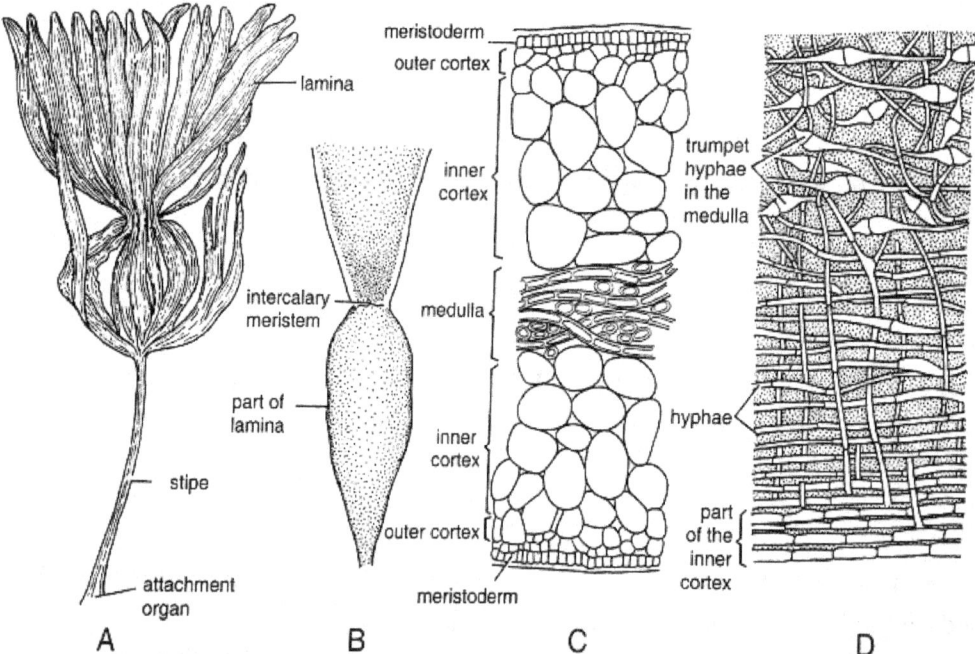

Fig. 1 A-D. *Laminaria*: Thallus structure; A. External structure, B. Portion of thallus showing intercalary meristem, C. Transverse section of stipe, D. Transverse section of lamina.

It is the outermost region, consists of one or two layers of small cubical cells containing chromatophores. It remains surrounded by a layer of mucilage and serves the function of assimilation.

ii) Cortex:

It lies just below the meristoderm (between meristoderm and central medulla). The cortex is usually differentiated into outer and inner cortex. The outer cortex is much wider and composed of elongated cells with pointed ends and the inner cortex composed of longer cells with square ends. Mucilage canals or tracts are present in the cortex. The cortical cells serve the function of both storage and mechanical support.

iii) Medulla:

It is the innermost layer, composed of loosely arranged cells and the spaces between them are filled with mucilage. Cross-connections are developed due to fusion of papillose outgrowths of the longitudinal walls. Due to regular cross-connections in the inner layer of cortex and also in the medulla, the layers appear as an entangled network.

The medulla is characterised by the presence of elongated hyphal structure, the 'trumpet hyphae', formed due to swelling at the septa which is larger at one side than the other. The transverse wall is perforated and shows a close resemblance with the sieve plates of sieve tubes. In older hyphae, the pores are sometimes blocked by callus pads.

2. Blade (Lamina):

Like stipe, the blade also shows the following three main regions in transverse section :

(i) Meristoderm,

(ii) Cortex (outer and inner) and

(iii) Medulla.

Only the elements of the medulla are drawn out due to greater surface area.

Reproduction in Laminaria:

The plant reproduces by all the three means: vegetative, asexual and sexual.

1. Vegetative Reproduction:

The vegetative reproduction takes place by stolons (*L. sincelarii, L. longipes* etc.). The stolons develop horizontally from the holdfast, grow upwardly into plantlets.

2. Asexual Reproduction:

It takes place by haploid zoospores developed within the club-shaped unilocular sporangia, borne inside the sori. These haploid zoospores are also called meiozoospores or gonozoospores. Sori develop extensively on both the surface of the blade and nearly cover the entire area. Numerous sterile hyphae, the paraphyses are present intermingled with the sporangia.

Development:

Sporangia develop from the superficial cells (meristoderm) of the blade. Initially these cells elongate and divide transversely into upper paraphysis initial and lower basal cell. The paraphysis initial elongates and forms an elongated thread with a club-shaped upper region. During this time, the basal cell broadens and thus the paraphyses occupy a limited part towards its basal region. The upper region of the paraphysis contains numerous chromatophores and fucosan vesicles. Each one at its apex develops a cup-like gelatinous thickening. By the side of paraphysis, the basal cells enlarge and differentiate into sporangia by transverse wall. During growth, the sporangia make their way and remain between the paraphyses. The nucleus of the sporangium undergoes first meiotic division, following successive mitotic divisions, thereby 32, sometimes 64 haploid nuclei are developed.

Each nucleus with some cytoplasm and chromatophore gets differentiated into a unit, which later gets differentiated into zoospore. Zoospores are without eye spot (Henry and Cole, 1982), with two lateral flagella (biflagellate) of unequal length (longer one tinsel and shorter one whiplash).

Dehiscence of the sporangia takes place by the lateral pressure of paraphyses on the sporangium. After liberation, the zoospores swim actively for some time. Out of 32 or 64 zoospores, 50% develop into male gametophyte and the remaining 50% into female gametophyte. After swimming for some time, the meio-zoospores settle down to germinate. After withdrawing flagella, it becomes rounded and secretes a wall. On germination, it develops a small protuberance, which increases in length and its apex becomes bulbous and appears almost like a dumb-bell. The bulbous region develops into a filamentous gametophyte, consisting of elongated cells. The male and female gametophyte develop male and female sex organ, respectively.

3. Sexual Reproduction (Fig.2)

It is of oogamous type. Sperms and egg produce inside the antheridium (male sex organ) and oogonium (female sex organ), respectively. The male plants become many celled before they develop antheridia, but the female plants develop oogonia at 2-3 celled stage.

Antheridia develop at the tip of lateral branches of many-celled male plants. Protoplast of each one-celled antheridium metamorphoses into a single pyriform or ellipsoidal antherozoid with two flagella (biflagellate) placed laterally. Sperms are liberated through the apical region of antheridium. The male plant degenerates after the gametes are released. Oogonia develop on 2-3 celled female plants. Oogonia are longer and thicker than the other cells of the female plant. The oogonial protoplast rounds off and becomes differentiated into an egg. At maturity, the egg is not extruded out, but remains attached at the apex of the oogonial wall.

Fertilisation:

After liberating from the antheridium, the antherozoid swims in water. On reaching the oogonium, it unites with egg and forms zygote.

Germination of Zygote:

Soon after the formation of zygote, it develops into a sporophytic plant. Initially, the zygote undergoes successive transverse division and develops into a vertical row of 4-10 cells. Except the lowermost one, all other cells of the filament further divide by both vertical and transverse divisions, thus forming an expanded, thin and flat blade.

The lowermost cell elongates to form the first rhizoid, but later several rhizoids are developed. Further division takes place in the lower part of the blade, thus forming meristematic region. With further development, the lower and upper parts of the meristem get differentiated into stipe and blade respectively. Parthenogenetically, an egg may develop into a plant, as that of sporophytic plant, which is irregular in shape.

Alternation of Generations:

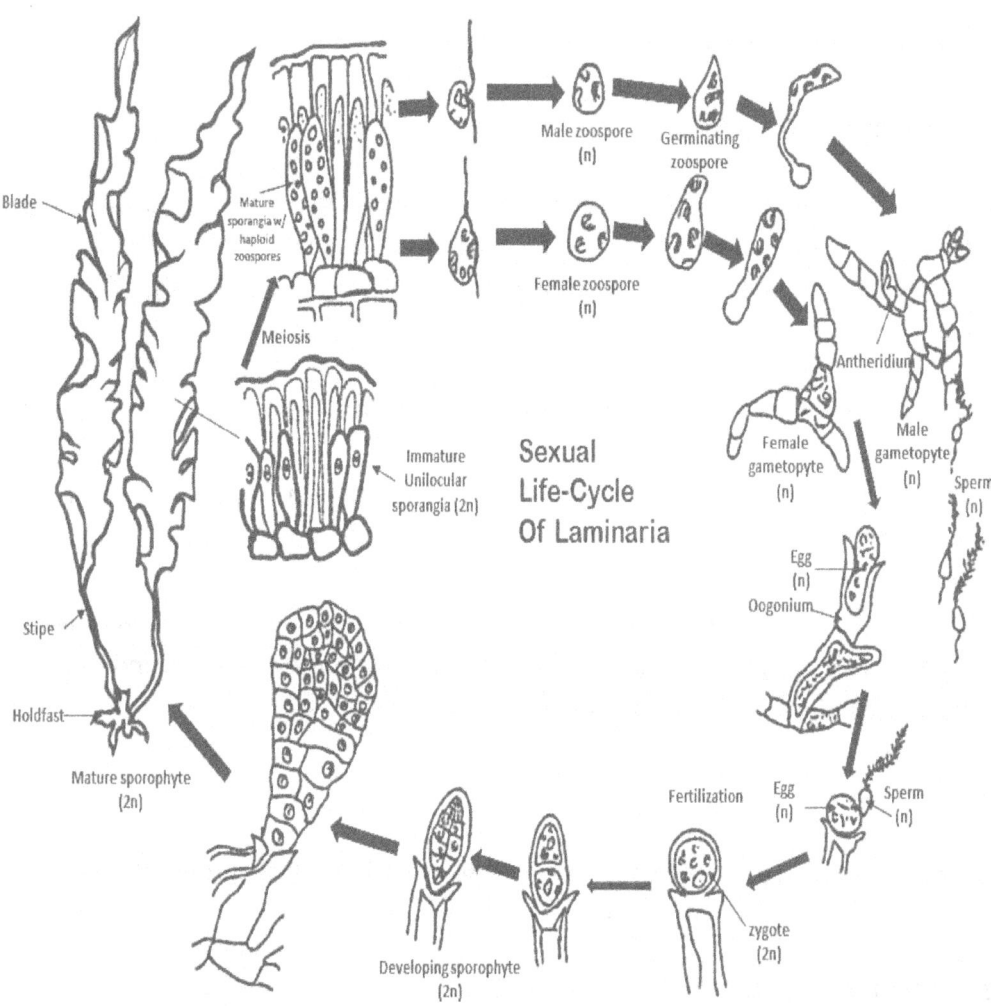

Fig.2. *Laminaria*: Life cycle

Laminaria exhibits a heteromorphic alternations, showing a prominent and well-developed sporophyte and a very small gametophyte. Asexual reproduction takes place in the sporophytic plant. The zoospores are developed inside the zoosporangium through meiosis and are thus called as zoomeiospores. Half of the zoomeiospores develop multicellular male thalli, where antherozoids are produced inside the antheridium. The other half develop into female thalli, those develop egg inside oogonium. The entire process of sexual reproduction is unique of its kind. The egg does not retain within oogonium, but it still remains attached with the oogonial wall, indicating a transitional stage between the primitive and advanced nature. There is spontaneous germination of zygote without rest, while it still remains attached with the female gametophyte. This is found to be an important feature in the life cycle of Laminaria.

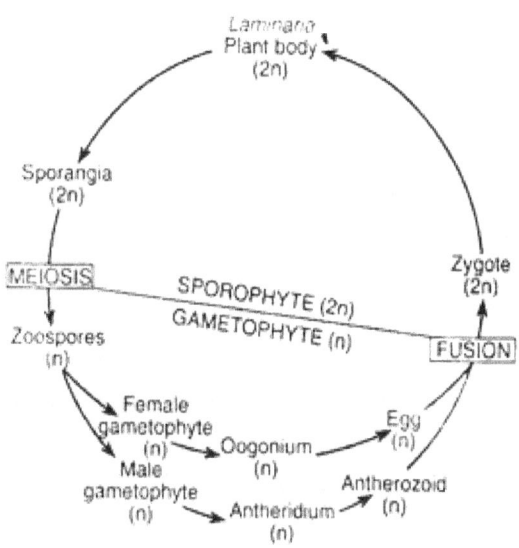

Fig.3. *Laminaria*: Graphic Life cycle

****

24.

Rhodophycae: Special Characteristics, Classification, Economic importance

It is a large group of algae consisting of about 831 genera and over 5; 250 species. They are commonly known as red algae due to the presence of a water soluble red pigment, r- phycoerythrin. The r-phycoerythrin is, however, present sufficiently and completely to mask the chlorophyll a, giving the characteristic red colouration. More than 98% members are marine and the rest grow in fresh water. The fresh water members grow in stagnant water (e.g., Asterocystis, Compsopogon etc.) as well as in flowing water (e.g., *Lamanea, Thorea, Batrachospermum* etc.). The marine species have the ability to live at greater depth (even at 30-90 meters) than the other members of different classes. They also exhibit a high degree of parasitism and epiphytism. The parasitic members show great reduction in their size and pigmentation. Some parasitic members are Ceramium condicola on *Codium fragile, Polysiphonia lanosa* on *Ascophyllum nodosum* etc. The epiphytic members like Rhodochorton, Ceratocolax etc. grow on other members of Rhodophyceae. Porphyridium, a unicellular member, is terrestrial and grow on damp soil. Marine members commonly grow in sublittorial zone, but a few members like *Rhodocorton, Corallina* and *Bostrychia* grow in intertidal zone.

Special Characteristics of the rhodophycae are mentioned below:

1. Maximum number of the algae are marine (except few are fresh water).

2. Some are color less and parasitic in nature that is they are unable to make their food themselves.

3. Rhodophycae do not have any motile stage and Thallophyta- in nature.

4. There structure can be unicellular, filamentous, parenchymatous and pseudoparenchymatous. Chloroplasts commonly discoid, without chloroplasts, endoplasmic reticulum.

5. Cell wall contains cellulose and phycocolloid like agar, carrageenan and funori.

6. Food is reserved as floridian starch formed in the cytoplasm, outside the chloroplasts. Mitosis closed, centrioles lacking, polar rings are present. Cleavage incomplete in most algae, presence of pit connection and pit plugs in most.

7. Rhodophycae contain female sex organ called carpogonium. It is flask shaped with long receptive neck called trychogyne.

8. Red algae contains male sex organ called spermatangium. Male sex organ contains non-flagellate gametes called spermatia.

9. External fertilization takes place. Post fertilization changes are complicated.

10. Some algae deposit calcium carbonate on their surface. This algae is known as coralline algae. Example of coralline algae is Corallina. This algae help in developing coral reefs along with corals.

11. Chloroplasts (chromatophores) have monothylakoid lamellae. Photosynthetic pigments include chlorophyll a, chlorophyll d (doubtful), carotenoids and phycobilins (phycoerithrin, phycocyanin, allophycocyanin). Red is due to phycoerithrin.

12. Life cycle is isoor heteroteromorphic, generally extended and triphasic.

Classification of Rhodophyceae:

The rhodophyceae has been divided primarily into two subclasses, one having only one Order under it, while the other possesses six orders.

A brief outline of the classification is given below:

I. Subclass Bangioideae:

Plants simple, filaments never aggregated; no pit-connections between the cells; very little specialization of sex organs; direct division of the zygote gives rise to carposporangia; terrestrial, freshwater, and marine.

Order Bangiales:

Plants ranging from unicellular to parenchymatous form; accessory reproduction by means of monospores; life cycle haplobiontic. Examples—*Compsopogon, Porphyra,* etc.

II. Subclass Florideae:

Plants usually filamentous with aggregation of filaments forming pseudo-parenchymatous thalli; cells with pit-connections; sex organs clearly differentiated ; gonimoblast filaments bearing carposporangia formed directly or indirectly from zygote; mainly marine.

Order 1:

Nemalionales:

Plants filamentous; construction of thallus uniaxial or multiaxial; accessory reproduction with the help of monospores or tetraspores; life cycle haplobiontic; some forms freshwater.

Examples:

Nemalion, Batrachospermum, etc.

Order 2:

Gelidiales:

Plants filamentous; uniaxial; generally with cruciate tetrasporangia; life cycle probably diplobiontic.

Example:

Gelidium.

Order 3:

Cryptonemiales:

Plants filamentous; uniaxial or multiaxial; tetraspores zonate or cruciate; some special fertile areas (conceptacles, nemathecia) noted in some cases; life cycle diplobiontic.

Examples:

Cryptosiphonia, Corallina, Lithothamnion, etc.

Order 4:

Gigartinales:

Plants filamentous; uniaxial or multiaxial; tetraspores zonate or cruciate; life cycle diplobiontic.

Examples:

Agardhiella, Iridea, etc.

Order 5:

Rhodymeniales:

Plants filamentous; multiaxial; tetraspores tetrahedral or cruciate; life cycle diplobiontic.

Examples:

Rhodymenia, Gastrodonium, Champea, etc.

Order 6:

Ceramiales:

Plants filamentous; uniaxial; tetraspores generally tetrahedral; carpogonial branch 4-celled; life cycle diplobiontic.

Examples:

Ceramium, Callithamnion, Polysiphonia, etc.

Economic importance of red algae are:

• Used as edible food – Some red algae like *Porphyra* or Laver, *Rhodymenia, Chondrus* or Irish moss (*Gracilaria, Gelidium*) are used as food or food processing.

• Extraction of bromine – At first bromine was extracted from red algae called *Rhodomela*.

• Used in microbiological media preparation - *Gelidium, Gracilaria* etc are used to yield agar which is used in culture media, media stabilizer and media thickener.

• Used as agrophyte – *Gelidella acerosa*. Is used as an agrophyte.

• Uses as adhesive – funori acts as an adhesive agent.

• Clearing agent – Carrageenin acts as a clearing agent.

• Emulsifiers – *Chondrus* or Irish moss act as emulsifiers.

• Anti-bacterial agent - *Polysiphonia* kills microbes as it act as anti-bacterial agent.

****

25.

Nemalionales: *Batrachospermum*

The order Nemalionales includes 250 species, belonging to 35 genera. Except for few freshwater forms like *Batrachospermum, Lemanea, Thorea* and *Tuomeya,* most of the of the members of this order are marine. The thallus is made up of uniaxial or multiaxial filaments. Each cell of the filament has one or more discoid chromatophores. These algae reproduce asexually by means of monospores, produced singly in monosporangia. The diploid phase is represented by zygote only. The first division of the zygote is meiotic and the resultant nuclei eventually give rise to haploid carpospores. Thus these algae show **haplobiontic life-cycle**.

The order Nemalionales includes eight families:

1. Acrochaetiaceae
2. Batrachospermaceae
3. Lemaneaceae
4. Naceariaceae
5. Bonnemaisoniaceae
6. Thoreaceae
7. Nelminthocladiaceae
8. Chaetangiaceae

Family. Batrachospermaceae

This family includes algae whose plant body is of soft, thick, gelatinous branched thallus differentiated into structures resembling nodes and internodes. Branches arise in whorls from the nodes. Asexual reproduction is by monospores and sexual reproduction oogamous by the development of carpogonia and spermatangia. Zygotic nucleus, divides meiotically immediately after development. This is followed by the formation of gonimoblast filaments from which carposporangia and carpospores are developed in a cystocarp.

Batrachospermum

Systematic position:

Class: Rhodophyceae

Order: Nemalionales

Family: Batrachospermaceae

Genus: *Batrachospermum*

Occurrence:

This is one of the fresh water forms of Rhodophyceae. This alga is found in slow running streams and on the banks of lakes and ponds. It is more commonly found in well aerated waters. The plants are blue-green, olive-green, violet and reddish in colour. The colour varies as a result of the differences in light intensity. The species which grow in deep water are reddish or violet in colour whereas the species growing in shallow water are olive-green in colour. The alga is also known as the 'frog spawn'. The plants are mucilaginous, moniliform or beaded in appearance to the naked eye. The plants may reach a length of twenty centimetres and may easily be collected from the slow running streams around Dehradun especially in winter season.

Thallus Structure

The plant is characterized by having a thallus body of delicate, freely-branched, filiform filaments, varying in size from a few to several centimetres in length. The branches are usually moniliform, and have a gelatinous consistency. Its colour varies from bluish-green to olive-green, but may be of violet or reddish tint when it grows in deep water under shady places.

The older portions of the thallus are made up of large cells arranged in an axial row bearing numerous vertical branches, which are either simple or divided in a forked manner, and the lateral branches are repeatedly forked. The lateral branches usually occur in the form of globose clusters at the nodes imparting to the plant its characteristic beaded appearance and these may lie quite close to or a little distance apart from one another.

From the basal cell of the lateral branches threads also grow downwards covering the external surface of the large axial row of cells, ultimately forming a complete cortex-like sheath. The protoplast of each cell is uninucleate, and contains several discoid or elongated chromatophores, each with a pyrenoid.

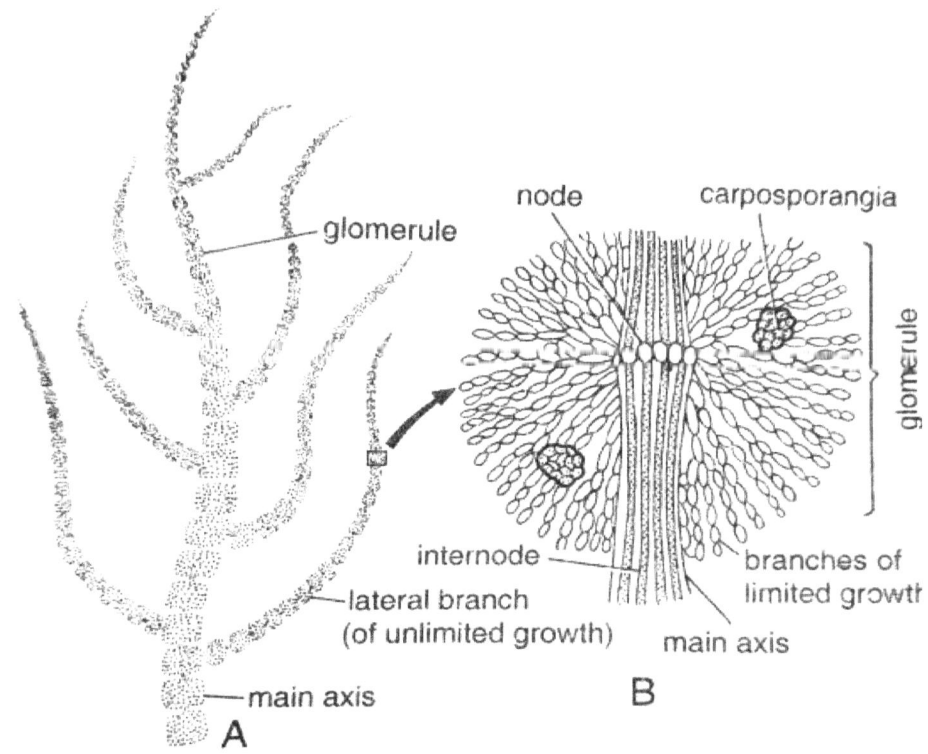

Fig. 1 A-B. *Batrachospermum*: Thallus structure.

Growth

Formation of branches of limited growth: Growth; in main element occurs by apical single cell. Transverse division occurs in the cell. It cut off cells towards the posterior side. Each of these coils cuts off of four small cells. These cells become initials for the side branches. These initials divide many times. Closter of small branches arises from these groups of lateral cells. It gives the plant beaded pattern. Each cluster of side branch is called glomerule. These branches produce whorls. Growth of central axis: The cells of the central axis elongates very much. Therefore, lateral cells separate from each other. Thus they form node like structure on the axis.

Formation of pseudo cortex: The cells at the nodes produce Filaments towards downside. They closely surround the central cells up to next node. Thus they form loose covering around the central axis. This loose covering is called **pseudo cortex.**

Formation of braches of unlimited growth: One or more cells on each node may acts as apical cell. This cell produces lateral branches of unlimited growth like main axis.

Asexual reproduction: In several species of *Batrachospermum* the short branches of the filaments of Chantransia stage produce monospores. These monospores again produce Chantransia stage, and again the apical cells of this stage produce new plants.

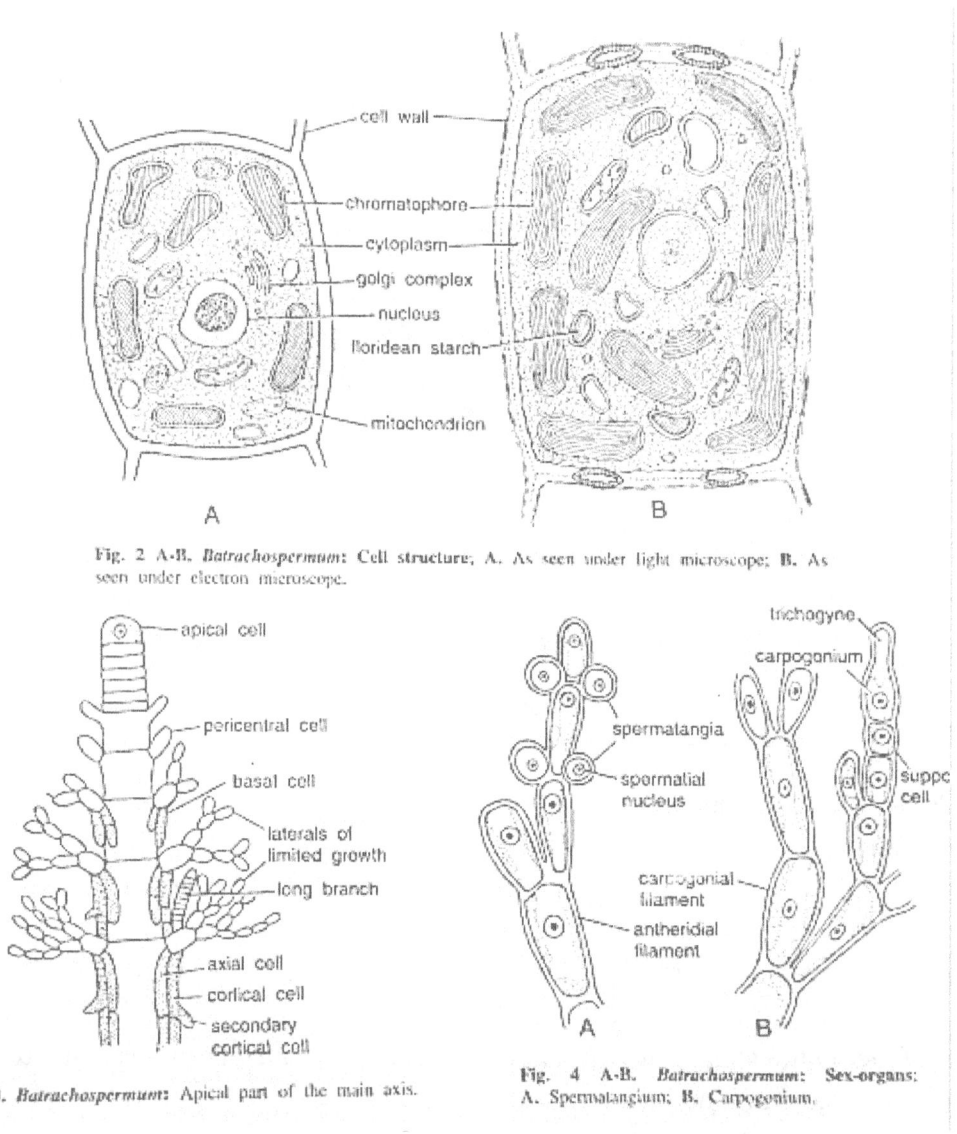

Fig. 2 A-B. *Batrachospermum*: Cell structure; A. As seen under light microscope; B. As seen under electron microscope.

Fig. 3. *Batrachospermum*: Apical part of the main axis.

Fig. 4 A-B. *Batrachospermum*: Sex-organs: A. Spermatangium; B. Carpogonium.

Sexual reproduction: The sexual reproduction is advanced oogamous and takes place by means of male and female sex organs known as antheridia and carpogonia respectively. The plants may be monoecious or dioecious.

Development of antheridium: The antheridium develops from a uninucleate, colourless antheridium mother cell. From each antheridium mother cell one to four antheridia are developed. The antheridia are produced in clusters at the apical points of the short laterals. In the beginning they appear as protuberances arising sub terminally and successively from

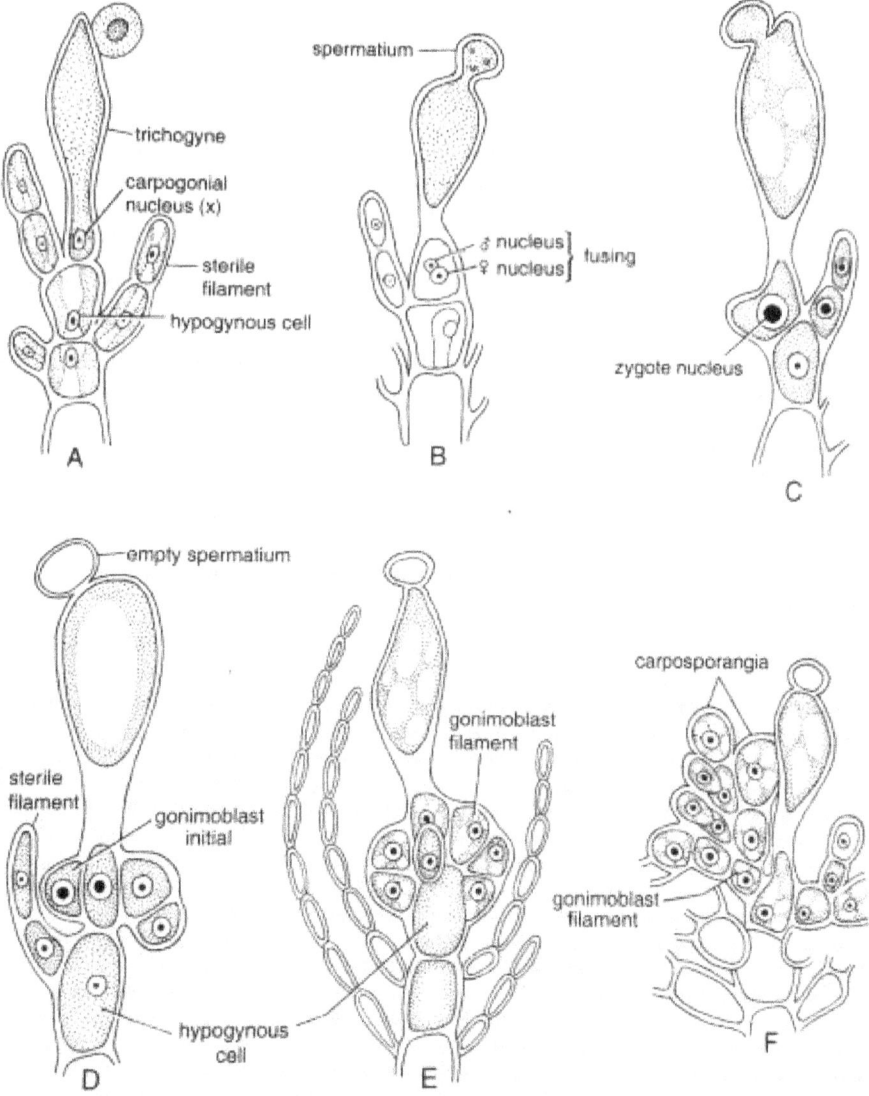

Fig. 5 A-F. *Batrachospermum*: Fertilization and post-fertilization changes; A-B. Fertilization, C. Zygote, D-F. Post-fertilization changes.

different sides of the mother cell. Later on, the small protuberances sub terminally become spherical.

Each antheridium contains a single spermatium on its maturity. The non-motile, spermatium liberates through a slit formed in the wall of antheridium. The spermatia remain floating in the water. In *Batrachospermum,* the nucleus divides into two as soon as it contacts the trichogyne. So at the time of the fertilization spermatium contains two nuclei and sometimes known as spermatium complex.

Development of carpogonium: The carpogonia develop on the terminal ends of the short laterals. The terminal cell of the lateral divides into four cells. The uppermost cell develops into the carpogonium. The carpogonium consists of a swollen basal portion which contains an egg and known as carpogonium and an elongated receptive part, the trichogyne. The carpogonium is flask-like. In the majority of Florideae the cytoplasm of the carpogonium is colourless, but in *Batrachospermum* it bears a pale plastid.

Fertilization: The non-motile spermatia float in the water. Many spermatia approach the trichogyne. One of the spermatia attaches itself to the trichogyne. The wall of contact dissolves and one of the two nuclei of the spermatium passes through this hole into the trichogyne, reaching in the basal swollen part of the carpogonium where it fuses with the female egg and develops into the zygote. Thereafter the trichogyne shrivels down up to the constriction in between trichogyne and carpogonium. Simultaneously a cross wall develops at this juncture.

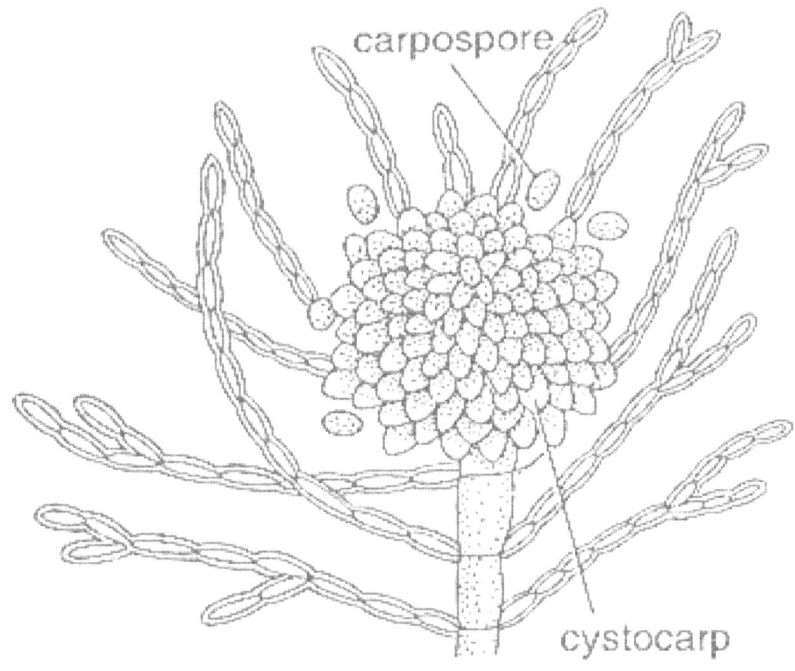

Fig. 6. *Batrachospermum*: Mature cystocarp (carposporophyte).

Germination of zygote: The diploid nucleus of the zygote divides meiotically producing two haploid nuclei. Thereafter one of the two nuclei migrates into the lateral protrusion of the zygote. A wall separates this protrusion from the rest of the zygote and this way the gonimoblast initial is formed, other daughter nucleus divides repeatedly several times, forming a large number of gonimoblast initials. The gonimoblast initials divide again and again and a gonimoblast filament develops from each initial. The gonimoblasts become branched and terminal cells of these develop into carposporangia. Each carposporangium produces a single rounded, haploid carpospore. The structure having gonimoblast filaments, carposporangia and carpospores is known as cystocarp or carposporophyte. At maturity of the carposporangia, the walls split off and the carpospores are liberated. Each carpospore gives rise to the juvenile branched filamentous body. The juvenile form of *Batrachospermum* resembles an alga known as Chantransia, and therefore, known as the Chantransia stage. The terminal cells of these plants act as apical cells and develop into new *Batrachospermum* plants.

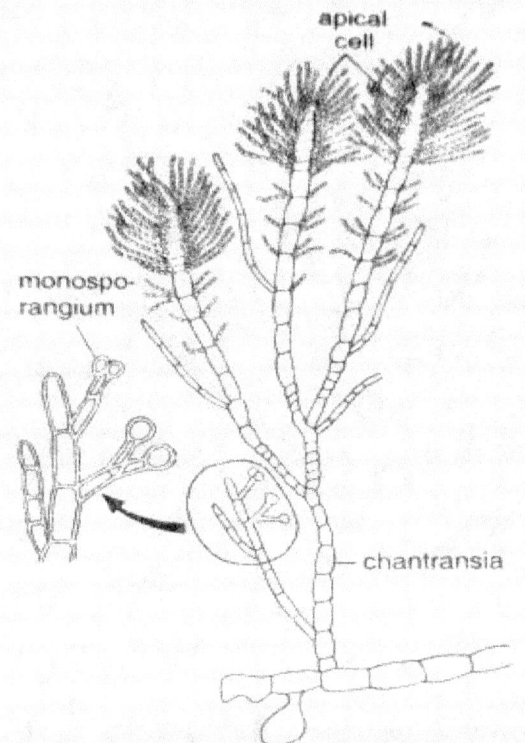

Fig. 7. *Batrachospermum*: Juvenile stage (*Chantransia* stage).

****

26.

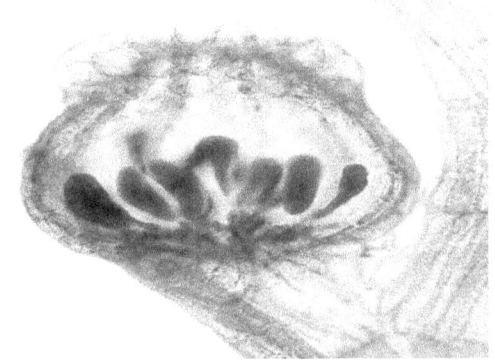

Ceramiales: *Polysiphonia*

The order ceramiales is represented by about 160 genera and 900 species. The algae of this group are mostly marine. The thalli are complex multi-axial or polysiphonous. After fertilization an auxiliary cell is cut off from the supporting cell of the carpogonium. The 2n zygote nucleus divides by mitotic division and one of the diploid nuclei enters supporting cell through auxiliary cell. From supporting cell arise a number of gonimoblast filaments. The terminal cell of gonimoblast filament bears carposporangium which forms a diploid carpospore. The complete structure is called as carposporophyte. The carpospore on germination forms tetra sporophyte. The tetra sporangia on tetra sporophytes form haploid tetra spores. The tetra spores on germination form haploid gametophytic plants. Thus, the life cycles are triphasic and diplobiontic.

The order Ceramiales is divided into four families:

1. Ceramiaceae e.g., *Ceramium*

2. Delesseriaceae e.g., *Delesseria*

3. Rhodomelaceae e.g., *Polysiphonia*

4. Dasyaceae e.g., *Dasya*

Family. Rhodomelaceae

The algae of family Rhodomelaceae are polysiphonous. The thallus consists of central siphon and pericentral siphons. The cells of central siphon are uninucleate. The male and female reproductive structures develop on special reproductive branches called as trichoblasts. After fertilization carposporophytes and tetra sporophytes are formed. The life cycles are diplobiontic.

Polysiphonia

Systematic Position

Class: Rhodophyceae

Order: Ceramiales

Family: Rhodomelaceae

Genus: *Polysiphonia*

Occurrence

Polysiphonia is a marine alga. It is widely distributed in the coastal area in the littoral zone. Most of the species of *Polysiphonia* prefer still waters and appear brownish-red to purple-red. The species are either annual or perennial and the thallus shows bushy appearance. Most of the 150 species of the alga occurs in the eastern coast of North America and around the Great Britain. In India, it is found along the western coast. A few species are also found in Pacific Ocean. Some species are epiphytic, lithophytic or semi-parasitic. *P. ferulaceae* is epiphytic on *Gelidium sp.* which is another red alga. Many species grow on the rocks. *P. fastigata* is a semiparasite on Ascophyllum sp. *P. ureceolata* is an epiphytic on *Laminania*.

Thallus Structure

The species of *Polysiphonia* are freely branched with polysiphonous construction. The thallus of the alga is basically filamentous. Due to repeated branching the thallus appears as a bushy structure. The thallus is attached to a substratum by means of holdfast. The plants attain a height of a few to several centimetres. The thallus is hetrotrichous. It is differentiated into two types of branches-prostrate and vertical. Prostrate branches are spread out on the substratum to which they are attached with the help of unseptate unicellular rhizoids. The rhizoids are flattened structures for better anchorage and are termed haptera. The rhizoids arise from the peripheral cells of the branch. The erect vertical fi laments arise from the upwards tips of the prostrate fi laments, or from its any other part. The fi laments consist of elongated siphon like cells arranged above one another in vertical rows and hence the name *Polysiphonia.* The dome-shaped apical cell cuts off segments proximally which produces lateral branches before dividing longitudinally into central and pericentral cells. In *Polysiphonia,* the pericentral cells are of the same length as the axial cells, so that the primary cortication of the axial fi lament is carried out by a smaller number of cells. The number of pericentral cells varies from 4 to 24, and their number and arrangement is of taxonomic importance. The young thallus of the alga in only 6–7 cells in height and are uniaxial. The young plants grow by means of an apical cell. In older parts of the thallus the pericentral cells undergo pericinal divisions to form a multilayered parenchymatous cortex. The cells of the cortex are smaller than the pericentral cells. The central siphons as well as the pericentral siphons are interconnected by means of pit-connections. The primary pit connections between these siphons are developed at the time of cell divisions. The Secondary pit connections develop between pericentral cells and cortical cells. In several species of Polysiphonia two types of branches arise from the main axis on the lateral sides just 2–5 cells behind the apical cell. The branches are short and long. The short branches are limited growth and are termed trichoblasts. They are uniseriate i.e., monosiphonous. They are arranged spirally around the main fi lament. In the perennial species they fall off annually before the onset of the winter season. The trichoblasts are forked in a dichotomous manner. The long branchesarise sometimes directly from the central siphon but more often from the basal cell of the trichoblasts. The basal cell of a trichoblast cuts off a small lateral protuberance by means of an oblique septum. This outgrowth acts as an apical all for the growth of the Long Branch, which has a row of central siphons surrounded by pericentral cells.

Fig.1. *Polysiphonia:* Habit

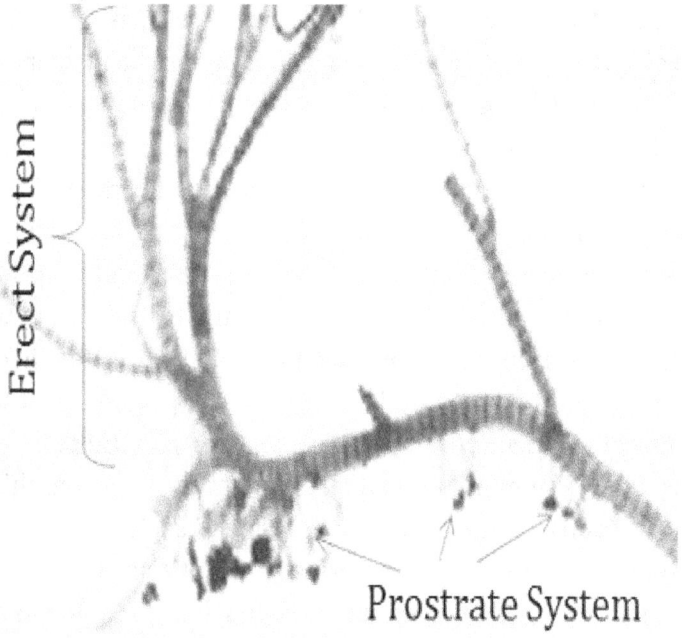

Prostrate & Erect System of Thallus

Fig.2.*Polysiphonia:* Thallus position

Reproduction:

In the life cycle of Polysiphonia occurs three distinct stages – gametophyte, carposporophyte and tetrasporophyte.

1. Gametophytes: (Fig.3.d) The gametophyte plants of *Polysiphonia* are concerned with the production of gametes which involves oogamous type of sexual reproduction. *Polysiphonia* is doiecious i.e., male and female sex organs are borne on different gametophyte plants called the male and female gametophytes respectively.

(a) Male gametophyte – It bears the male sex organs called spermatangia. The spermatangia are borne in dense clusters and are closely packed forming a compact cone shaped structure on short monoecious branches near the apices of the male plants. The branches are called the male trichoblasts. The male trichoblast usually consists of two basal cells constituting the stalk. The stalks usually forks into two branches. In some species both the branches may become fertile. In others, one branch may become fertile and the other one sterile. The sterile branch may develop into a repeatedly forked sterile axis. The entire contents of the spermatangium is metamorphosed into a single male nucleus i.e., spermatium.

(b) Female gametophyte – It bears the female sex organs called the carpogonia. The carpogonium is somewhat a flask-shaped structure. It consists of a swollen basal portion and a long tubular portion also known as trichogyne. The female gamete is the uninucleate protoplast. The trichogyne simply functions as a receptive organ. The carpogonium is situated at the tip of a short lateral curved branch consisting of 4 cells, known as carpogonial branch or filament. The basal cell of the carpogonial filament is known as supporting cell. It cuts off two sterile filament initials– one towards the base and the second laterally. The basal sterile cell remains undivided while the lateral sterile filament immediately divides to form a 2-celled lateral sterile filament. This is the structure of the carpogonium at the time of fertilization. The two pericentral cells now divide and redivide to form the envelope of the carpogonium which is known aspericarp.

(c) Fertilization – The liberated spermatia are carried by the currents of sea water. As they come in vicinity of the carpogonium, one of them adheres to the trichogyne. At the point of contact, the wall between the two dissolves. The male nucleus then enters the trichogyne and moves down and finally reaches the female nucleus, and finally fuses with it to accomplish the fertilization. After fertilization, a series of development takes place which results in the formation of cystocarp. The cystocarp consistsof the placental elements, gonioblast filaments bearing carposporangia and the surrounding sheath known as pericarp.

2. Carposporophyte: The diploid portion of the cystocarp constitutes the carposporophyte. It consists of the zygote and the gonioblasts bearing the carposporangia. It remains attached to the female plant of *Polysiphonia* and is responsible for the production of carpospores. The carposporophyte (**Fig.3.h),** is surrounded and protected by a two-layered pericarp, opened at the tip. At maturity the pericarpcontains a diploid carpospores which float out through the opening or ostiole and are carried away by water currents. On coming in contact with the solid object, the carpospores germinates to produce a diploid tetrasporophytic generation.

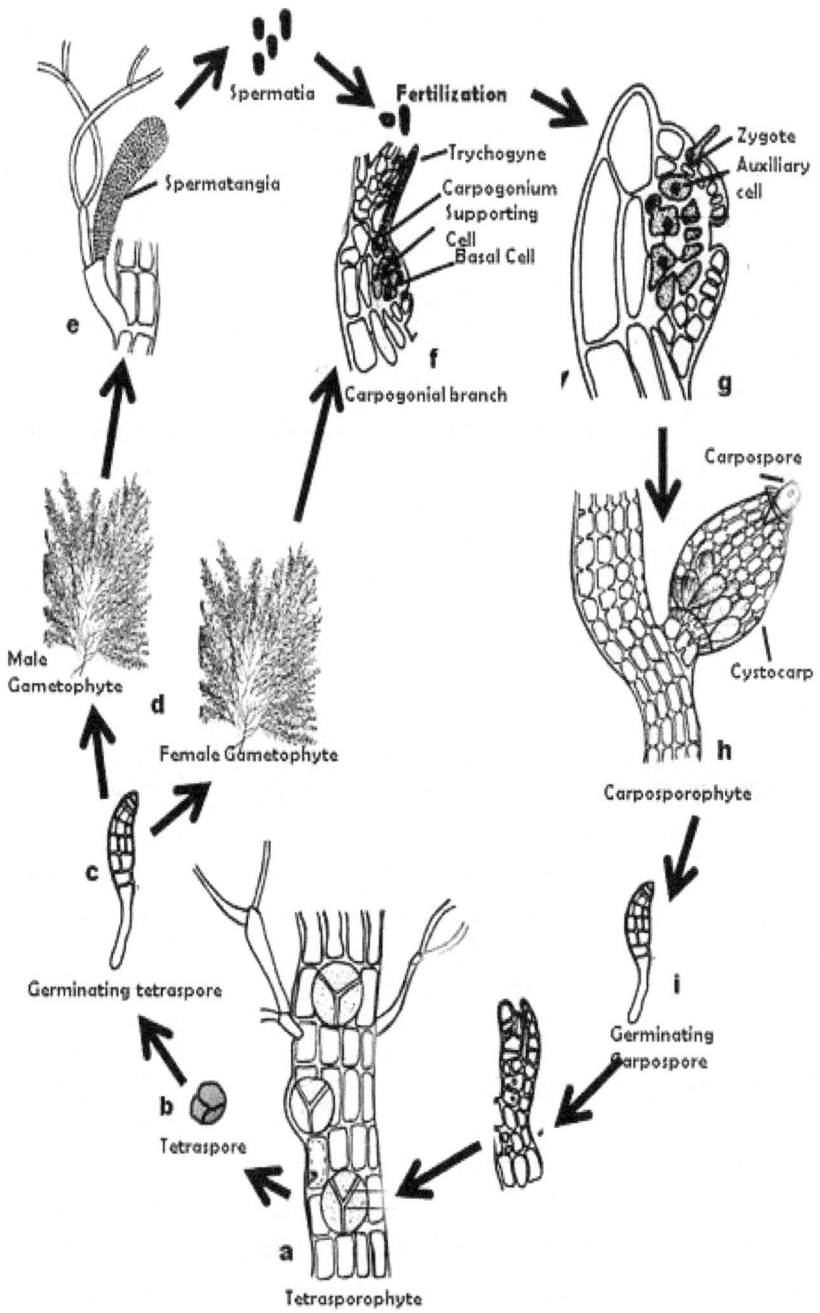

Fig.3. *Polysiphonia:* life history illustrating different stages of development

3. Tetrasporophyte: (Fig.3.a) It is a free living individual. The thallus in its vegetative structure exactly resembles the gametophytic plants. It consists of a central siphon encircled with the placental siphons. The branching of the thallus is lateral. The branches arise from the apical region of the central filament. The tertasporophyte produces tetrasporangia from the pericentral cells in the apical region of the branch. The diploid nucleus of tetrasporangium by meiotic division produces four haploidtetraspores (n). When the tetraspores attain maturity, the sporangial wall ruptures and the tetraspores escape. On attachment with thesubstratum, the tetraspores germinate to produce a haploid gametophytic plant again.

****

27.

Origin and Evolution of Sex in Algae

Primarily there are two methods of reproduction in plant, i.e., asexual and sexual. In asexual type of reproduction, no sex is involved and there is no fusion of any kind of gametes or cells. This type of reproduction is being effected by special cells known as spores. Each such spore is capable to develop into a new plant. On the other hand, in sexual reproduction the two cells or the gametes unite together, and the zygote is resulted, which develops into a new plant. Normally, the individual gametes are incapable to produce new plants without fusion.

The most primitive algae, i.e., the members of Cyanophyceae (Myxophyceae), e.g., *Gloeocapsa, Chroococcus,* etc., reproduce by means of fusion, whereas *Oscillatoria, Nostoc, Anabaena,* etc., reproduce vegetatively, by means of a group of few cells called hormogonia, which later on give rise to new plants by further division. Some members of Cyanophyceae reproduce by means of arthrospores. This shows that all the blue-green algae reproduce asexually and the sexual reproduction is altogether absent. In rest of algae, whether unicellular or filamentous, both asexual and sexual methods of reproduction prevail. It is thought that probably the gametes have been originated from the motile asexual spores or zoospores. Except few cases, the sexual reproduction has not replaced the asexual method but has been added as a supplementary method. The zoospores usually resemble the gametes except for their size. For example, in algae such as *Chlamydomonas, Ulothrix, Cladophora,* etc., it is thought that the zoospores were probably produced before the gametes originated. In above mentioned and many other filamentous forms, the asexual reproduction takes place by means of uninucleate and biflagellate zoospores which are formed by repeated division contents of the ordinary vegetative cell. The zoospores are produced, in favourable conditions. In adverse conditions, the gametes are formed in the vegetative cells which earlier produced zoospores in favourable conditions. The gametes are quite similar to zoospores but for their size and behaviour. In forms like *Chlamydomonas debaryanum* the zoospores and gametes are quite similar.

Origin of sex:

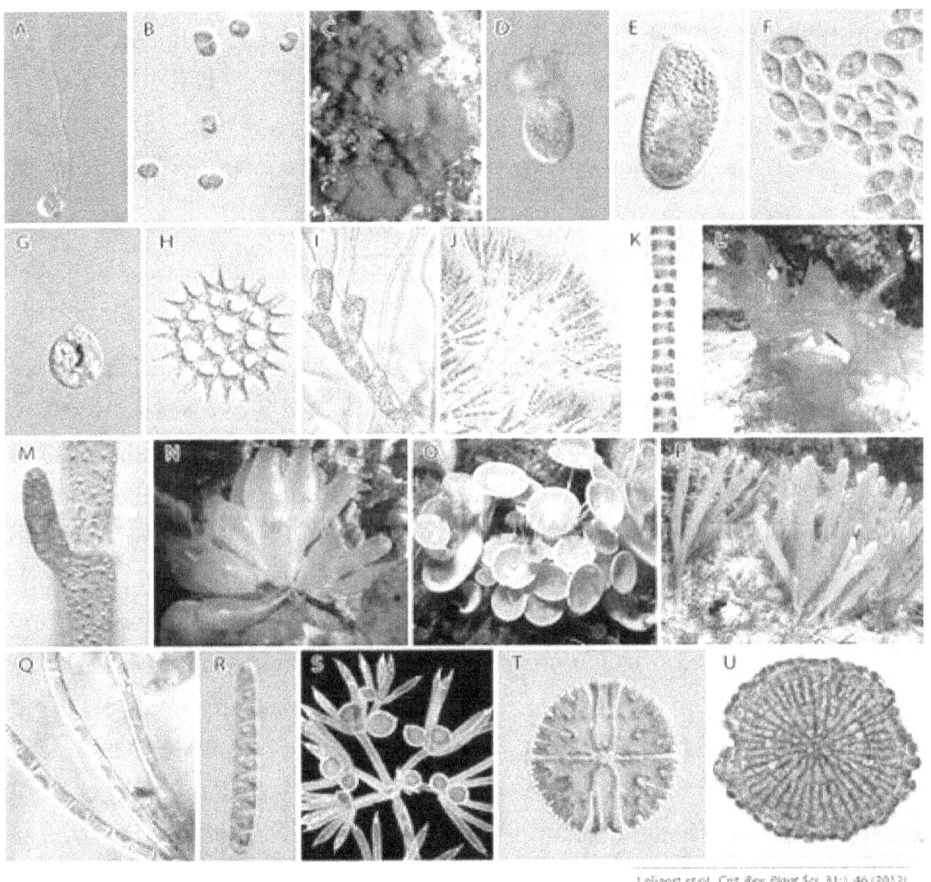

Fig. 1.Green algae. A: *Pterosperma*. B: *Nephroselmis*. C: *Palmophyllum*. D: *Tetraselmis* (Chlorodendrophyceae. E: *Chlorella* (Trebouxiophyceae, Chlorellales), F: *Oocystis* (Trebouxiophyceae, Oocystaceae), G: *Haematococcus* (Chlorophyceae, Chlamydomonadales), H: *Pediastrum* (Chlorophyceae, Sphaeropleales), I: *Bulbochaete* (Chlorophyceae, Oedogoniales. J: *Chaetophora* (Chlorophyceae, Chaetophorales), K: *Ulothrix* (Ulvophyceae, Ulotrichales), L: *Ulva* (Ulvophyceae, Ulvales), M: *Cladophora* (Ulvophyceae, Cladophorales), N: *Boergesenia* (Ulvophyceae, Cladophorales), O: *Acetabularia* (Ulvophyceae, Dasycladales), P: *Caulerpa* (Ulvophyceae, Bryopsidophyceae), Q: *Klebsormidium* (Klebsormidiophyceae), R: *Spirotaenia*, S: *Nitella* (Charophyceae), T: *Micrasterias* (Zygnematophyceae, Desmidiales), U: *Coleochaete* (Coleochaetophyceae), (photo by CFD).

In *Ulothrix*, two types of zoospores, i.e., the micro- and the macrozoospores, are produced and the gametes are produced in the same manner as the zoospores are formed. The microzoospores are quite similar to the gametes. At the end of the season, when the plants run in the shortage of food and other favourable conditions, the cells begin to produce gametes. Cholnosky, however, advocates that the gametes in Ulothrix variabilis develop from potential zoospores which fail to escape and undergo more divisions to give rise to the gametes. According to him, the gametes are quite similar except their size and number of flagella. In *Oedogonium,* the gametes and the zoospores are quite similar except their size.

One can conclude that the gametes are formed as a result of aging of cells and unfavourable conditions. It may be said that the gametes are the reduced type of zoospores. Such reduced zoospores cannot reproduce asexually. They can give rise to a new offspring only with the result of fusion and formation of the zygote. All these examples support that the gametes have been originated from zoospores by means of reduction.

Evolution of sex:

The members of Myxophyceae do not show any trace of sexuality. In other classes of algae, the sexuality has been established. Now to trace out the evolution of sex in algae, it becomes necessary to know about the nature of the gametes in beginning. It is thought, that in the beginning the gametes were of similar shape and size. These gametes were morphologically identical, but physiologically different, e.g., in *Ulothrix* and certain species of *Chlamydomonas,* the gametes are of plus and minus strains. This is known as isogamy. Thereafter anisogamy developed. In this process, the two uniting gametes are of different sizes, e.g., in *Chlamydomonas braunii* and *Pandorina*. In this condition, the number of divisions in one cell is more or less than the other cell, and therefore, the uniting gametes are unequal in size. Thereafter developed oogamy. The oogamy is the highly evolved condition of heterogamy. In oogamy, one cell does not divide at all it simply increases in size, accumulates sufficient food material and acts as female cell or egg. The other cell divides repeatedly producing small motile cells, acting as male gametes. However, in *Polysiphonia* the resultant male gametes are non-motile. Less evolved oogamy is found in *Chlamydomonas coccifera*, Volvox and Oedogonium.

The evolution of sex in algae has not taken place in any one phylogenetic line. This has taken place along several independent lines. The example of this statement is found in Volvocales, where Gonium is isogamous, *Pandorina* slightly anisogalnous, *Eudorina* and *Pleodorina* marked by anisogamy and *Volvox* oogamous. This *Gonium-Pandorina-Eudorina-Pleodorina-Volvox* series also shows progressive somatic differentiation. It has been assumed that it is correlated with the evolution from isogamy to oogamy through anisogamy. Whereas on the other hand, in another genus of Volvocales, i.e., *Chlamydomonas,* where species within the genus show all gradations from isogamy to oogamy, e.g., *Chlamydomonas snowiae* is isogamous, *C.braunii* is anisogamous and *C. coccifera* is oogamous.

In a few green algae, such as Spirogyra, *Vaucheria*, and *Chara* and in nearly all the brown and red algae a far more advanced condition has been reached. In Spirogyra, the two uniting gametes are produced inside the ordinary vegetative cells known as gametangia. The gametangia are not differentiated into male and female structures. Those which receive the contents are supposed to be female ones while those from which the contents pass out are supposed to be male ones. In *Vaucheria,* the gametes are borne in well-differentiated gametangia, known as antheridia and oogonia. The antheridia produce antherozoids and the oogonium contains an egg.

In *Chara,* there is further advancement. Here the sex organs are not developed all over the plant body, but are confined to nodes on the branches of limited growth. The male and female sex organs are highly developed and specialized organs known as globule and nucule respectively. The spermatogenous filaments develop within globule while the nucule possesses an egg. In brown algae, *Fucus* and *Sargassum* possess fertile conceptacles on the receptacles which contain male and female sex organs. The oogonia may be liberated even before fertilization, and the fertilization takes place in the water. *Fucus* contains eight eggs in its oogonium whereas in *Sargassum* seven eggs degenerate but one. The highest evolution of sex in algae has been seen in the red algae. In *Polysiphonia,* the male spermatia are nonmotile. After their liberation, the spermatia are carried away to the female by means of water currents. From these examples, it becomes quite clear, that there is no evolutionary sequence in the evolution of sex in algae. It is thought that it developed and evolved independently in green, brown and red algae. In conclusion, we can say that the sex has been developed in response to circumstances and gametes have been developed from zoospores by means of reduction. The isogamy gave rise to oogamy through anisogamy. During oogamy, the female sex organ increases in size and stores food material to face adverse weather conditions, whereas, the male sex organ gives rise to numerous male gametes which ensure fertilization.

The most important feature of sexual reproduction is the union of two gametic nuclei, which results in the formation of a diploid zygote nucleus. Sooner or later reduction division takes place which provides a means of reshuffling of paternal and maternal chromosomes brought together during the act of fertilization. Asexual reproduction results in organic similarity while sexual reproduction in diversity. The sexual reproduction is not primarily meant for the multiplication of individuals. It is actually meant for reproduction of heritable variations that accelerate the process of evolution.
****

28. Economic Importance of Algae

Algae constitute an important group of plants which mostly occur in aquatic (both freshwater and marine) habitats. They are commonly known as pond scum and sea weeds. Their controlled growth can provide positive help in tackling some present day problems such as disposal of sewage and industrial wastes, conservation of water and soil, and improvement of quality and yield of food products. However, their uncontrolled growth causes pond epidemics, water fouling and fouling of marine vessels. At present many research organizations in India (Indian Agricultural Research Institute, New Delhi; Central Salt and Marine Chemicals Research Institute, Bhavnagar) and abroad are actively engaged in developing marine products.

Useful or Positive Importance

(a) Algae Used as Food:

Algae as food

Fig.1.Porphyra

Porphyra, Ulva, Chlorella, Alaria, Rhodymenia, Chondrus, Laniinaria and *Nostoc* are the most commonly used edible forms. People of China, Japan, etc. have long been using seaweeds and other algae as a source of food. Moreover indirectly algae provide food to man through their position in food chain. In both fresh and marine water, algae are ingested by lower animals which in turn are eaten by fish, shellfish and sea mammals. Thus, algae can be fruitfully utilized in fish culture. So there has greet economic importance of algae as food in some area.

(b) Algae Used as Fodder:

The blue green algae as well as green algae and a few red algae form the food for various fishes and aquatic animals. Some of the marine algae are used as fodder for Cattle by the Americans, French and Danish people. The people living in the American Pacific coasts use pieces of sea weeds from which kelp is prepared for poultry.

For this purpose small industries for cultivation of *Archophyllum, Fucus, Laminaria,* etc., have been raised. *Pelvetia*, a brown sea weed, increases the milk-content of the cattle and sometimes its butter and fat. A few red algae, like *Rhodymenia,* are taken by the cattle. *Sargassum* is a good cattle food.

Fig.2. *Fucus*

(c) Algae Used in Agriculture:

Some blue green algae, like *Nostoc, Anabaena* etc., increases the fertility of soil by nitrogen fixation, thus thereby helping the production of crops. The people of France, Ceylon, etc., use a few marine algae as fertilisers.

Chlorella is used to purify the soil. It liberates oxygen by photosynthesis which decomposes the rotten substances. The brown algae, like *Pelargophycus, Macracystis, Nereocystis,* etc. furnish us with potash, which is used as a manure.

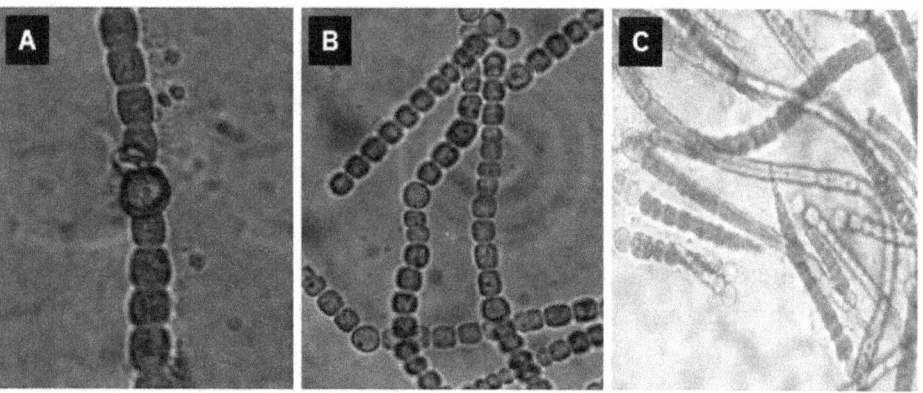

Fig.3. Nitrogen fixing cyanobacteria *(A) Nostoc sp.; (B) Anabaena sp. (C) Calothrix sp*

(d) Algae Used in Industry:

The algae are used in various kinds of industries.

These are as follows:

(i) Agar Industry:

Fig.4. *Gelidium and agar agar*

Agar is a transparent substance obtained from the red algae, like *Gelidium, Gracillaria*, etc. It is solid at normal temperature but becomes liquid above 95°C. Agar has diverse uses. Agar is used for the preparation of culture media for fungi and bacteria in the laboratory as well as a sizing material in the textile industry, as a laxative, and as a solidifying agent in the preparation of desserts.

(ii) Algin Industry:

Algin is a gelatinous or jelley like substance obtained from the cell-wall of some red algae. It is used in different commercial purposes, like the preparation of shampoos, cosmetics, hair dressings, shaving creams and other articles of toilet, shoe polishes, lubricating jellies, ice-creams and desserts. It is also used in paints and rubber industry.

(iii) Kelp Industry:

The ash left behind, after burning the brown algae, like *Laminaria, Sargassum*, etc., is known as kelp. It chiefly contains potassium and iodine. Formerly, these algae were used for the production of the ingredients but it has been discarded at present.

Fig.4. Kelp/Kombu harvesting on the beach

(iv) Diatomaceous Earth Industry:

The diatomaceous earth is a kind of earth deposited at the bottom of the sea, after the death and decay of Diatoms. The largest diatomaceous earth deposit has been discovered in Santamaria oil-fields in California.

The uses of diatomaceous earth are-much varied, viz., as a filtering material, as a substance used in insulation of boilers, blast furnaces, etc., as a cementing material, as a mild abrasive in metal polishes and tooth-pastes, as an absorbent for liquid nitroglycerine (the explosive material of the dynamite), as a material for painting the ship to protect it against the attack of sea-barnacles.

Fig.5. Diatomaceous Earth Grinding Mills | Williams Crusher

(e) Algae Used in Medicine:

A large number of medicines are obtained from algae. The brown algae like *Laminaria, Sargassum,* etc., are good sources of iodine. These algae also yield medicines for treatment of goitre and gland diseases. Antibiotics are also prepared from some of these algae. Chloranthus, an antibiotic is obtained from *Chlorella.* The red algae like *Gelidium, Gracillaria*, etc., yield agar, which is a laxative and used in the preparation of medicinal pills and ointments.

Antibiotics and Medicines

- Some algae yield antibiotics e-g, Chlorellin Is obtained from green alga Chlorella, that inhibits the growth of certain bacteria.
- Brown algae is used in manufacture of various goiter medicines

(f) Algae Used in the Purification of Water:

Fig.6. Algae Used in the Purification of Water

The green algae play an important role in the purification of water in which they grow. They utilise the carbon dioxide released by the various animals during their respiration and give off oxygen during photosynthesis and thus the water is purified.

(g) Algae Used in Aviation:

The aviator prior to their flight store lumps of algae, such as Chlorella, in the craft along with Oxygen-cylinders. The oxygen of the cylinders are utilised by the aviators when the plane rises to high altitude where the air is much rarefied. In case, there be a shortage or stoppage of oxygen supply from the cylinders the algae give off necessary oxygen for carrying on respiration.

(h) Algae as a Source of Petroleum and Gas:

The minute marine algae about the solar energy store it up as a potential energy inside their body. When these algae are taken by the marine animals it gets in their body and is stored up as such. After the death of these animals their dead bodies are deposited in heaps at the bottom of the sea. Subsequently the dead bodies are decomposed by bacteria in absence of oxygen and being subjected to a great hydrostatic pressure of sea water. Their organic substances containing the energy are gradually converted into oil and gas. These are utilised by man for various purposes.

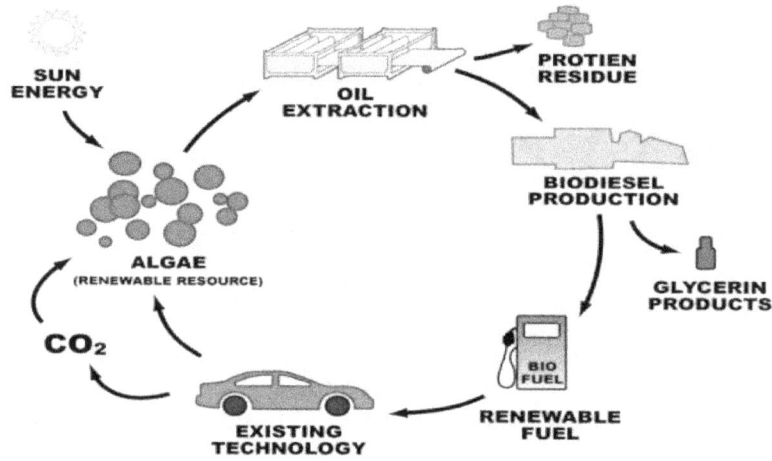

Fig.7. 'Algae Biodiesel' – Tremendous potential for next-generation Green Energy

B. Some of the major harmful effects of Algae

1. Harmful to living stock:

The algae are harmful to humans in several ways. Volvocales, Chlorococcales, Myxophyceae and several others occur in such a great abundance in water, that they colour the whole water either green or blue green and cause the death of fishes. The algae block the gills of the fishes and they respire during night and make the respiration of fishes difficult by complete depletion of oxygen. Sometimes the algae are found in water so abundantly that they make difficult to drink water of livestock. Some blue green algae have been reported poisonous and they directly cause the death of living stock who drink this contaminated water.

2. Blocking of photosynthesis:

The epiphytic algae which are found upon other plants and trees block photosynthesis and indirectly harm the trees and plants.

3. Parasitic algae:

The well-known disease 'red rust of tea' is not caused by any parasitic fungus but an algal form *Cephaleuros virescens*. This causes havoc to tea plants in Assam tea gardens.

Besides, this parasitic form attacks several other plants, e.g., *Mangifera,* Rhododendron. Coffea, etc. The heavy losses are caused to tea and coffee by this parasitic algal form.

Fig.8. Cephaleuros Species, the Plant-Parasitic Green Algae

4. Mechanical injury:

Sometimes the filamentous forms of algae are found in such a great abundance and net-like behaviour, that many fishes and other aquatic animals may get perish in these tangles, and direct death is inflicted upon them.

5. Contamination of water supply:

Many blue-green, green and other algae contaminate the water of city reservoirs. This contamination develops a foul odour in the water and makes the water unhygienic. The algae also form some mucilaginous secretions which are the seats of harmful bacteria and other pathogens causing several human and animal diseases.

6. Fouling of ships:

Some algae are attached to the ships, and this is called fouling of ship. The fouling retards the speed of the ship. To avoid this nuisance the ships are periodically dried up and painted with copper paint.

7. Deterioration of exposed fabrics:

Commonly in rainy season, if the wet fabrics are exposed, within a few days a blue green alga appears on it and makes the cloth black spotted and weak. This was a serious problem during the Second World War. This algal growth is usually followed by bacterial infection and the fibres are completely destroyed.

****

29.

Culturing of algae

General principles

A culture can be defined as an artificial environment in which the algae grow. In theory, culture conditions should resemble the alga's natural environment as far as possible; in reality many significant differences exist, most of which are deliberately imposed.

A culture has three distinct components: a) a culture medium contained in a suitable vessel; and b) the algal cells growing in the medium; and c) air, to allow exchange of carbon dioxide between medium and atmosphere.

Extensive measures must be taken to keep pure unialgal cultures chemically and biologically clean. Chemical contamination may have unquantifiable, often deleterious, and therefore undesirable effects on algal growth. Biological contamination of pure algal cultures by other eukaryotes and prokaryotic organisms in most cases invalidates experimental work, and may lead to the extinction of the desired algal species in culture throughout-competition or grazing. In practise, it is very difficult to obtain bacteria-free (axenic) cultures, and although measures should be taken to minimize bacterial numbers, a degree of bacterial contamination is often acceptable.

For an entirely autotrophic alga, all that is needed for growth is light + CO_2 + H_2O + nutrients + trace elements. By means of photosynthesis the alga will be able to synthesize all the biochemical compounds necessary for growth. Only a minority of alga seem, however, to be entirely autotrophic; many are unable to synthesize certain biochemical compounds (certain vitamins, for example) and will require these to be present in the medium. This condition is known as auxotrophy.

Based on their growth characteristics, two kinds of cultures can be defined.

- In limited volume (batch) cultures, resources are finite. When the resources present in the culture medium are abundant, growth occurs according to a sigmoid curve, but once the resources have been utilised by the cells, the cultures die unless supplied with new medium. In practise this is done by subculturing, i.e. transferring a small volume of existing culture to a large volume of fresh culture medium at regular intervals.

- In continuous cultures, resources are potentially infinite: cultures are maintained at a chosen point on the growth curve by the regulated addition of fresh culture medium. In practise, a volume of fresh culture medium is added automatically at a rate proportional to the growth rate of the alga, while an equal volume of culture is removed.

Physical parameters

Temperature

The temperature at which cultures are maintained should ideally be as close as possible to the temperature at which the organisms were collected; polar organisms ($<10°C$); temperate ($10-25°C$); tropical ($>20°C$). An intermediate value of $18-20°C$ is most often employed. Temperature controlled incubators usually use constant temperature (transfers to different temperatures should be conducted in steps of $2°C$/week), although some models permit temperature cycling. In temperate regions ambient room temperature is generally acceptable for culturing purposes.

Light

Natural light is usually sufficient to maintain cultures in the laboratory. Cultures should never be exposed to direct sunlight (which may cause photopigment damage), and should therefore be placed next to a north-facing window (in the northern hemisphere).

Artificial lighting by fluorescent bulbs is often employed for culture maintainance and experimental purposes. Light intensity should range between 0.2-50% of full daylight (= $1660 \mu E/s/m^2$), with 5-10% (c. $80-160 \mu E/s/m^2$) most often employed. Light quality (spectrum) depends on type of bulb used (see manufacturer's technical data), the most common types being 'cool white' or 'daylight' bulbs. Light intensity and quality can be manipulated with filters. Many microalgal species do not grow well under constant illumination, and hence a light/dark (LD) cycle is used (maximum 16:8 LD, usually 14:10 or 12:12).

Mixing

Mixing of microalgal cultures may be necessary under certain circumstances: when cells must be kept in suspension in order to grow (particularly important for heterotrophic dinoflagellates); in concentrated cultures to prevent nutrient limitation effects due to stacking of cells and to increase gas diffusion. It should be noted that in the ocean cells seldom experience turbulence, and hence mixing should be gentle. The following methods may be used: bubbling with air (may damage cells); plankton wheel or roller table (about 1 rpm); gentle manual swirling. Most cultures do well without mixing, particularly when not too concentrated, but when possible gentle manual swirling (once each day) is recomended.

Types of culture vessel

Culture vessels should have the following properties: nontoxic (chemically inert); reasonably transparent to light; easily cleaned and sterilized; provide a large surface to volume ratio (depending on organism).

Certain materials which could potentially be used for culture vessels may leach chemicals which have a deleterious effect on algal growth into the medium. The use of chemically inert materials is particularly important when culturing oceanic plankton and during isolation. A list of materials which should be safe, inhibitory, or toxic to algal cultures is given by Stein (1973). Recommended materials for culture vessels and media preparation include:

- Teflon (very expensive, used for media preparation);
- Polycarbonate (expensive and becomes cloudy and cracks with repeated autoclaving);
- Polystyrene (cheaper alternative to Teflon and polycarbonate, not autoclavable);
- Borosilicate glass (has been shown to inhibit growth of some species).

Materials which should generally be avoided during microalgal culturing include all types of rubber, and PVC. The following types of culture vessel are most commonly used for microalgal culturing (but may not prove suitable for all species):

- Erlenmeyer flasks (glass or polycarbonate) with cotton, glass, polypropylene, or metal covers.

- tubes (glass or polycarbonate) with cotton, glass, polypropylene or metal caps.
- polystyrene tissue culture flasks, purchased as single-use sterile units (eg. Iwaki, Nunc, Corning).

Cleaning/sterilization of culture materials

Cleaning procedure

- scrub (abrasive brushes not appropriate for most plastics) and soak with warm detergent (not domestic detergents, which leave a residual film on cultureware – use laboratory detergent such as phosphate-free Decon);
- rinse extensively with tap water;
- soak in 10% HCl for 1 day-1 week (not routinely necessary, but particularly important for new glass and polycarbonate material);
- rinse extensively with distilled and finally bidistilled water;
- leave inverted to dry in a clean, dust free place.

Sterilization

Sterilization can be defined as a process which ensures total inactivation of microbial life (NB. not the same as desinfection, which is defined as an arbitrary reduction of bacterial numbers). The primary purpose of sterilization is to prevent contamination by unwanted organisms, but it may also serve to eliminate unwanted chemicals. There are several sterilisation methods and the choice depends on the purpose and material used:

- **Gas** (ethylene oxide, often used for disposable plastic material);
- **Moist heat - Autoclave or pressure cooker** (1-2 Bar, 121°C, in pure saturated steam).

  For sterility the steam must penetrate the material (wrap in material that allows access of steam - kraft paper or aluminium). Autoclave steam may introduce chemical contaminants; glass and polycarbonate vessels should be autoclaved containing a small amount of bidistilled water which is poured out (thus diluting contaminants) under sterile conditions immediately prior to use. Never close vessels (risk of implosion); use cotton wool bungs, or leave screw caps slightly open.

- **Dry heat**

Oven (at least 2 hours at 160°C). Cover neck of vessel with aluminium foil to maintain sterility on removal from oven.

Flame (heating to incandescence for metal)

Problems: few materials stand high temperatures (glass, teflon, silicone, metal, cotton).

- **Radiation**

    X and γ - radiation (industrial applications for disposable plastics that cannot stand more than 60°C, not always effective)

    UV radiation 240-280 nm (not often used for culture materials).

Transfer protocol

The following procedures should always be used when preparing media or transferring cultures:

- work in clean place or preferably in a laminar flow cabinet (cabinet must be turned on at least 30 minutes before transfer; if equipped with UV lamps, leave on overnight prior to use).
- Clean working surface with 70% alcohol (ethanol/isopropanol) prior to and after use.
- Clean hands with disinfecting soap and rinse with 70% alcohol prior to all operations.
- When not using a laminar flow cabinet (and to be safe even when using a cabinet), sterilise (flame) the neck of vessel of origin before and after transfer (not possible with some plastic vessels, which must, therefore, be opened in a laminar flow cabinet).
- Pipettes must be clean and sterile; use autoclavable tips for repeating pipettes (eg. Gilson), pre-wrapped sterile single use plastic/glass pipettes, or if using non-sterile glass pipettes (with cotton plugs), sterilise in the flame before use.

The frequency of culture transfer depends on species and culture conditions (it is advisable to follow growth for each unfamiliar species to get a feeling). Transfers should be conducted as population growth slows, preferably before population growth ceases (stationary phase). Culture transfers (usually 1-5 % of culture volume into fresh sterile medium) can be conducted directly (pouring, not used when vessel neck is flame sterilized), or more often with a pipette.

Culture media

Media for the culture of marine phytoplankton consist of a seawater base (natural or artificial) which may be supplemented by various substances essential for microalgal growth, including nutrients, trace metals and chelators, vitamins, soil extract and buffer compounds.

Seawater base

The quality of water used in media preparation is very important.

Natural seawater can be collected near shore, but its salinity and quality is often quite variable and unpredictable, particularly in temperate and polar regions (due to anthropogenic pollution, toxic metabolites released by algal blooms in coastal waters). The quality of coastal water may be improved by ageing for a few months (allowing bacterial degradation of inhibitory substances), by autoclaving (heat may denature inhibitory substances), or by filtering through acid-washed charcoal (which absorbs toxic organic compounds). Most coastal waters contain significant quantities of inorganic and organic particulate matter, and therefore must be filtered before use (e.g. Whatman no. 1 filter paper).

The low biomass and continual depletion of many trace elements from the surface waters of the open ocean by biogeochemical processes makes this water much cleaner, and therefore preferable for culturing purposes.

Collect seawater from the front of the boat (or pump from subsurface) to avoid contamination. Seawater can be stored in polyethylene carboys, and should be stored in cool dark conditions.

Artificial seawater, made by mixing various salts with deionized water, has the advantage of being entirely defined from the chemical point of view, but they are very laborious to prepare, and often do not support satisfactory algal growth. Trace contaminants in the salts used are at rather high concentrations in artificial seawater because so much salt must be added to achieve the salinity of full strength seawater.

Some of the more successful artificial seawater media that have been developed include the ESAW medium of Harrison *et al.* (1980), and the AK medium of Keller *et al.* (1987).

Nutrients, trace metals, and chelators

The term 'nutrient' is colloquially applied to a chemical required in relatively large quantities, but can be used for any element or compound necessary for algal growth.

Table 1. The average concentrations of constituents of potential biological importance found in typical seawater (after Brand, 1990).

Element	Average Molar concentration (range in brackets)
Group A	
Na^+	4.7×10^{-1}
K^+	1.02×10^{-2}
Mg^{2+}	5.3×10^{-2}
Ca^{2+}	1.03×10^{-2}
Cl^-	5.5×10^{-1}
SO_4^{2-}	2.8×10^{-2}
HCO_3^-	2.3×10^{-3}
BO_3^{3-}	4.2×10^{-4}
Group B	
Br^-	8.4×10^{-4}
F^-	6.8×10^{-5}
IO_3^-	4.4×10^{-7}
Li^+	2.5×10^{-5}
Rb^+	1.4×10^{-6}

Sr^{2+}	8.7×10^{-5}
Ba^{2+}	1×10^{-7}
MoO_4^{2-}	1.1×10^{-7}
VO_4^{3-}	2.3×10^{-8}
CrO_4^{2-}	4×10^{-9}
AsO_4^{3-}	2.3×10^{-8}
SeO_4^{2-}	1.7×10^{-9}
Group C	
NO_3^-	3×10^{-5} (10^{-8} to 4.5×10^{-5})
PO_4^{3-}	2.3×10^{-6} (10^{-7} to 3.5×10^{-6})
Fe^{3+}	1×10^{-9} (10^{-10} to 10^{-7})
Zn^{2+}	6×10^{-9} (5×10^{-11} to 10^{-7})
Mn^{2+}	5×10^{-10} (2×10^{-10} to 10^{-6})
Cu^{2+}	4×10^{-9} (5×10^{-10} to 6×10^{-9})
Co^{2+}	2×10^{-11} (10^{-11} to 10^{-10})
SiO_4^{4-}	1×10^{-4} (10^{-7} to 1.8×10^{-4})
Ni^{2+}	8×10^{-9} (2×10^{-9} to 1.2×10^{-8})

Group A: concentrations of these constituents exhibit essentially no variation in seawater, and high algal biomass cannot deplete them in culture media. These constituents do not, therefore, have to be added to culture media using natural seawater, but do need to be added to deionized water when making artificial seawater media.

Group B: also have quite constant concentrations in seawater, or vary by a factor less than 5. Because microalgal biomass cannot deplete their concentrations significantly, they also do not need to be added to natural seawater media. Standard artificial media (and some natural seawater media) add molybdenum (as molybdate), an essential nutrient for algae, selenium (as selenite), which has been demonstrated to be needed by some algae, as well as strontium, bromide and flouride, all of which occur at relatively high concentrations in seawater, but none of which have been shown to be essential for microalgal growth.

Group C: all known to be needed by microalgae (silicon is needed only by diatoms and some chrysophytes, and nickel is only known to be needed to form urease when algae are using urea as a nitrogen source). These nutrients are generally present at low concentrations in natural seawater, and since microalgae take up substantial amounts, concentrations vary widely (generally by a factor of 10 to 1000). All of these nutrients (except silicon and nickel in some circumstances) generally need to be added to culture media in order to generate significant microalgal biomass.

Nitrate is the N source most often used in culture media, but ammonium can also be used, and indeed is the preferential form for many algae since it does not have to be reduced prior to amino acid synthesis, the point of primary intracellular nitrogen assimilation into the organic linkage. Ammonium concentrations greater than 25µM are, however, often reported to be toxic to phytoplankton, so concentrations should be kept somewhat low.

Inorganic (ortho) phosphate, the P form preferentially used by microalgae, is most often added to culture media, but organic (glycerol)phosphate is sometimes used, particularly when precipitation of phosphate is anticipated (when nutrients are autoclaved in the culture media rather than separately, for example). Most microalgae are capable of producing cell surface phosphatases which allow them to utilise this and other forms of organic phosphate as a phosphorus source.

The trace metals which are essential for microalgal growth are incorporated into essential organic molecules, particularly a variety of coenzyme factors which enter into photosynthetic reactions. Of these metals, the concentrations (or more accurately the biologically available concentrations) of Fe, Mn, Zn, Cu and Co (and sometimes Mo and Se) in natural waters may be limiting to algal growth. Little is known about the complex relationships between chemical speciation of metals and biological availability. It is thought that molecules which complex with metals (chelators) influence the availability of these elements. Chelators act as trace metal buffers, maintaining constant concentrations of free ionic metal. It is the free ionic metal, not the chelated metal, which influences microalgae, either as a nutrient or as a toxin. Without proper chelation some metals (such as Cu) are often present at toxic concentrations, and others (such as Fe) tend to precipitate and become unavailable to phytoplankton. In natural seawater, dissolved organic molecules (generally present at concentrations of 1-10mg l^{-1}) act as chelators. The most widely used chelator in culture media additions is EDTA (ethylenediaminetetraacetic acid), which must be present at high concentrations since most complexes with Ca and Mg, present in large amounts in seawater. EDTA may have an additional benefit of reducing precipitation during autoclaving. High concentrations have, however, occasionally been reported to be toxic to microalgae. As an alternative the organic chelator citrate is sometimes utilised, having the advantage of being less influenced by Ca and Mg. The ratio of chelator:metal in culture medium ranges from 1:1 in f/2 to 10:1 in K medium. High ratios may result in metal deficiencies for coastal phytoplankton (ie. too much metal is complexed), and many media therefore use intermediate ratios.

In today's aerobic ocean, iron is present in the oxidized form as various ferric hydroxides and thus is rather insoluble in seawater. While concentrations of nitrogen, phosphorus, zinc and manganese in deep water are similar to plankton elemental composition, there is proportionally 20 times less iron in deep water than is apparently needed, leading to the suggestion that iron may be the ultimate geochemically limiting nutrient to phytoplankton in the ocean (Brand, 1986). Very little is known about iron in seawater or phytoplankton uptake mechanisms due to the complex chemistry of the element. Iron availability for microalgal uptake seems to be largely dependent on levels of chelation. It is highly recommended that iron be added as the chemically prepared chelated iron salt of EDTA rather than as iron chloride or other iron salts; the formation of iron chelates is relatively slow, and iron hydoxides will form first in seawater, leading to precipitation of much of the iron in the culture medium.

Apparently as a result of the extreme scarcity of copper in anaerobic waters, copper did not begin to be utilised by organisms until the earth became aerobic and copper increased in abundance. Consequently copper does not seem to be an obligate requirement, algae either not needing it, or needing so little that free ionic copper concentrations in natural seawater are sufficient to maintain maximum growth rates (Brand, 1986). Copper may indeed be toxic, particularly to more primitive algae, and hence copper, if added to culture media at all, should be kept at low concentrations.

The provision of manganese, zinc and cobalt in culture medium should not be problematical since even fairly high concentrations are not thought to be toxic to algae.

Vitamins

Roughly _ of all microalgal species tested have been shown to have a requirement for vitamin B_{12}, which appears to be important in transferring methyl groups and methylating toxic elements such as arsenic, mercury, tin, thallium, platinum, gold, and tellurium (Brand, 1986), around 20% need thiamine, and less than 5% need biotin.

It is recommended that these vitamins are routinely added to seawater media. No other vitamins have ever been demonstrated to be required by any photosynthetic microalgae.

Soil extract

- Prepared by heating, boiling, or autoclaving a 5 to 30% slurry of soil in fresh water or seawater and subsequently filtering out the soil.

Soil extract has historically been an important component of culture media. The solution provides macronutrients, micronutrients, vitamins, and trace metal chelators in undefined quantities, each batch being different, and hence having unpredictable effects on microalgae. With increasing understanding of the importance of various constituents of culture media, soil extract is less frequently used. Soil extract should only be used on a non-experimental basis.

Buffers

The control of pH in culture media is important since certain algae grow only within narrowly defined pH ranges, and in order to prevent the formation of precipitates. Except under unusual conditions, the pH of natural seawater is around 8. Because of the large buffering capacity of natural seawater (due to a bicarbonate buffering system, HCO_3 being present at c. 2.2M) it is quite easy to maintain the pH of marine culture media. The buffer system is overwhelmed only during autoclaving, when high temperatures drive CO_2 out of solution and hence cause a shift in the bicarbonate buffer system and an increase in pH, or in very dense cultures of microalgae, when enough CO_2 is taken up to produce a similar effect. As culture medium cools after autoclaving, CO_2 reenters solution from the atmosphere, but certain measures must be taken if normal pH is not fully restored:

- The pH of seawater may be lowered prior to autoclaving (adjustment to pH 7-7.5 with 1M HCl) to compensate for subsequent increases.
- Certain media recipes include additions of extra buffer, either as bicarbonate, Tris (Tris-hydroxymethyl-aminomethane), or glycylglycine to

supplement the natural buffering system. Tris may also act as a Cu buffer, but has occasionally been cited for its toxic properties to microalgae. Glycylglycine is rapidly metabolized by bacteria and hence can only be used with axenic cultures. These additions are generally not necessary if media are filter sterilized, unless very high cell densities are expected.

- The problem of CO_2 depletion in dense cultures may be reduced by having a large surface area of media exposed to the atmosphere relative to the volume of the culture, or by bubbling with either air (CO_2 concentration c.0.03%) or air with increased CO_2 concentrations (0.5 to 5%). Unless there is a large amount of biomass taking up the CO_2, the higher concentrations could actually cause a significant decline in pH. When bubbling is employed, the gas must first pass through an in line 0.2µm filter unit (eg. Millipore Millex GS) to maintain sterile conditions. For many microalgal species, aeration is not an option since the physical disturbance may inhibit growth or kill cells.

Media preparation

The salinity of the seawater base should first be checked (30-35‰ for marine phytoplankton), and any necessary adjustments (addition of fresh water/evaporation) made before addition of enrichments.

Always use reagant grade chemicals and bidistilled (or purer) water to make stock solutions of enrichments. Gentle heating and/or magnetic stirring of stock solutions can be used to ensure complete dissolution. When preparing a stock solution containing a mixture of compounds, dissolve each individually in a minimal volume of water before mixing, then combine and dilute to volume.

Seawater, stock solutions of enrichments and the final media must be sterile in order to prevent (or more realistically minimize) biological contamination of unialgal cultures. Several methods are available for sterilization:

- **autoclaving** is the most widely used technique for sterilizing culture media, and is the ultimate guarantee of sterility (including the destruction of viruses). A commercial autoclave is best, but pressure cookers of various sizes are also suitable. Sterility requires 15 minutes at 1-2 Bar pressure and a temperature of 121°C in the entire volume of the liquid (ie. longer times for larger volumes of liquid; approximately 10 min for 100ml, 20 min for 2l, 35 min for 5 l). Bottles containing media should be no more than _ full, and should be left partially open or plugged with cotton wool or covered with aluminium foil.

Ensure the heating elements are covered with distilled water, and the escape valve should not be closed until a steady stream of steam is observed. After autoclaving, the pressure release valve should not be opened until the temperature has cooled to below 80°C.

Autoclaving is a process which has many effects on seawater and its constituents, potentially altering or destroying inhibitory organic compounds, as well as beneficial organic molecules. Because of the steam atmosphere in an autoclave, CO_2 is driven out of the seawater and the pH is raised to about 10, a level which can cause precipitation of the iron and phosphate added in the medium. Some of this precipitate may disappear upon re-equilibriation of CO_2 on cooling, but both the reduced iron and phosphate levels, and the direct physical effect of the precipitate may limit algal growth. The presence of EDTA and the use of organic phosphate may reduce precipitation effects. Addition of 5% or more of distilled water may also help to reduce precipitation (but may affect final salinity). The best solution, however, if media are autoclaved, is to sterilize iron and phosphate (or even all media additions) seperately and add them aseptically afterwards.

Autoclave steam may contaminate the media (i.e. with trace metals from the autoclave tubing). Autoclaving also produces leaching of chemicals from the medium receptacle into the medium (silica from glass bottles, toxic chemicals from plastics). Autoclaving in well-used Teflon or polycarbonate vessels reduces leaching of trace contaminants.

Autoclaving will cause evaporation of water, and hence an increase in salinity (usually of c. 1‰). Distilled water can be added prior to autoclaving to compensate for this increase.

- **Pasteurization** (heating to 90-95°C for 30 minutes) of media in Teflon or polycarbonate bottles is a potential alternative, reducing the problems of trace metal contamination and alteration of organic molecules inherent with autoclaving. Pasteurization does not, however, completely sterilize the seawater; it kills all eukaryotes and most bacteria, but some bacterial spores probably survive. Heating to 90-95°C for at least 30 minutes and cooling, repeated on two successive days ('**tyndallisation**') may improve sterilization efficiency; it is assumed that vegetative cells are killed by heat and heat resistant spores will germinate in the following cool periods and be killed by subsequent heating.

- **Ultraviolet radiation** can be used to sterilize seawater, but very high intensities are needed to kill everything in the seawater (1200 W lamp, 2-4h for culture media in quartz tubes). Such intense UV light necessarily alters and destroys the organic molecules in seawater and generates many long lived free radicals and other toxic reactive chemical species (Brand, 1986). Seawater exposed to intense UV light must, therefore, be stored for several days prior to use to allow the level of these highly reactive chemical species to decline.
- Sterilization by commercial **microwave** apparatus is another option. Microwave sterilization has not, as yet, been widely employed in culture media preparation due to uncertainties about sterilization efficiency. Trials should be conducted before use of this method to ensure sterility of seawater.
- **Sterile filtration** is probably the best method of sterilizing seawater without altering the chemistry of the seawater, as long as care is taken not to contaminate the seawater with dirty filter apparatus. Sterilization efficiency is, however, to some extent reduced compared with heat sterilization methods. Membrane filters of 0.2µm porosity are generally considered to yield water free of bacteria, but not viruses. 0.1µm filters can be used, but the time required for filtration of large volumes of culture media may be excessively long. The filtration unit must be sterile: for small volumes (<50ml) pre-sterilized single use filter units for syringe filtration (e.g. Millipore Millex GS) can be used; for volumes up to 1 litre reusable autoclavable self-assembly filter units (glass or polycarbonate) with 47mm cellulose ester membrane filters (e.g. Millipore HA) can be used with suction provided by a vacuum pump; for larger volumes an in-line system with peristaltic pump and cartridge filters may be the best option.

Filter units (particularly disposable plastic systems), and the membrane filters themselves can also leak toxic compounds into the filtrate. The first portion of filtrate (e.g. 5% of volume to be filtered) should be discarded to alleviate this problem.

Most stock solutions of culture medium additions can be sterilized separately by autoclaving, although vitamin stock solutions are routinely filtered through 0.2µm single use filter units (e.g. Millipore Millex GS), since heat sterilization will denature these organic compounds. Filter sterilization of all additions may reduce uncertainties about stability of the chemical compounds and contamination from autoclave steam, but absolute sterilization is not guaranteed. Stock solutions can be stored in ultraclean sterile glass, polycarbonate or Teflon tubes/bottles. In order to minimize effects of any microbial contaminations, all stock solutions should be stored in a refrigerator at 4°C, except vitamin stocks which are stored frozen at –20°C and thawed immediately prior to use.

Some culture medium recipes

The recipes of 3 media which have proved successful for the culture of coccolithophores are given. In practise dilutions of f/2 and K medium (eg. 10%-->f/20, K/10) are sufficient to maintain good coccolithophorid growth. A variety of alternative marine culture media recipes are given by Stein (1975), and on the web pages of the major culture collections (eg. CCMP, Utex).

f/2-Si (Guillard, 1975)

To 996ml of sterile seawater aseptically add:

Quantity	Compound	Stock solution (sterile)	Final conc. in medium
1.0ml	$NaNO_3$	75.0g/litre H_2O	884µM
1.0ml	$NaH_2PO_4.H_2O$	5.0g/litre H_2O	36µM
1.0ml	f/2 trace metal solution	(see recipe below)	(see below)
1.0ml	f/2 vitamin solution	(see recipe below)	(see below)

f/2 trace metal solution

To 950ml distilled H_2O add:

Quantity	Compound	Stock solution	Final conc. in medium
3.15g	$FeCl_3.6H_2O$	-	11.7µM

Quantity	Compound	Stock solution	Final conc. in medium
4.36g	Na$_2$EDTA.2H$_2$O	-	12µM
1.0ml	CuSO$_4$.5H$_2$O	9.8g/litre H$_2$O	0.04µM
1.0ml	Na$_2$MoO$_4$.2H$_2$O	6.3g/litre H$_2$O	0.03µM
1.0ml	ZnSO$_4$.7H$_2$O	22.0g/litre H$_2$O	0.08µM
1.0ml	CoCl$_2$.6H$_2$O	11.9g/litre H$_2$O	0.05µM
1.0ml	MnCl$_2$.4H$_2$O	178.2g/litre H$_2$O	0.9µM

Make up to 1 litre with distilled H$_2$O, sterilize (autoclave or filter) and store in fridge.

f/2 Vitamin solution

To 950ml distilled H$_2$O add:

Quantity	Compound	Stock solution	Final conc. in medium
1.0ml	Vit. B$_{12}$ (cyanocobalamin)	0.5g/litre H$_2$O	0.37nM
1.0ml	Biotin	5.0mg/litre H$_2$O	2.0nM
100.0mg	Thiamine HCl	-	0.3µM

Make up to 1 litre with distilled H2O, filter sterilize into plastic vials and store in freezer.

K(-Si) (Keller *et al.*, 1987)

To 994 ml of sterile seawater aseptically add:

Quantity	Compound	Stock solution (sterile)	Final conc. in medium
1.0ml	$NaNO_3$	75.0g/litre H_2O	884µM
1.0ml	NH_4Cl *	0.535g/litre H_2O	10µM
1.0ml	Na_2glycerophosphate **	2.16g/litre H_2O	10µM
1.0ml	TRIS-base (pH7.2) ***	121.1g/litre H_2O	1mM
1.0ml	K trace metal solution	(see recipe below)	(see below)
1.0ml	f/2 vitamin solution	(see recipe below)	(see below)

* should not be autoclaved (volatile when heated)

** Inorganic orthophosphate can be substituted (particularly if not autoclaving)

*** can be omitted (particularly if not autoclaving, or if cell density will not be very high)

K trace metal solution

To 950ml distilled H_2O add:

Quantity	Compound	Stock solution	Final conc. in medium
4.3g	(Na)FeEDTA	-	11.7µM
37.22g	$Na_2EDTA.2H_2O$	-	100µM
0.5ml	$CuSO_4.5H_2O$	9.8g/litre H_2O	0.02µM
1.0ml	$Na_2MoO_4.2H_2O$	6.3g/litre H_2O	0.03µM
1.0ml	$ZnSO_4.7H_2O$	22.0g/litre H_2O	0.08µM

1.0ml	$CoCl_2 \cdot 6H_2O$	11.9g/litre H_2O	0.05µM
1.0ml	$MnCl_2 \cdot 4H_2O$	178.2g/litre H_2O	0.9µM
1.0ml	H_2SeO_3	1.29mg/litre H_2O	0.01µM

Make up to 1 litre with distilled H_2O, sterilize (autoclave or filter) and store in fridge.

f/2 Vitamin solution

To 950ml distilled H_2O add:

Quantity	Compound	Stock solution	Final conc. in medium
1.0ml	Vit. B_{12} (cyanocobalamin)	0.5g/litre H_2O	0.37nM
1.0ml	Biotin	5.0mg/litre H_2O	2.0nM
100.0mg	Thiamine HCl	-	0.3µM

Make up to 1 litre with distilled H2O, filter sterilize into plastic vials and store in freezer.

BWM (blue water medium) Brand (unpublished)

To 995.5 ml sterile seawater, aseptically add:

Quantity	Compound	Stock solution (sterile)	Final conc. in medium
1.0ml	$NaNO_3$	8.5g/litre H_2O	100µM
1.0ml	NH_4Cl *	0.535g/litre H_2O	10µM

1.0ml	Na₂glycerophosphate *	2.16g/litre H₂O	10µM
1.0ml	BWM trace metal solution	(see recipe below)	(see below)
0.5ml	BWM vitamin solution	(see recipe below)	(see below)

* should not be autoclaved (volatile when heated)

** can be substituted for inorganic orthophosphate

BWM trace metal solution

To 950ml distilled H$_2$O add:

Quantity	Compound	Stock solution	Final conc. in medium
0.367g	(Na)FeEDTA	-	1µM
3.72g	Na$_2$EDTA.2H$_2$O	-	10µM
1.0ml	CuSO$_4$.5H$_2$O	0.25g/litre H$_2$O	0.001µM
1.0ml	ZnSO$_4$.7H$_2$O	28.75g/litre H$_2$O	0.1µM
1.0ml	CoCl$_2$.6H$_2$O	2.38g/litre H$_2$O	0.01µM
1.0ml	MnCl$_2$.4H$_2$O	19.8g/litre H$_2$O	0.1µM
1.0ml	H$_2$SeO$_3$ *	1.29g/litre H$_2$O	0.01µM

* optional

Make up to 1 litre with distilled H$_2$O, sterilize (autoclave or filter) and store in fridge.

BWM Vitamin solution

To 950ml distilled H₂O add:

Quantity	Compound	Stock solution	Final conc. in medium
1.0ml	Vit. B_{12} (cyanocobalamin)	13.55g/litre H_2O	0.01µM
1.0ml	Biotin	2.5mg/litre H_2O	0.001µM
33.3mg	Thiamine HCl	-	0.1µM

Make up to 1 litre with distilled H2O, filter sterilize into plastic vials and store in freezer.

Some aquatic microalgae grow well on solid substrate.

A 3% high grade agar can be used for the solid substrate. The agar and culture medium should not be autoclaved together, because toxic breakdown products can be generated. The best procedure is to autoclave 30% agar in deionized water in one container and nine times as much seawater base in another. After removing from the autoclave, sterile nutrients are added aseptically to the water, which is then mixed with the molten agar. After mixing, the warm fluid is poured into sterile petri dishes, where it solidifies when it cools. The plate is inoculated by placing a drop of water containing the algae on the surface of the agar, and streaking with a sterile implement. The plates are then maintained under standard culture conditions. This method may be particularly effective for cleaning cultures infected with bacteria, clean colonies of the algal species being isolated from the plate into fresh liquid medium.

Considerations in the selection of culture media

Two approaches to selection of media composition:

- In theory it is best to work on the principle that if the alga does not need the addition of any particular chemical substance to the culture media (i.e. if it has no observable positive effect on growth rate), don't add it.
- In practise it is often easier to follow well known (and presumably, therefore, well tried) media recipes, and safer to add substances 'just in case' (providing they have no observable detrimental effect on algal growth).

When choosing a culture medium, the natural habitat of the species in question should be considered in order to determine its environmental requirements:

Table 2. Considerations in the selection of culture media (after Brand, 1986).

Habitats	Eutrophic, coastal, variable environment, well-mixed, nutrient rich	Oligotrophic, oceanic, predictable environment, stratified, nutrient poor
Organisms	*r*-selected, rapid growth rate, wide environmental tolerance, highly autotrophic, high biomass/low diversity communities	*K*-selected, slow growth rate, narrow environmental tolerance, tendency towards auxotrophy/mixotrophy/photoheterotrophy, low biomass/high diversity communities
	diatoms	
	hetero- coccolithophores holo-	
	dinoflagellates	
Culture media	f/2, GPM, IMR, ESAW	K, BWM ??????

To date, culture techniques have been quite successful in culturing the *r*-selected species from eutrophic habitats, but quite poor at culturing the *K*-selected species from oligotrophic habitats. The media recipes currently available should be adequate for the culture of diatoms, some coccolithophorids (particularly coastal heterococcolithophorids) and few coastal dinoflagellates, the exact choice for a particular species therefore dependant on trial and error (taking into account the above discussion). K and BWM media are perhaps the best available options for successful culturing of oceanic coccolithophorids and dinoflagellates, but further advances, including the reduction in concentration of many nutrients, and the possible use of organic nutrients, should also be considered.

- It must be remembered that in culturing in general there are (within limits) no right and wrong methods; culture media have only developed by culturers trying out various additions (usually based on theoretical considerations), and hence innovation is actively encouraged.

Pure, axenic cultures

If biological contaminants appear in a culture, the best remedy is to isolate a single cell from the culture with a micropipette, and try to establish a new, clean clonal culture. Alternatively the culture can be streaked on an agar plate (see below) in the hope of attaining a colony free of contaminants. Neither of these methods work well, however, for eliminating bacteria that attach firmly to the surface of microalgae. Placing a test tube of microalgal culture in a low-intensity 90 kilocycles/sec ultrasonic water bath for varying lengths of time (a few seconds to tens of minutes) can sometimes physically separate bacteria without killing the algae, making it easier to obtain an axenic culture by micropipette isolation. Often, however, to achieve an axenic culture, antibiotics must be used. Best results appear to occur when an actively growing culture of algae is exposed to a mixture of penicillin, streptomycin and gentamycin for around 24 hours. This drastically reduces the growth of bacteria while allowing the microalgae to continue to grow, increasing the chances of obtaining an axenic cell when using micropippette or agar streaking isolation. Different algal species tolerate different concentrations of antibiotics, so a range of concentrations should be used (generally 50-500 mg/l). Other antibiotics that can be used include chloramphenicol, tetracycline, and bacitracin. Antibiotic solutions should be made with distilled water and filter-sterilized (0.2μm filter units) into sterile tubes, and should be stored frozen until use. Another approach is to add a range of antibiotic concentrations to a number of subcultures and then select the culture that has surviving algal cells but no surviving bacteria or other contaminants. Sterility of cultures should be checked by microscopic examination and by adding a small amount of sterile bacterial culture medium (e.g. 0.1% peptone) to a microalgal culture and observing regularly for bacterial growth. Absence of bacterial growth does not, however, ensure that the microalgal culture is axenic, since the majority of bacteria do not respond to standard enrichments. In reality there is no way of demonstrating that a microalgal culture is completely axenic. In practise, therefore, axenic usually means 'without demonstrable unwanted prokaryotes or eukaryotes'. Some microalgal cultures may die when made axenic, probably due to the termination of obligate symbiotic relationships with bacteria.

Equipment checklist

- a north-facing window (in the northern hemisphere) in a room in which temperature is not too high (normally <25°C, depending on type of algae being grown), supplementary fluorescent lights, or; a reach-in (fridge-like) incubator, with controllable temperature and illumination, or; a cupboard type reach-in, or walk-in incubator (very expensive).

- Bench space (preferably in a clean lab), and/or; a laminar flow cabinet. 70% ethanol, a bunsen burner, sterile glass/plastic pipettes (2ml, 10ml, 25ml), pipette bulb.

- Detergent (laboratory, not domestic), 10% HCl, and distilled water for cleaning culture ware.
- Culture vessels; borosilicate glass Erlenmeyer (or other shape) flasks with cotton wool bungs / glass covers, or pre-wrapped sterile polystyrene culture flasks.
- seawater (for natural based media), reagent grade chemicals and bidistilled water for media enrichments, an analytical balance (1mg sensitivity), spatula, hot-plate magnetic stirrer, sterile glass / polystyrene / polycarbonate tubes/bottles for media enrichments, media recipients – 1 litre glass/polycarbonate/Teflon bottles, fridge/freezer, pH meter, salinity meter.
- equipment sterilization: either; an autoclave / pressure cooker (for moist heat sterilization) or; an oven (for dry heat sterilization)

- media sterilization either; an autoclave / pressure cooker or; a microwave oven or; filter apparatus (see discussion above)

****

30.

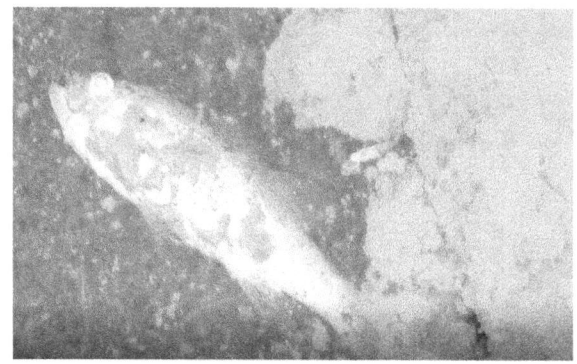

Toxins of Algae

Toxins are all compounds that are either synthesized by the algae or formed by the composition of metabolic products and hence represent an intrinsic characteristic of the organism. Of the millions of species of microalgae those that produce specific toxins scarcely exceed a hundred. These occur in both salt and freshwaters, and while most are planktonic some are benthic or floating at water surface. Toxins can attract particular attention when they cause the death of livestock that has drunk water containing them or fish and shellfish in the sea, or humans that consume these. Toxins have been divided into different classes based on the syndromes associated with exposure to them, such as paralytic shellfish poisoning (PSP), diarrheic shellfish poisoning (DSP), neurotoxic shellfish poisoning (NSP), ciguatera fish poisoning (CFP), and amnesic shellfish poisoning (ASP). Algae which seem to be directly producer of toxic substances mostly belong to three taxonomic groups: Cyanophyta, Haptophyta, and Dinophyta. In addition to these there are some groups which include one or two toxic members. Species of Chattonella and Heterosygma, belonging to the Raphidophyceae, form toxic red tides in Japanese waters and a few diatoms of the genus Peusdonitzschia produce domoic acid, a low molecular amino acid causing amnesic shellfish poisoning. Among the 50 freshwater existing cyanobacteria genera, 12 are capable of producing toxins. While blue-green algae have significant taste and odor constituents, representing a moldy smell, their toxic metabolites have no taste, odor, or color. The risk of exposure to algal toxins may come from drinking water, recreational water, dietary supplements, or residue on produce irrigated with contaminated water and consumption of animal tissue. Avoiding cyanobacteria toxins is not as easy as avoiding a harmful algal bloom as toxins may be present in fish, shellfish and water even after the bloom has dissipated. Cyanobacterial toxins are responsible for a variety of health effects such as skin irritations, respiratory ailments, neurological effects, and carcinogenic effects. The three major classes of these compounds are:

❖ Cyclic peptides (*nodularins, microcystins*). *Nodularia*, a well-known cyanobacterium, produce nodularins that are primarily a concern in marine and brackish waters thus creating a risk to recreational swimmers. The 65 variants of microcystins, however, are isolated from freshwaters worldwide and are produced by *Microcystis* (the most commonly identified cyanobacteria in human and animal poisonings), Anabaena, and other algae. They are very stable in the environment and resistant to heat, hydrolysis, and oxidation. Both toxins have an affinity for the liver. Other symptoms of exposure to microcystins may range from weakness, loss of appetite, vomiting, and diarrhea to cancer.

❖ Alkaloids (anatoxin, saxitoxin). Anatoxins may affect the nervous system, skin, liver, or gastrointestinal tract. These neurotoxins can cause symptoms of diarrhea, shortness of breath, convulsions and death, in high doses, due to respiratory failure. Saxitoxins are the cause of paralytic shellfish poisonings in humans consuming contaminated shellfish. There are no reports of similar poisonings via the drinking water route.

❖ Lipopolysaccharides (endotoxins). A similar cell wall toxin as found in Salmonella bacteria, but appears to be less toxic.

Summary of Toxic Algae and the Corresponding Metabolites

Scientific Name	Class	Toxin
Nodularia spp.	Cyanophyceae	Nodularin
Microcystis spp.	Cyanophyceae	Microcystin
Chondrus armata	Floridophyceae	ASP domoic acid
Prymnesium parvum	Haptophyceae	Fish toxin
Phaeocystis pouchetii	Haptophyceae	PUA
Chrysochromulina polylepis	Haptophyceae	Ichtytoxic glycolipids
Alexandrium spp.	Dinophyceae	PSP
Dinophysis spp.	Dinophyceae	DSP
Prorocentrum lima	Dinophyceae	DSP
Gymnodinium breve	Dinophyceae	NSP
Ptychodiscus brevis	Dinophyceae	NSP
Gambierdiscus toxicus	Dinophyceae	Ciguatoxin/maitotoxin
Pseudonizschia spp.	Bacillariophyceae	ASP domoic acid
Thalassiosira rotula	Bacillariophyceae	PUA
Skeletonema costatum	Bacillariophyceae	PUA

Algal toxins can cause diarrhoea, vomit, tingling, paralysis and other effects in humans, mammals or fish. Algal toxins are produced by various algae and are found both in seawater and fresh water. The algal toxins can be retained in shellfish or contaminate drinking water. They have no taste or smell, and are not eliminated by cooking or freezing.

M. O. P. Iyengar, Professor F.E. Fritsch

Mandayam Osuri Parthasarathy Iyengar (15 December 1886–10 December 1963) was a prominent Indian botanist and phycologist who researched the structure, cytology, reproduction and taxonomy of Algae. He is known as the "father of Indian phycology" or "father of algology in India". He was the first President of Phycological Society of India. He primarily studied spirogyra. A series of outstanding contributions on algae resulted, studies on colonial Volvocales of South India, on *Fritschiella* and *Ecballocystopsis* to mention some.

Iyengar was born in Madras where his father M.O. Alasingrachariar worked as an attorney. The wealthy family was known for achievements in many walks of lives. After studies at the Hindu High School, he went to Presidency College, obtaining a BA degree in 1906 and an MA in 1909. He then became a curator in the Government Museum at Madras and became a lecturer in the Teacher's College in 1911. He became a professor of botany in the Presidency College in 1920 and worked on algae aside from teaching. He worked in the UK in 1930 along with Professor F.E. Fritsch at the Queen Mary College from where he received a PhD. Iyengar was an active sportsman and swimmer. He rescued two of his students from drowning in the Pamban in 1925. He was also a billiards champion in Madras.

Born	15 December 1886
Died	10 December 1966
Nationality	India
Known for	Pioneering Algal Studies in India
Scientific career	
Fields	Phycologist
Institutions	Madras University

Professor F.E. Fritsch, 1879-1954

The book for which Fritsch may be best known is 'The Structure and Reproduction of the Algae'. He also revised, indeed largely rewrote, G.S. West's 'British Freshwater Algae'. His other publications include several reviews of ecological, taxonomic, classificatory, morphological and evolutionary aspects of phycology.

The career of Felix Eugen Fritsch took him to the University of Munich, University College London, and the Royal Botanic Gardens (Kew). Following this he started a new Botany department at what is now Queen Mary College, University of London. He became professor in 1924 and retired in 1948.

For many years he had been concerned about the lack of a British freshwater biological station, and later highlighted this unsatisfactory situation in his presidential address to the British Association for the Advancement of Science. In 1929 the Freshwater Biological Association (FBA) was founded, and in 1931 it was decided that work would start in three rented rooms at Wray Castle in the English Lake District. The staff consisted of two recent graduates and a boy, with a student to do research. From these humble beginnings arose one of the best and most famous freshwater biological organisations in the world and one with a strong phycological section. Fritsch was chairman of the FBA's Council until his death and came 'nearer than anyone else to being it's the FBA's) founder' (Anon. 1955). As his wife said –'it was his favourite child'. Another institution owes its existence in large part to Fritsch: The Culture Collection of Algae and Protozoa, started by Dr E.G. Pringsheim. In 1938, Dr Pringsheim arrived in England from the German University in Prague, a refugee from Nazism. Fritsch helped him and played a major part in getting him to the University of Cambridge. It was as a result of his advocacy that the Botany Department made space for Pringsheim's culture collection, soon to be enlarged by British isolates. Later, the University took over responsibility for the collection until NERC (Natural Environmental Research Council) made special provision for it. The freshwater and marine parts of this are now housed at the Dunstaffnage Marine Laboratory at Oban under the management of the Scottish Association for Marine Science. In 1912, Fritsch started to put illustrations of freshwater algae onto foolscap sheets of paper. These were cut out of, or traced from, published papers. As the collection grew, so it became ever more useful to students and visiting phycologists. We, his research students, called it his scrap collection. He would work on it when too tired to do other things, and enlarged and improved the Collection in his retirement. At his death there were about 20,000 such illustrations. The Fritsch Collection of Illustrations of Freshwater Algae now contains millions of illustrations, and a microfiche edition is available.

****

aerial	Growing in a habitat above the ground or water
akinete	A thick-walled spore that can survive harsh conditions and functions as an asexual resting stage
algaenans	Hydrocarbon polymers that help some algal cells resist decay
anisogamy	Sexual reproduction involving two types of gametes of different sizes or shapes
antheridium	A cell that produces male gametes
apical cell	Cell located at the end of a filament or thallus
aplanospore	A nonflagellate spore that may be genetically capable of producing flagellated cells under the right conditions; produced by divisions of the parent cell
araphid diatom	Diatom lacking a raphe structure on either valve
areolae	Pores or perforations in the siliceous valves of diatoms
asexual reproduction	Reproduction involving the cells of a single parent
autocolony	A miniature version of the parent colony produced asexually by one of its cells; a coenobium
autospore	A nonflagellate spore
axial	A chloroplast located in the center of a cell along its longitudinal axis
basal	Concerning the base of a thallus or filament
benthic	Inhabiting the bottom of an aquatic environment
biflagellate	Cells with two flagella
bloom	Visible or nuisance algal growth often associated with nutrient-rich conditions
blue-green algae	See cyanobacteria
brackish	Slightly salty waterbodies
calyptra	Thickened or enlarged tip of a cyanobacterial filament

charophyceans or charophyte green algae	The group of green algae most closely related to the land plants, includes the zygnematalean algae (*Spirogyra* and the desmids) as well as several other genera and the stoneworts
chlorophyll-a	Primary photosynthetic pigment in algae and higher plants
chlorophyll-b	Photosynthetic pigment found in higher plants, green algae, and some euglenoids
chlorophyll-c	Photosynthetic pigment found in chrysophytes, synurophytes, diatoms, tribophytes, dinoflagellates, cryptophytes, and brown algae; includes c1 and c2 components
chlorophyceans	Green algae
chloroplast (plastid)	Membrane-bound organelle involved in photosynthesis
chrysophytes	Mixotrophic algae with golden-brown pigmentation
chytrids	Group of small, colorless fungi that may parasitize algal cells
chromatophore	An organelle that contains pigments; i.e. a plastid
cingulum (pl. cingula)	In the dinoflagellates, a transverse groove that usually holds the transverse flagellum; another term for girdle band
coenobium (pl. coenobia)	A type of colony with a genetically-predetermined morphology and number of cells; each cell of the parent colony can produce an autocolony that is an exact replica of the parent colony
coenocytic	Multinucleate and lacking tranverse cell walls; siphonous
conjugation	In the zygnematalean algae, sexual reproduction involving nonflagellate gametes that combine in a special tube or structure; see also lateral and scalariform conjugation
contractile vacuole	A membrane-bound vesicle that expands and contracts to expel excess water
cross walls	Transverse cell walls
cyanelles	The blue-green plastids of the glaucophytes that originally resulted from an endosymbiotic event
cyanobacteria	Photosynthetic bacteria with a characteristic blue-green color, commonly known as the blue-green algae
cytokinesis	Division of the cytoplasm during cell division
desmid	A group of unicellular or filamentous green algae; see also placoderm desmids and saccoderm desmids

detritus	Dead or decaying organic matter
diatoms	Photosynthetic microalgae with siliceous cell walls
dioceious	Organisms in which male and female (or + and -) gametes are produced on different individuals
dinoflagellates	Unicellular algae that are often covered in thick, armoured plates
double false branching	Growth pattern that occurs when a filament breaks apart (often at a dead cell or heterocyst) and both pieces emerge through the mucilage sheath and continue to grow
embryophytes	The group consisting of bryophytes and vascular plants
endosymbiont	An organism that lives inside another, benefitting both individuals
epilithic	Growing on the surface of rock or stone
epiphytic	Growing on the surface of plants or other algae
eutrophic	Waters rich in dissolved nutrients (especially nitrogen and phosphorus); leads to accelerated growth of algae and plants that depletes oxygen levels and reduces biodiversity
euglenoids	Organisms that are unicellular or colonial, colorless or pigmented, and may move with an undulating, shape-changing motion (see metaboly)
eyespot	A light-sensitive, red-pigmented spot usually found in the cell anterior of some flagellated algal cells or reproductive cells; sometimes called a *stigma*
false branching	Single false branching occurs when a filament breaks apart (often at a dead cell or heterocyst) and one piece emerges through the mucilage sheath and continues to grow; see also double false branching
flagellum (pl. flagella)	A long, whiplike organelle that usually protrudes from a cell and is used for motility
fragmentation	Type of asexual reproduction where the parent thallus breaks into pieces to form new individuals; may be accidental or genetically programmed
frustule	The siliceous components of a diatom cell; i.e. the valves and girdle bands
fucoxanthin	A brown carotenoid pigment used in photosynthetis, found in the golden-brown and brown algae

gametes	Reproductive cells
gas vesicles	In cyanobacteria, cylindrical structures used to increase cell bouyancy
gas vacuoles	Sometimes synonymous with gas vesicles, also refers to a group of gas vesicles
girdle band or cingulum (pl. cingula)	In diatoms, the rings or bands of silica encircling the valves where they overlap. In desmids, the short wall segments oriented transversely at the central portion of the cell
glaucophytes	A group of algae that contain blue-green plastids called cyanelles, sometimes called Glaucocystophytes
golden-brown algae	Algal groups with golden pigmentation; i.e. the chrysophytes, synurophytes, haptophytes, and diatoms
gone	In the desmids, a cell derived from zygote germination that may not have the typical morphology of that species
grana (pl.)	Stacks of thylakoids found in the discoid chloroplasts of land plants and some green algae
green algae	Morphologically diverse group of algae with green-pigmented chloroplasts
hair cells	Elongate, usually colorless cells or extensions of cells found at branch apices, increases surface area for nutrient and gas exchange and may protect the cells from herbivory
halophilic	Inhabiting a salty environment
haploid	Having a single set of chromosomes
heterocyst	In some cyanobacteria, a large, thick-walled cell involved in nitrogen fixation
heteropolar	Filaments with distinct basal and apical components or cells that are assymetrical along the transverse axis
heterotrophic	Organisms that feed on organic matter from external sources, and are osmotrophic or phagotrophic
holdfast cells	Structure that attaches a thallus or filament to the substrate
hormogonium (pl. hormogonia)	In filamentous cyanobacteria, short, usually motile segments that break off from the main filament and are used for asexual reproduction and dispersal
hypnospore	A thick-walled resting cyst
hypnozygote	A thick-walled, nonmotile resting zygote that remains dormant for a period of time until

	germinating when conditions improve; found in green algae, dinoflagellates, and some other flagellated organisms
intercalary	Located in the middle of a thallus, not on the end
isogamy	Sexual reproduction involving two types of gametes that function differently but are morphologically identical
isopolar	Symmetrical along the tranverse axis, having the same morphology at both ends
isthmus	In desmids, the constricted region between the two semicells that usually contains the nucleus
lateral conjugation	Conjugation where two cells on the same filament reproduce within a conjugation tube
littoral zone	Region near the shore of lakes, ponds, or large rivers
lorica	A protective covering surrounding flagellated cells of some algal groups
meristem	Portion of a thallus where the cells are able to divide and produce new growth
metaboly	Shape-changing motion of the euglenoid genera that does not involve flagellar movements
metaphytic	Growing among the floating microscopic community in the littoral zone that is often associated with aquatic plants
mixotrophic	Cells with both photosynthetic and heterotrophic or phagotrophic nutrition
monoecious	Organisms in which male and female (or + and -) gametes are produced on the same individual
monophyletic	A group of organisms that share a single common ancestor
multicellular	Composed of more than one cell
multinucleate	Having many nuclei
necridic cells (necridia)	Dead cells in cyanobacteria that disrupt the main filament to produce false branching or hormogonia
oligotrophic	Waters low in nutrients and organic production, typically rich in biodiversity

oogamy	Sexual reproduction involving the fusion of a small, flagellated male gamete with a larger, nonflagellated female gamete
oogonium (pl. oogonia)	A female reproductive structure capable of producing one or more eggs
osmotrophic	Organisms that feed on dissolved organic molecules absorbed from the outside environment
parietal	Near the outer or peripheral surface of a cell or thallus; commonly used to describe chloroplasts situated along the cell wall
pellicle	The surface covering of euglenoid cells composed of proteinaceous strips in a helical pattern
periphytic	Growing attached on the surfaces of plants, other algae, or other substrates
phagotrophic	Organisms that feed by ingesting particulate organic carbon or intact cells
phototaxis	Movement in response to light
photosynthetic	Organisms that use sunlight as an energy source to synthesize carbohydrates from carbon dioxide and water
phylogenetic	Concerning the relationships of evolutionary development
phytoplankton	Floating or swimming microscopic algae and cyanobacteria
placoderm desmids	Unicellular or pseudofilamentous desmids with cell walls made up of two parts of different ages separated by a central isthmus that are covered in pores
plastid	See chloroplast
polar nodule	In cyanobacteria, thickened regions where the heterocyst cell wall attaches to a vegetative cell
polyeders	In *Hydrodictyon* and *Pediastrum*, polygonal cells produced from zygote germination that give rise to new coenobia
polyphyletic	A group of organisms descended from more than one ancestor, may be more closely related to other organisms outside of the taxonomic group
protoplasm, protoplast	The living substance of a cell
pseudofilament	Loose chain of individual cells held together by mucilage or attached at the end walls, rather than sharing a common cell wall as in a true filament
pyrenoid	A proteinaceous structure associated with algal chloroplasts that often forms storage compounds

quadriflagellate	Cells with four flagella
raphe	In diatoms, a slit in the valve face that allows the cell to move along a substrate
rhizoid	Rootlike extensions usually used to attach a thallus to the substrate
saccoderm desmids	Desmids with a homogeneous cell wall that lacks pores; are not differentiated into two semi-cells
scalariform conjugation	Conjugation involving two filaments that allign laterally and form conjugation tubes between adjacent cells in a ladder-like arrangement
semicell	In placoderm desmids, one-half of a cell
sexual reproduction	Reproduction involving the cells of a single parent
silica deposition vesicle	Involved in the formation of the siliceous valves of diatoms and siliceous scales of chrysophytes and synurophytes
siphonous	See coenocytic
striae	Term used to described elongated markings on the cell walls of desmids and diatoms
subaerial	Growing on or near the earth's surface
sulcus	In dinoflagellates, a longitudinal groove in the ventral side of the cell that holds the longitudinal flagellum
suture	In dinoflagellates, the regions between thecal plates
syngamy	The fusion of two gametes during fertilization
synurophytes	Motile flagellates covered by siliceous scales; the scaled chrysophytes
terminal	Located at the end of a thallus
thallus	General form of an alga that, unlike a plant, is not differentiated into stems, roots, or leaves
thylakoid	Flattened membranous sacs that form the chloroplast membranes
tribophyte algae	See yellow-green algae
trichome	A filament of a blue-green alga or cyanobacterium
trichogyne	In some charophycean green algae and in the red algae, the elongated portion of the female reproductive structure that recieves the male gametes

turbidity	Measure of the cloudiness of a waterbody due to suspended particles
unicellular	Composed of a single cell
uniseriate	Composed of a single row of cells.
valve	In the diatoms, the two siliceous halves of the cell wall
vegetative cells	Cells involved in growth, nutrition, or asexual reproduction, but not sexual reproduction
vesicles	SDV, gas
yellow-green algae	Rare algae with yellow-green pigmentation; the tribophyte algae
zoospore	A flagellated spore
zygospore	A thick-walled resting spore formed from the fusion of gametes or a fertilized oogonium

Glossary of Terms for Algae

References

- Adelman WJ, Fohlmeister JF, Sasner JJ, Ikawa M (1982) Toxicon 20:513PubMedCrossRefGoogle Scholar
- Afreen S, Fatma T (2013) Laccase production and simultaneous decolourization of synthetic dyes by cyanobacteria. Inter. J. Innovative Research Sci. Eng. Tech. 2:3563–3568Google Scholar
- Ali Laila KM, Mostafa Soha SM (2009) Evaluation of potassium humate and *Spirulina platensis* as a bio-organic fertilizer for sesame plants grown under salinity stress. The 7th International Conference of Organic Agriculture. 13–15 December, Egypt. J Aquat Res 871369388Google Scholar
- Aly MHA, Abd El-All Azza AM, Mostafa Soha SM (2008) Enhancement of sugar beet seed germination, plant growth, performance and biochemical compounds as contributed by algal extracellular products. J Agric Sci Mansoura Univ 33(12):8429 8448Google Scholar
- Anderson LK, Toole CM (1998) A model for early events in the assembly pathway of cyanobacterial phycobilisomes. Mol Microbiol 30:467–474PubMedCrossRefGoogle Scholar
- Anwer R, Khrusheed S, Fatma T (2012) Detection of immunoactive insulin in *spirulina*. J Appl Phycol 24:583–591CrossRefGoogle Scholar
- Amal ZH, Soha SMM, Hamdino MIA (2010) Influence of different cyanobacterial application methods on growth and seed production of common Bean under various levels of mineral nitrogen fertilization. Nat Sci 8(11):202–212Google Scholar
- Arment AR, Carmicchael WW (1996) Evidence that microcystin is a thio-template product. J Phycol 32:591–597CrossRefGoogle Scholar
- Ayehunie S, Belay A, Baba A, Ruprecht RM (1998) Inhibition of HIV-1 replication by an aqueous extract of *Spirulina platensis* (*Arthrospira platensis*). J Acquir Immune Defic Syndr Hum Retrovirol 18:7–12PubMedCrossRefGoogle Scholar
- Bapat AV, KIyer R, Rao SP (1996) Effect of cyanobacterial extract on somatic embryogenesis in tissue culture of sandalwood (Santalum album). J Med Aromatic Plants Sci 18(1):1014Google Scholar
- Bastia AK, Satpathy DP, Adhikary SP (1993) Heterotrophic growth of several filamentous blue-green algae. Arch Hydrobiol 70:65–70Google Scholar
- Becker W (2004) The nutritional value of microalgae for aquaculture. In: Richmond A (ed) Microalgae for aquaculture. Handbook of microalgal culture. Blackwell, Oxford, pp 380–391Google Scholar
- Bergman B, Rasmussen U, Rai AN (2007) Cyanobacterial associations. In: Elmerich C, Newton WE (eds) Associative and endophytic nitrogen-fixing bacteria and cyanobacteria associations. Kluwer Academic Publishers, DordrechtGoogle Scholar
- Bhaleya P, Mathur T, Pandya M, Fatma T, Rattan A (2006) Fitoterapia 77:233–235CrossRefGoogle Scholar

- Bhaskar US, Gopalswamy G, Raghu R (2005) A simple method for differential extraction and purification of C-phycocyanin from *Spirulina platensis* Geitler. Indian J Exp Biol 43:277–279PubMedGoogle Scholar
- Blinkova LP, Gorobets OB, Baturo AP (2001) Biological activity of Spirulina.jjj. Zh Mikrobiol Epidemiol Immunobiol 2:114–118PubMedGoogle Scholar
- Bolch CJS, Orr PT, Jones GJ, Blackburn SI (1999) Genetic, morphological, and toxicological variation among globally distributed strains of *Nodularia* (cyanobacteria). J Phycol 35:339–355CrossRefGoogle Scholar
- Boone DR, Castenholz RW (eds) (2001) Bergeys manual of systematic bacteriology, hylum. BX Cyanobacteria. Springer, New York, pp 437–599Google Scholar
- Brand F (1903) Morphologische-physiologische betrachtungen uber cyanophycean. Belh Bot Cbl 15:31–64Google Scholar
- Brandl H, Gross RA, Lenz RW, Fuller RC (1990) Plastic bacteria? Progress and prospects for polyhydroxy-alkanoate production in bacteria. Adv Biochem Eng Biotechnol 41:77–93PubMedGoogle Scholar
- Brijvir S (1992) Thesis: studies on the biology of alkalophilic cyanobacteriaGoogle Scholar
- Burja AM, Banaigs BE, Abou-Mansour, Burgess JG, Wright PC (2001) Marine cyanobacteria-a prolific source of natural products. Tetrahedron 57:9347–9377CrossRefGoogle Scholar
- Campbell J, Edward Stevens SJR, Balkwill LD (1982) Accumulation of polyb-hydroxybutyrate in *Spirulina platensis*. J Bacteriol 149:361–363PubMedCentralPubMedGoogle Scholar
- Canabaeus L (1929) Uber die Heterocysten und Gasvakuolen der Blaualgen und ihre Beziehung zueinander. In: Kolkowitz R (ed) Pflanzenforschung. Fischer, Jena, pp 1–48Google Scholar
- Carmichael WW, Drapeau C, Anderson DM (2000) Harvesting of *Aphanizomenon flos-aquae* Ralfs ex Born. & Flah. Var. flos-aquae (Cyanobacteria) from Klamath Lake for human dietary use. J Appl Phycol 12:585–595CrossRefGoogle Scholar
- Carpenter EJ (2002) Marine cyanobacterial symbiosis. Biol Environ 102:15–18Google Scholar
- Cerniglia CE, Gibson TD, Van Baalen C (1980a) Oxidation of naphthalene by the cyanobacteria *Oscillatoria* sp., strain JCM. J Gen Microbial 116:485–494Google Scholar
- Cerniglia CE, Van Baalen C, Gibson TD (1980b) Metabolism of naphthalene by cyanobacteria and microalgae. J Gen Microbiol 116:495–500Google Scholar
- Chadd HE, Newman J, Man NH, Carr NG (1996) Identification of iron superoxide dismutase and a copper/zinc superoxide enzyme activity within marine cyanobacterium *synechococcus* sp. WH 7803. FEMS Microbiol Lett 138:161–165PubMedCrossRefGoogle Scholar

- ❖ Chen GQ, Wu Q (2005) The application of polyhydroxyalkanoates as tissue engineering materials. Biomaterials 26:6565–6578PubMedCrossRefGoogle Scholar
- ❖ Chiaramonti D (2007) Bioethanol: role and production technologies. In: Ranalli P (ed) Improvement of crop plants for industrial end uses. Springer, Dordrecht, pp 209–251CrossRefGoogle Scholar
- ❖ Cohen Y, Jørgensen BB, Revsbech NP, Poplawski R (1986) Adaptation to hydrogen sulfide of oxygenic and anoxygenic photosynthesis among cyanobacteria. Appl Environ Microbiol 51:398–407PubMedCentralPubMedGoogle Scholar
- ❖ Couto SR (2007) Decolouration of industrial azo dyes by crude laccase from Trametes hirsute. J Hazard Material 148:768–770Google Scholar
- ❖ Darley WM (1982) Algal biology. A physiological approach. Blackwell Scientific, OxfordGoogle Scholar
- ❖ Darnell DW, Greene B, Henzl MT, Hosea JM, Mopherson RA (1986) Selective recovery of gold and other metal ions from an algal biomass. Environ Sci Technol 20:206–208CrossRefGoogle Scholar
- ❖ De Caire GZ, De Mule MCZ, Doallo S, Halperin RD, Halperin L (1976) Accion de extractos algales acuosos y etereos de *Nostoc muscorum* Ag. (N° 79a). Bol Soc Agric Bot 17:289–300Google Scholar
- ❖ De Caire GZ, De Cano MS, De Mule MCZ, Halperin DR, Galvagno MA (1987) Action of cell-free extracts and extracellular products of *Nostoc muscorum* on growth of *Sclerotinia sclerotiorum*. Phyton 47:43–46Google Scholar
- ❖ De Caire GZ, De Cano SM, De Mule ZCM, De Halperin RD (1990) Antimycotic products from the Cyanobacterium *Nostoc muscorum* against *Rhizoctonia solani*. Phyton 51:1–4Google Scholar
- ❖ De Mule MCZ, De Caire GZ, Doallo S, De Halperin DR, Halperin L (1977) Accion de extractos algales acuosos y etereos de *Nostoc muscorum* Ag. (n°79a). Efecto sobre el desarrollo del hongo *Cunninghamella blakesleana* en el medio de Mehlich. Bol Soc Arg Bot 18:121–128Google Scholar
- ❖ Elhai J, Wolk CP (1988) Conjugal transfer of DNA to cyanobacteria. Methods Enzymol 167:747PubMedCrossRefGoogle Scholar
- ❖ Ellis BE (1977) Degradation of phenolic compounds by fresh water algae. Plant Sci Lett 8:213–216CrossRefGoogle Scholar
- ❖ BR Vashishta. Botany (for Degree Students). S. Chand, 1995.
- ❖ FE Fritsch. & 1945 structure and reproduction in algae vol. i& ii, 1935.
- ❖ FE Fritsch. reprinted 1948. The structure and reproduction of the algae, 1, 1935.
- ❖ Gilbert Morgan Smith et al. Fresh-water algae of the united states. 1933.

- OP Sharma. Algae. series on diversity of microbes and cryptogams, 2011. Edward G Bellinger and David C Sigee. Freshwater algae: identification and use as bioindicators. JohnWiley & Sons, 2015.
- Brand L. E. (1986). Photosynthetic picoplankton. *Canadian Bulletin of Fisheries and Aquatic Sciences* **214**, 205-233.
- Brand L. E. (1990). The isolation and culture of microalgae for biotechnological applications. In: *Isolation of biotechnological organisms from nature*, Labeda D. P. (ed.). McGraw-Hill, New York, pp. 81-115.
- Guillard R. R. L. (1975). Culture of phytoplankton for feeding marine invertebrates. In: *Culture of marine invertebrate animals*, Smith W. L. and Chanley M. H. (eds.). Plenum, New York, pp. 29-60.
- Guillard R. R. L. and Keller M. D. (1984). Culturing dinoflagellates. In: *Dinoflagellates*, Spector D. L. (ed.). Academic, New York, pp. 391-442.
- Harrison PJ, Waters RE & Taylor FJR (1980). A broad spectrum artificial seawater medium for coastal and open ocean phytoplankton. *J. Phycol.* **16**, 28-35.
- Keller M.D., Selvin R. C., Claus W and Guillard R. R. L. (1987). Media for the culture of oceanic ultraphytoplankton. *J. Phycol.* **23**, 633-638.
- Novarino G. (1998) general aspects of microalgal cultivation. In, Report and abstracts of CODENET First Annual Meeting.
- Provasoli L., McLaughlin J. J. A. and Droop M. R. (1957). The development of artificial media for marine algae. *Archiv für Mikrobiologie.* **25**, 392-428.
- Stein J. R. (1973). *Handbook of phycological methods: Culture methods and Growth measurements*. Cambridge University Press, London and New York. 448pp.

  o **Web Source**
- https://www.biologydiscussion.com
- https://guidancecorner.com/nostoc/
- https://biologyeducare.com/volvox/
- https://www.biologydiscussion.com/algae/classification-of-chaetophorales-with-diagram-algae
- http://domainbiology.blogspot.com/2017/10/bacillariophyta-definition.html
- https://www.biologydiscussion.com/algae/diatoms-characteristics-occurrence-and-reproduction
- https://www.botanylibrary.com/algae/class-rhodophyceae/rhodophyceae-features-and-reproduction-algae-botany/14569
- https://www.biologydiscussion.com/algae/useful-notes-on-the-origin-and-evolution-of-sex-in-algae
- http://ina.tmsoc.org/CODENET/culturenotes.htm
- https://en.wikipedia.org
- https://biologyboom.com

www.ingramcontent.com/pod-product-compliance
Lightning Source LLC
Chambersburg PA
CBHW080450220526
45465CB00006B/2219